ATOMIC DYNAMICS IN LIQUIDS

ATOMIC DYNAMICS IN LIQUIDS

N. H. March
Coulson Professor of Theoretical Chemistry,
University of Oxford

M. P. Tosi
Professor of Theoretical Physics,
University of Trieste

DOVER PUBLICATIONS, INC.
New York

Published in Canada by General Publishing Company, Ltd., 30 Lesmill Road, Don Mills, Toronto, Ontario.
Published in the United Kingdom by Constable and Company, Ltd., 3 The Lanchesters, 162–164 Fulham Palace Road, London W6 9ER.

This Dover edition, first published in 1991, is an unabridged and unaltered republication of the work first published by The Macmillan Press Ltd., London, in 1976.

Manufactured in the United States of America
Dover Publications, Inc., 31 East 2nd Street, Mineola, N.Y. 11501

Library of Congress Cataloging-in-Publication Data

March, Norman H. (Norman Henry), 1927–
Atomic dynamics in liquids / N. H. March, M. P. Tosi.
p. cm.
Includes bibliographical references and index.
ISBN 0-486-66598-4 (pbk.)
1. Liquids. 2. Fluid dynamics. 3. Hydrodynamics. I. Tosi, M. P. II. Title.
QC145.2.M37 1991
530.4′2—dc20 90-21861
CIP

CONTENTS

Preface

This book presents the subject of dynamics of atoms in liquids in a form which should be useful for established research workers in the field of the physics and chemistry of the liquid state. It should also be useful to research students entering this field but, of course, quite a bit of background will then be necessary for an understanding of this work. Specifically, we have assumed that the reader will have successfully completed a basic course on statistical mechanics and thermodynamics and, since the book deals with quantal as well as classical fluids, a basic course in quantum mechanics. However, only in the advanced part of Chapter 9 on Critical Phenomena is detailed knowledge of quantum field theory assumed.

Some of the material in this book has formed the basis of advanced level courses given by the authors. One of us in particular (M.P.T.) has drawn upon it for lectures given at the University of Rome and at the Postgraduate School at L'Aquila, organised by the Consiglio Nationale delle Ricerche for condensed matter physicists. Supplemented by some material from Landau and Lifshitz's *Statistical Physics*, courses of from 10 to 20 lectures to first and second year postgraduate students have been given at different times on portions of this book.

There are several points relating to the presentation which need brief comment. First, though we have called the work *Atomic Dynamics in Liquids*, some attention is also paid to the dynamics of electrons in Chapter 7. Secondly, we have thought it helpful to include, albeit in rather elementary terms, an account of the recent progress in the understanding of liquid helium three, and of critical phenomena. We are conscious that neither of us has contributed to these fields and that great use has been made of the presentation developed by other workers (particularly A. J. Leggett on helium three, who of course is in no way responsible for any of our final presentation here).

Though we felt it important to include a final chapter on the liquid surface, this is now a subject for a book in its own right and we are aware that the choice of material is idiosyncratic. Fortunately, in C. A. Croxton's book *The Liquid State*, there is a useful source for readers who wish to take this field somewhat further.

Many colleagues have influenced our thinking in this area over the past decade. To them, and to all authors and publishers who have given us permission to include figures and tabular material from the original sources, we are most grateful.

Finally, we shall be grateful if readers who find the book useful would write to tell us where we might make improvements in the future. It is too much to hope that a work on this scale, and of such a degree of complexity, could be completely free from error.

London and Rome, 1976

N. H. M.
M. P. T.

ATOMIC DYNAMICS IN LIQUIDS

CHAPTER 1

STATIC STRUCTURE AND THERMODYNAMICS

The most basic characteristic of a liquid is that it possesses short-range order, as opposed to the long-range periodicity of a crystalline solid. Since the structure of a crystal is determined experimentally by observing the Bragg reflections of X-rays, it is natural to seek a quantitative description of the liquid structure via the intensity I of X-rays scattered through an angle 2ϑ say, from the liquid. If we introduce the usual variable $k=4\pi \sin \vartheta/\lambda$, λ being the X-ray wavelength, then the liquid structure factor $S(k)$ is defined by

$$S(k)=I/Nf^2 \tag{1.1}$$

where N is the total number of atoms in the liquid, assumed monatomic. In eqn (1.1), $f(k)$ is the atomic scattering factor, i.e. the Fourier transform of the electron density in the atom. $f(k)$ falls from a value Z, the atomic number, at $k=0$ to zero at large k, the asymptotic behaviour being proportional to k^{-4}.

1.1 Definition of radial distribution function $g(r)$ and structure factor $S(k)$

To gain insight into the relation between the liquid structure factor $S(k)$ in (1.1) and the atomic arrangement, one now uses the Debye theory of the diffraction of X-rays by a liquid (see for example Gingrich, 1943 or March, 1968 for a detailed discussion), which yields

$$I=Nf^2\left[1+\sum_{\mathrm{n}}{}' \frac{\sin kR_{\mathrm{nm}}}{kR_{\mathrm{nm}}}\right] \tag{1.2}$$

Here the R_{nm} denote the interatomic distances in the fluid, while the prime denotes the exclusion of the term $R_{\mathrm{nm}}=0$ from the summation. Clearly, eqn (1.2) is revealing an interference pattern between scattered X-rays from pairs of atoms. Therefore, it is helpful to introduce the pair distribution function $g(r)$, which, following Zernike and Prins (1927), is defined by setting $\rho g(r)4\pi r^2\,\mathrm{d}r$ equal to the total number of atoms in a spherical shell of radius r and thickness $\mathrm{d}r$ centred on a chosen atom at the origin of coordinates. Here, $\rho=N/V$ is the average number density of N atoms in volume V.

Replacing the summation in eqn (1.2) by an integration, and omitting the

X-ray scattering from a uniform density of electrons (merely a delta function at $k=0$ in the limit of an infinite liquid sample), we find from eqns (1.1) and (1.2) that

$$S(k)=1+\rho\int_0^\infty 4\pi r^2[g(r)-1]\frac{\sin kr}{kr}\,\mathrm{d}r \tag{1.3}$$

Hence, the liquid structure factor is related by Fourier transform to the atomic arrangement around a given atom at the origin. By noting that $\sin kr/kr$ is the s wave ($l=0$ term) in the expansion of the plane wave $\exp(i\boldsymbol{k}\cdot\boldsymbol{r})$ in spherical waves, $S(k)$ in (1.3) can be rewritten as

$$S(k)=1+\rho\int[g(r)-1]\exp(i\boldsymbol{k}\cdot\boldsymbol{r})\,\mathrm{d}\boldsymbol{r} \tag{1.4}$$

In a crystalline solid, $g(r)$ in (1.4) would not be isotropic but would be a sum of delta functions (neglecting atomic vibrations) centred on the crystal lattice sites. Evidently, eqn (1.4) would lead then to non-zero $S(\boldsymbol{k})$ only when $\boldsymbol{k}$ was a reciprocal lattice vector. These Bragg reflections can be thought of, roughly, as blurred out and spherically averaged in the liquid, leading to a continuous function $S(k)$ reflecting directly the short-range order via $g(r)$.

We could view eqns (1.3) or (1.4) as the forms from which X-ray intensity could be predicted from a calculated pair distribution function $g(r)$. But a more fruitful approach to date has been to invert the Fourier transform and hence to derive $g(r)$ from the measured $S(k)$ via

$$\begin{aligned} g(r)&=1+\frac{1}{8\pi^3\rho}\int[S(k)-1]\exp(-i\boldsymbol{k}\cdot\boldsymbol{r})\,\mathrm{d}\boldsymbol{k}\\ &=1+\frac{1}{2\pi^2\rho r}\int_0^\infty[S(k)-1]k\sin kr\,\mathrm{d}k \end{aligned} \tag{1.5}$$

Of course, the determination of $g(r)$ from the measured $S(k)$ is subject to some problems and the kind of errors which can arise have been discussed by Paalman and Pings (1963)

The static structure factor $S(k)$ for liquid thallium near its freezing point, as determined by neutron scattering (see Enderby and March, 1966), is recorded in *figure 1.1*, while *figure 2.10* (page 33) shows the radial distribution function $g(r)$ in a Lennard–Jones fluid simulating liquid argon, as determined by computer simulation† by Verlet (1968).

It is important to stress here that a great deal of the theory of liquids is centred round the pair distribution function $g(r)$, and its dynamic generalisation given in chapter 3. Therefore, we turn immediately to discuss how some important thermodynamic properties are related to $g(r)$.

†For a short discussion, see chapter 2.

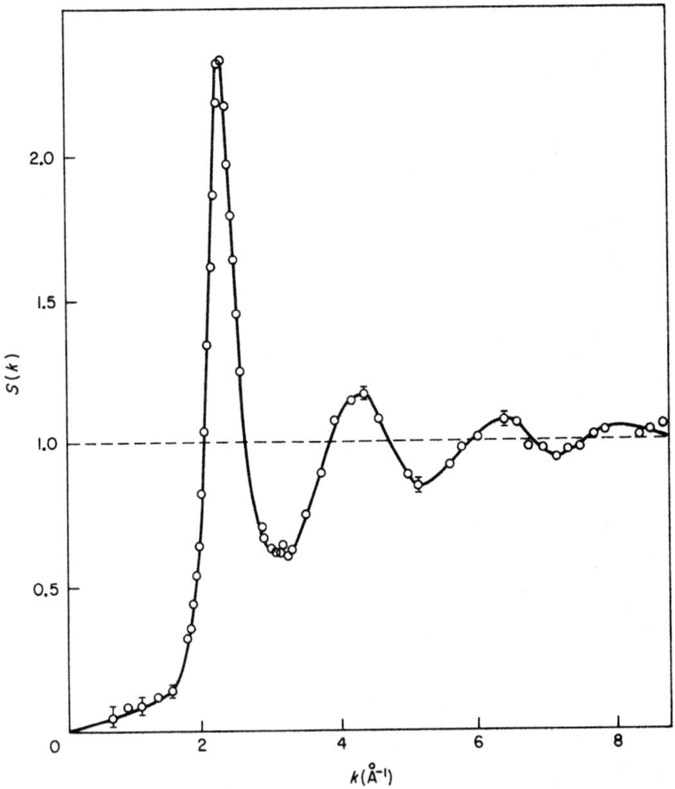

Figure 1.1 Static structure factor $S(k)$ for liquid thallium at 600 K, as determined by neutron scattering (see Enderby and March, 1966)

1.2 Internal energy and equation of state with pair forces

In this section expressions will be derived for the internal energy E and the equation of state. The assumption will be made that the intermolecular potential energy Φ can be decomposed into a sum of pair potentials

$$\Phi(\boldsymbol{R}_1 \ldots \boldsymbol{R}_N) = \sum_{i<j} \phi(R_{ij}) \tag{1.6}$$

The objectives will be to obtain E and the pressure p in terms of the pair distribution function $g(r)$ and the potential $\phi(r)$.

In a classical monatomic liquid at temperature T, the kinetic energy per degree of freedom is $\frac{1}{2}k_B T$ and hence

$$E = \tfrac{3}{2}Nk_B T + \langle \Phi \rangle \tag{1.7}$$

The mean potential energy $\langle\Phi\rangle$ is simply

$$\langle\Phi\rangle = Z^{-1}\int\cdots\int \exp\left(-\Phi/k_{\mathrm{B}}T\right)\Phi\,\mathrm{d}\boldsymbol{R}_1\cdots\mathrm{d}\boldsymbol{R}_{\mathrm{N}} \tag{1.8}$$

where Z is the partition function. From the assumption (1.6) of pair forces, the sum consists of $N(N-1)/2$ terms, all of which make equal contributions to $\langle\Phi\rangle$. Hence we find

$$\langle\Phi\rangle = \frac{N(N-1)}{2}\iint \phi(R_{12})\left[\frac{\int\cdots\int \exp\left(-\Phi/k_{\mathrm{B}}T\right)\mathrm{d}\boldsymbol{R}_3\cdots\mathrm{d}\boldsymbol{R}_{\mathrm{N}}}{Z}\right]\mathrm{d}\boldsymbol{R}_1\,\mathrm{d}\boldsymbol{R}_2 \tag{1.9}$$

But from the definition of the distribution functions (see also equation (2.3) of section 2.1) we can introduce the radial distribution function and write

$$\begin{aligned}\langle\Phi\rangle &= \frac{N(N-1)}{2}\iint \phi(R_{12})\frac{(N-2)!}{N!}\rho^2 g(R_{12})\,\mathrm{d}\boldsymbol{R}_1\,\mathrm{d}\boldsymbol{R}_2\\ &= \tfrac{1}{2}\rho^2 V\int_0^\infty \phi(r)g(r)4\pi r^2\,\mathrm{d}r\end{aligned} \tag{1,10}$$

one of the integrations yielding immediately the volume V of the fluid. Hence we have finally (cf. H. S. Green, 1952)

$$E = \tfrac{3}{2}Nk_{\mathrm{B}}T + \frac{N\rho}{2}\int_0^\infty \phi(r)g(r)4\pi r^2\,\mathrm{d}r \tag{1.11}$$

The potential energy term exhibited in this equation for the internal energy could have been written down directly on physical grounds, since the number of molecules on average within a distance between r and $r+\mathrm{d}r$ of a given molecule is $\rho g(r)4\pi r^2\,\mathrm{d}r$ and the factor $\frac{1}{2}$ avoids counting interactions twice over.

Similarly, the equation of state can be obtained. To do so, we make use of the classical virial theorem, which relates the average of the kinetic energy, $\bar{K}$ say, to the virial of the forces. The virial of the pressure p is $3pV$, yielding, for a perfect gas

$$2\bar{K} = 3pV \tag{1,12}$$

When there is a force $\boldsymbol{F}_{\mathrm{i}}$ acting on the ith molecule at $\boldsymbol{R}_{\mathrm{i}}$, the average of $-\sum \boldsymbol{R}_{\mathrm{i}}\cdot\boldsymbol{F}_{\mathrm{i}}$ has to be calculated, the summation extending over all the molecules. For central pair forces, this becomes again the average of $N(N-1)/2$ terms, one term being

$$\int R_{12}\frac{\partial\phi(R_{12})}{\partial R_{12}}\left[\int\frac{\exp\left(-\Phi/k_{\mathrm{B}}T\right)}{Z}\mathrm{d}\boldsymbol{R}_3\cdots\mathrm{d}\boldsymbol{R}_{\mathrm{N}}\right]\mathrm{d}\boldsymbol{R}_1\,\mathrm{d}\boldsymbol{R}_2 \tag{1.13}$$

Expressing this once more in terms of the radial distribution function yields, since $\bar{K}=\frac{3}{2}Nk_{\mathrm{B}}T$, as used in the calculation of E,

$$3pV=3Nk_{\mathrm{B}}T-\frac{N(N-1)}{2}\int R_{12}\frac{\partial\phi(R_{12})}{\partial R_{12}}\frac{(N-2)!}{N!}\rho^2 g(R_{12})\,\mathrm{d}\boldsymbol{R}_1\,\mathrm{d}\boldsymbol{R}_2 \quad (1.14)$$

or

$$p=\rho k_{\mathrm{B}}T-\frac{\rho^2}{6}\int r\frac{\partial\phi}{\partial r}g(r)\,\mathrm{d}\boldsymbol{r} \quad (1.15)$$

In principle then, from an assumed law of force and knowledge of the corresponding radial distribution function, the internal energy E and the fluid pressure p can be estimated from eqns (1.11) and (1.15).

It is important to note at this point that while equation (1.15) requires knowledge of both $g(r)$ and $\phi(r)$, there is another route to the equation of state of a fluid, which is more fundamental in that it does not require the assumption (1.6) of pair potentials. However, to derive this further relation we must go from the canonical ensemble used above to the grand canonical ensemble.

1.3 Relation of liquid structure factor at $k=0$ to compressibility

Let us consider an open region, i.e. one in which particles can come and go freely, drawn in a system of infinite extent. We shall now show that the fluctuation in the number of particles in this region is given by the volume integral of $g(r)-1$, which is specifically the isothermal compressibility of the liquid. Another interesting example of such a relation between fluctuations and thermodynamic quantities yields the specific heat c_v, which is discussed in appendix 2.1.

One reason for the interest in the above relation between the volume integral of the radial distribution function, or equivalently, from eqn (1.3), the long wavelength limit of $S(k)$, and the compressibility (first derived by Ornstein and Zernike, 1914), is because of the difficulty of extending the diffraction experiments, referred to in section 1.1, to small scattering angles.

Consider a member of the grand canonical ensemble in which the open region, of volume V, contains exactly N particles. For a specified configuration of the particles, $\boldsymbol{R}_i$ say, the singlet density $\rho(\boldsymbol{r}_1)$ at point $\boldsymbol{r}_1$ in this region, and the density $\rho^{(2)}(\boldsymbol{r}_1\boldsymbol{r}_2)$ of pairs of particles at points $\boldsymbol{r}_1$ and $\boldsymbol{r}_2$ are respectively given by

$$\rho(\boldsymbol{r}_1)=\sum_{i=1}^{N}\delta(\boldsymbol{R}_i-\boldsymbol{r}_1) \quad (1.16)$$

and

$$\rho^{(2)}(\boldsymbol{r}_1\boldsymbol{r}_2)=\sum_{i\neq j=1}^{N}\delta(\boldsymbol{R}_i-\boldsymbol{r}_1)\delta(\boldsymbol{R}_j-\boldsymbol{r}_2) \quad (1.17)$$

It follows immediately from these definitions that

$$\int_V \mathrm{d}\boldsymbol{r}_1 \rho(\boldsymbol{r}_1) = N \tag{1.18}$$

$$\int_V \int_V \mathrm{d}\boldsymbol{r}_1\, \mathrm{d}\boldsymbol{r}_2 \rho^{(2)}(\boldsymbol{r}_1 \boldsymbol{r}_2) = N^2 - N \tag{1.19}$$

The distribution functions for single particles and for pairs of particles are obtained by averaging the respective densities over phase space and over all numbers N of particles with the probability distribution of the grand canonical ensemble,

$$w_{GC} = \exp\left[(\Omega + N\mu - H_N)/k_B T\right] \tag{1.20}$$

where $\Omega = -pV$, μ is the chemical potential, and H_N is the Hamiltonian of the set of N particles. In a fluid the averaged densities have the form $\langle \rho(\boldsymbol{r}_1) \rangle = \langle N \rangle / V = \rho$ and $\langle \rho^{(2)}(\boldsymbol{r}_1 \boldsymbol{r}_2) \rangle = \rho^2 g(r_{12})$, where ρ is the bulk number density of the fluid and $g(r_{12})$ is the radial distribution function, dependent only on the scalar distance r_{12}. By taking the average of eqn (1.19) it therefore follows that

$$\lim_{k \to 0} S(k) \equiv 1 + \rho \int \mathrm{d}V \left[g(r_{12}) - 1\right] = \frac{\langle N^2 \rangle - (\langle N \rangle)^2}{\langle N \rangle} \tag{1.21}$$

which is the desired relation between the structure factor at long wavelengths and the particle number fluctuations.

Next, we recall from fluctuation theory that the particle number fluctuations are related to thermodynamic properties of the system. The grand partition function is

$$\exp(-\Omega/k_B T) = \sum_{N=0}^{\infty} \exp\left[N\mu - F(N,T,V)\right]/k_B T \tag{1.22}$$

where $F(N,T,V)$ is the Helmholtz free energy of a member of the grand ensemble containing N particles. By differentiating Ω and the average number of particles,

$$\langle N \rangle = \sum_{N=0}^{\infty} N \exp\left[\Omega + N\mu - F(N,T,V)\right]/k_B T \tag{1.23}$$

with respect to the chemical potential we find

$$\left(\frac{\partial \Omega}{\partial \mu}\right)_{T,V} = -\langle N \rangle \tag{1.24}$$

and

$$\left(\frac{\partial N}{\partial \mu}\right)_{T,V} = \frac{1}{k_B T}\left[\langle N\rangle\left(\frac{\partial \Omega}{\partial \mu}\right)_{T,V} + \langle N^2\rangle\right] \tag{1.25}$$

whence it follows that

$$\langle N^2\rangle - (\langle N\rangle)^2 = k_B T\left(\frac{\partial N}{\partial \mu}\right)_{T,V} \tag{1.26}$$

Finally, by using the thermodynamic identity

$$\left(\frac{\partial \mu}{\partial N}\right)_{T,V} = -\frac{1}{\rho^2}\left(\frac{\partial p}{\partial V}\right)_{T,N} \equiv \frac{1}{\rho^2 V K_T}$$

where K_T is the isothermal compressibility, we arrive at the Ornstein–Zernike relation,

$$\lim_{k\to 0} S(k) = \rho k_B T K_T \tag{1.27}$$

As we shall see in section 4.3.2, the above relation admits a dynamic interpretation, in that it is a consequence of the fact that the long wavelength excitations of the fluid are sound waves. In this connection we should then stress that, in the case of a fluid of charged particles, the fluctuations in the particle number must be such as to preserve the electrical neutrality of the open region. Charge separation involves the excitation of the plasmon mode, which contributes to the k^2 term in the expansion of $S(k)$ at small k (see section 7.1.2).

In summary, the tools for describing the short-range order in a monatomic liquid, $S(k)$ and $g(r)$ have been introduced. These, plus the pair potential $\phi(r)$, have been shown to be closely related to the thermodynamic properties of the liquids and, in particular, formulae have been derived for the internal energy, pressure and compressibility. Of course, a first principles approach to the theory of the liquid state should ideally allow all physical properties to be calculated from an assumed force law. The next chapter is concerned with this problem.

CHAPTER 2

CALCULATION OF LIQUID STRUCTURE FROM A LAW OF FORCE

The radial distribution function $g(r)$, by itself, gives a valuable but nevertheless limited picture of liquid structure, even for a monatomic fluid. Finer details of the structure of such a one-component fluid are contained in higher order correlation functions. These describe the relative distribution in space of groups of three or more atoms.

Though such a description would work in principle for a molecular fluid such as N_2 or Br_2, it is often useful in such a case to include the molecular unit, from the outset, in the structural distribution functions. Thus, in such cases the structure could be expressed not only by giving the relative distributions of the centres of mass for groups of two or more molecules, but also distribution functions describing the relative orientations of the molecules in space. However, such a description would get into some difficulty if we could heat the fluid sufficiently for an appreciable fraction of the molecules to be dissociated, whereas the original (though more cumbersome) description of the monatomic fluid by multi-particle correlation functions will always work for liquids with only one type of atom.

As we shall see in detail later, the structure of a liquid mixture has to be expressed in terms of partial distribution functions for the various types of atomic or molecular constituents.

In this chapter the multi-particle correlation functions for a monatomic classical liquid will be introduced and their relation to its radial distribution function discussed. Two essential points then emerge. First, a theoretical determination of the radial distribution function itself from an assumed pair force law requires detailed knowledge of the three-particle, or triplet, correlation function $g^{(3)}$. Indeed, as discussed below, the various theories of liquid structure can be classified according to the approximations they make to relate $g^{(3)}$ and g. Secondly, measurements of the structure factor, when the fluid is subject to an external perturbation, such as high pressure, allow experimental information to be obtained on some properties of the triplet function.

Before defining three-particle and higher order correlation functions, which will permit a first principles calculation of the liquid structure from a force law, we want to emphasise that there is an experimental route by which the structure can be obtained from the forces. This is the method of 'computer simulation', which has had a major impact on our understanding of the liquid state. Since it is not the object in this book to discuss details of experiments, we shall confine ourselves to the role of computer simulation in throwing light on the validity of the approximate theories of liquid structure discussed below. We stress that the merit of the experimental data supplied by computer simulation for the structure, is that we are given precisely that structure which comes from a pair force law. In real liquids, of course, additional questions as to the role of many-body forces arise and prevent a completely clear-cut comparison with structural theories. In the more powerful of the two methods available for computer simulation, that of molecular dynamics, one simply tells the computer to solve Newton's equation of motion for a system of particles in a box, interacting via a given pair force law. To avoid surface effects and give equal weight to all particles, it is customary to impose periodic boundary conditions. As a starting point of the calculation one can, for example, place the particles on a face-centred cubic lattice and give them random displacements (for details see Croxton, 1974 or Schofield, 1973). While this method is essential for dynamical properties (e.g. calculating the diffusion constant), Monte Carlo methods suffice for a determination of the liquid structure from a force law. We shall confront, below, the assumptions underlying the approximate theories of structure with the predictions of such computer experiments.

2.1 Multi-particle correlation functions

The procedure followed in section 1.3 in defining the singlet and the pair distribution function in the grand canonical ensemble may be generalised without difficulty to distribution functions for sets of s particles. We consider again a member of the grand ensemble in which an open region of volume V contains N particles, and write the density $\rho^{(s)}(\boldsymbol{r}_1, \ldots, \boldsymbol{r}_s)$ of groups of s particles at points $\boldsymbol{r}_1, \ldots, \boldsymbol{r}_s$, for a given configuration $\boldsymbol{R}_1, \ldots, \boldsymbol{R}_N$ of the N particles, as

$$\rho^{(s)}(\boldsymbol{r}_1, \ldots, \boldsymbol{r}_s) = \sum_{i_1, \ldots, i_s = 1}^{N}{}' \delta(\boldsymbol{R}_{i_1} - \boldsymbol{r}_1) \cdots \delta(\boldsymbol{R}_{i_s} - \boldsymbol{r}_s) \tag{2.1}$$

where the primed sum is restricted to the case where the indices $i_1, \ldots, i_s$ are all different. The corresponding distribution function is obtained by averaging the density (2.1) with the probability distribution of the grand canonical ensemble, given in eqn (1.20)

Let us carry out the average explicitly for the pair function, in order to see the structure of the general result. For a classical liquid we can carry out the integration over momentum space independently of the integration over the coordinates, when we have

$$\langle\rho^{(2)}(\boldsymbol{r}_1, \boldsymbol{r}_2)\rangle \equiv \rho^2 g(r_{12})$$
$$= \sum_{N=2}^{\infty} \frac{\exp[(\Omega + N\mu)/k_B T]}{(2\pi\hbar)^{3N} N!} \int \cdots \int \mathrm{d}p_1 \ldots \mathrm{d}p_N \, \mathrm{d}\boldsymbol{R}_1 \cdots \mathrm{d}\boldsymbol{R}_N$$
$$\times \exp(-H_N/k_B T) \sum_{i \neq j = 1}^{N} \delta(\boldsymbol{R}_i - \boldsymbol{r}_1)\delta(\boldsymbol{R}_j - \boldsymbol{r}_2)$$
$$= \sum_{N=2}^{\infty} \frac{\exp[(\Omega + N\mu)/k_B T)}{N!} \left(\frac{mk_B T}{2\pi\hbar^2}\right)^{3N/2} \int \cdots \int \mathrm{d}\boldsymbol{R}_1 \cdots \mathrm{d}\boldsymbol{R}_N$$
$$\times \exp(-\Phi_N/k_B T) \sum_{i \neq j = 1}^{N} \delta(\boldsymbol{R}_i - \boldsymbol{r}_1)\delta(\boldsymbol{R}_j - \boldsymbol{r}_2) \tag{2.2}$$

Since the particles are identical we can choose two particular values for the indices i and j ($i=1$ and $j=2$, say) and multiply the result by the number of ordered pairs, which is $N(N-1)$. We can therefore write

$$\rho^2 g(r_{12}) = \frac{1}{\Xi} \sum_{N=2}^{\infty} \frac{z^N}{(N-2)!} \int \cdots \int \mathrm{d}\boldsymbol{R}_3 \cdots \mathrm{d}\boldsymbol{R}_N \exp[-\Phi_N(\boldsymbol{r}_1, \ldots, \boldsymbol{R}_N)/k_B T] \tag{2.3}$$

In these expressions, Φ_N is the potential energy for a set of N particles, Ξ is the grand partition function,

$$\Xi \equiv \exp(-\Omega/k_B T) = \sum_{N=0}^{\infty} \frac{z^N}{N!} \int \cdots \int \mathrm{d}\boldsymbol{R}_1 \cdots \mathrm{d}\boldsymbol{R}_N \exp[-\Phi_N(\boldsymbol{R}_1, \ldots, \boldsymbol{R}_N)/k_B T) \tag{2.4}$$

and z is a thermodynamic property of the system,

$$z = \left(\frac{mk_B T}{2\pi\hbar^2}\right)^{3/2} \exp(\mu/k_B T) \tag{2.5}$$

known as its fugacity. This is to be determined as a function of temperature and density from the requirement that $\langle N\rangle/V$ in the thermodynamic limit ($V \to \infty$) coincides with the observed density of the fluid under the prevailing conditions of temperature and pressure.

The obvious extension of eqn (2.3) to the s-particle distribution function is

$$\langle\rho^{(s)}(\boldsymbol{r}_1, \ldots, \boldsymbol{r}_s)\rangle \equiv \rho^s g^{(s)}(\boldsymbol{r}_1, \ldots, \boldsymbol{r}_s)$$
$$= \frac{1}{\Xi} \sum_{N=s}^{\infty} \frac{z^N}{(N-s)!} \int \cdots \int \mathrm{d}\boldsymbol{R}_{s+1} \cdots \mathrm{d}\boldsymbol{R}_N$$
$$\exp[-\Phi_N(\boldsymbol{r}_1, \ldots, \boldsymbol{R}_N)/k_B T] \tag{2.6}$$

Of course, the properties of translational and rotational invariance of the fluid imply that $g^{(s)}$ depends explicitly on fewer variables than the s vectors $\boldsymbol{r}_1, \ldots, \boldsymbol{r}_s$

(for example, $g^{(3)}$ depends only on the lengths of two vectors and the angle between them).

Obviously we expect that if one of the s particles is taken to an infinite distance from the others, the s-particle function should reduce to the $(s-1)$-particle function. It is useful to isolate the s-particle correlations by subtracting from $g^{(s)}$ such asymptotic behaviours, which are already contained in the distribution functions of lower order. In particular, for a pair of particles it is useful to introduce the total correlation (or 'hole') function $h(r)$ as

$$h(r) = g(r) - 1 \tag{2.7}$$

Similarly, the total correlation function for triplets is defined as

$$t(\boldsymbol{r}_1, \boldsymbol{r}_2, \boldsymbol{r}_3) = g^{(3)}(\boldsymbol{r}_1, \boldsymbol{r}_2, \boldsymbol{r}_3) - g(r_{12}) - g(r_{23}) - g(r_{31}) + 2 \tag{2.8}$$

Such functions describe intrinsic s-body correlations in that they vanish if any one particle of the set is taken to infinity.

2.2 Force equation and theories of structure

A relation between the radial distribution function and higher order correlations is obtained by taking the gradient of $g(r_{12})$ with respect to the position of one particle. We assume that the particles interact via a two-body, central potential, $\phi(|\boldsymbol{R}_1 - \boldsymbol{R}_2|)$, and we write the potential energy Φ_N explicitly in the form

$$\Phi_N(\boldsymbol{R}_1, \ldots, \boldsymbol{R}_N) = \phi(R_{12}) + \sum_{i=3}^{N} \phi(R_{1i}) + \text{(terms independent of } R_1) \tag{2.9}$$

We then find from eqn (2.3)

$$\rho^2 \nabla_{r_1} g(r_{12}) = -\frac{1}{k_B T \Xi} \sum_{N=2}^{\infty} \frac{z^N}{(N-2)!} \int \cdots \int \mathrm{d}\boldsymbol{R}_3 \cdots \mathrm{d}\boldsymbol{R}_N \exp(-\Phi_N/k_B T) \left[\nabla_{r_1} \phi(r_{12}) + \sum_{i=3}^{N} \nabla_{r_1} \phi(|\boldsymbol{r}_1 - \boldsymbol{R}_i|) \right] \tag{2.10}$$

The sum in the square bracket contributes $(N-2)$ equal terms, whose value can be determined by taking $i=3$, say. By the use of eqn (2.6) for $s=3$ we then have

$$\nabla_{r_1} g(r_{12}) = -\frac{g(r_{12})}{k_B T} \nabla_{r_1} \phi(r_{12}) - \frac{\rho}{k_B T} \int \mathrm{d}\boldsymbol{r}_3 g^{(3)}(\boldsymbol{r}_1, \boldsymbol{r}_2, \boldsymbol{r}_3) \nabla_{r_1} \phi(r_{13}) \tag{2.11}$$

It is worth while to rewrite this equation by expressing $g(r_{12})$ in the form

$$g(r_{12}) = \exp[-U(r_{12})/k_B T] \tag{2.12}$$

where $U(r_{12})$ is playing the role of a 'potential of mean force' and is, of course, defined by equation (2.12). Then, eqn (2.11) immediately takes the form

$$-\nabla_{r_1} U(r_{12}) = -\nabla_{r_1}\phi(r_{12}) - \rho \int \mathrm{d}\boldsymbol{r}_3 \frac{g^{(3)}(\boldsymbol{r}_1, \boldsymbol{r}_2, \boldsymbol{r}_3)}{g(r_{12})} \nabla_{r_1}\phi(r_{13}) \tag{2.13}$$

In this form, the physical interpretation is quite clear. The left-hand side represents the total force on particle 1 when another atom is distant r_{12} from it and on the right-hand side this is split into two parts (see also Cole, 1967):

(1) From the direct interaction between particles 1 and 2.
(2) From the interaction between atom 1 and a third atom at $\boldsymbol{r}_3$, the force $\nabla_{r_1} \phi(r_{13})$ being simply weighted with the factor $g^{(3)}(\boldsymbol{r}_1, \boldsymbol{r}_2, \boldsymbol{r}_3)/g(r_{12})$ giving the probability of finding particle 3 at $\boldsymbol{r}_3$, given that there are certainly particles at $\boldsymbol{r}_1$ and $\boldsymbol{r}_2$.

Because of this physical interpretation we shall refer to eqn (2.13) as the 'force equation'. It provides us with a starting point for a theory of liquid structure, by which we mean a theory which allows the calculation of $g(r)$ from a given pair potential $\phi(r)$.

The presence of $g^{(3)}(\boldsymbol{r}_1, \boldsymbol{r}_2, \boldsymbol{r}_3)$, the three-atom correlation function, means that approximations have to be made. However, it is worth examining $g^{(3)}$ a little further in order to isolate which part is actually involved in the relation between ϕ and g. To do this, it is useful to take the scalar product of $(\boldsymbol{r}_2 - \boldsymbol{r}_1)$ with eqn (2.11). Then we find

$$-r_{12}\frac{\partial g(r_{12})}{\partial r_{12}} - \frac{r_{12}}{k_B T} g(r_{12}) \frac{\partial \phi(r_{12})}{\partial r_{12}}$$

$$= \frac{\rho}{k_B T} \int \mathrm{d}\boldsymbol{r}_3 g^{(3)}(\boldsymbol{r}_1, \boldsymbol{r}_2, \boldsymbol{r}_3) r_{12} \cos\vartheta \frac{\partial \phi(r_{13})}{\partial r_{13}} \tag{2.14}$$

where ϑ is the angle between $(\boldsymbol{r}_3 - \boldsymbol{r}_1)$ and $(\boldsymbol{r}_2 - \boldsymbol{r}_1)$. It will be convenient to simplify the notation by introducing $r_{12} = s$, $r_{13} = t$. Furthermore, $g^{(3)}$ will only depend on s, t and $\cos\vartheta$, for, in a homogeneous, isotropic fluid, it cannot depend on the choice of origin nor on the orientation of the triplet configuration in space. Exploiting this fact, we may write

$$g^{(3)}(\boldsymbol{r}_1, \boldsymbol{r}_2, \boldsymbol{r}_3) = g^{(3)}(s, t, \cos\vartheta)$$

$$= \sum_{l=0}^{\infty} Q_l(s, t) P_l(\cos\vartheta) \tag{2.15}$$

where the quantities $Q_l(s, t)$, as indicated, are now independent of $\cos\vartheta$, after the expansion in Legendre polynomials $P_l(\cos\vartheta)$. If we now insert (2.15) in (2.14), then because of the presence of $\cos\vartheta$ ($\equiv P_1(\cos\vartheta)$), only the $l=1$ term in (2.15) contributes (see for example Hutchinson, 1967). This shows that much less than complete knowledge of the three-body function is required in relating *structure* and *forces*. In particular, we could add any terms with $l=0$ to $g^{(3)}$ in

(2.14) and not affect the final equation. We shall use this fact below, in discussing approximate theories. Before doing so, it is of interest in view of the presence of $g^{(3)}$ in the force equation (2.13) to discuss how far the three-particle function is accessible to experimental determination.

2.3 Pressure dependence of structure factor $S(k)$ and three-atom correlation function

We have derived the force equation in section 2.2 by differentiating $g(r)$ with respect to r at constant density and temperature. Other relations between the pair distribution function and higher order correlation functions can be found by differentiating $g(r)$ with respect to density or temperature (Schofield, 1966). These relations imply the possibility of determining experimentally some properties of higher order correlations by diffraction experiments under different thermodynamic conditions, a possibility which has been exploited for liquid argon and liquid rubidium by Egelstaff, Page and Heard (1969 and 1971) using data from diffraction experiments under pressure.

Let us then consider, in particular, the density dependence of the pair function. In the grand ensemble the dependence of $g(r)$ on density at constant temperature is only in the fugacity z, and we can write

$$\left(\frac{\partial[\rho^2 g(r)]}{\partial\rho}\right)_T = V\left(\frac{\partial[\rho^2 g(r)]}{\partial\langle N\rangle}\right)_T = V\left(\frac{\partial[\rho^2 g(r)]}{\partial z}\right)_T \bigg/ \left(\frac{\partial\langle N\rangle}{\partial z}\right)_T \tag{2.16}$$

From eqn (2.3) we find

$$z\left(\frac{\partial[\rho^2 g(r)]}{\partial z}\right)_T = \Xi^{-1}\sum_{N=2}^{\infty}(N-\langle N\rangle)\frac{z^N}{(N-2)!}\int\cdots\int \mathrm{d}\boldsymbol{R}_3\cdots\mathrm{d}\boldsymbol{R}_N$$
$$\exp\left[-\Phi_N(\boldsymbol{r}_1,\ldots,\boldsymbol{R}_N)/k_BT\right] = \langle N\rho^{(2)}(\boldsymbol{r}_1,\boldsymbol{r}_2)\rangle - \langle N\rangle\langle\rho^{(2)}(\boldsymbol{r}_1,\boldsymbol{r}_2)\rangle \tag{2.17}$$

which evidently represents the fluctuation in the product of N with the pair density. After some manipulation the expression above is easily rewritten in terms of the three-atom function defined in eqn (2.6),

$$z\left(\frac{\partial[\rho^2 g(r)]}{\partial z}\right)_T = 2\rho^2 g(r) + \rho^3\int \mathrm{d}\boldsymbol{r}_3[g^{(3)}(\boldsymbol{r}_1,\boldsymbol{r}_2,\boldsymbol{r}_3) - g(r)] \tag{2.18}$$

From eqn (2.5) we also have

$$\frac{z}{V}\left(\frac{\partial\langle N\rangle}{\partial z}\right)_T = \frac{z}{V}\left(\frac{\partial N}{\partial\mu}\right)_T\left(\frac{\partial\mu}{\partial z}\right)_T = \frac{k_BT}{V}\left(\frac{\partial N}{\partial\mu}\right)_T = \rho k_BT\left(\frac{\partial\rho}{\partial p}\right)_T \tag{2.19}$$

and we arrive then at the result

$$\left(\frac{\partial[\rho^2 g(r)]}{\partial p}\right)_T = \left(\frac{\partial[\rho^2 g(r)]}{\partial \rho}\right)_T \Big/ \left(\frac{\partial p}{\partial \rho}\right)_T$$

$$= (k_B T)^{-1}\{2\rho g(r) + \rho^2 \int d\boldsymbol{r}_3 [g^{(3)}(\boldsymbol{r}_1, \boldsymbol{r}_2, \boldsymbol{r}_3) - g(r)]\} \qquad (2.20)$$

Finally, the use of eqn (1.27) allows us to rewrite the expression above in its most transparent form,

$$k_B T\left(\frac{\partial g(r)}{\partial p}\right)_T = \int d\boldsymbol{r}_3 [g^{(3)}(\boldsymbol{r}_1, \boldsymbol{r}_2, \boldsymbol{r}_3) - g(r)g(r_{23}) - g(r)g(r_{31}) + g(r)] \qquad (2.21)$$

Thus we see that the experimental study of the pressure dependence of the pair function gives information on the three-body correlations, integrated over all values of the coordinates of the third particle. The appearance of the various pair functions in the square bracket in eqn (2.21) implies that only the behaviour of the triplet function over the configurations in which the third particle is close to the other two particles is relevant.

A brief discussion of the temperature dependence of correlation functions and of the hierarchy of relations between s-particle and $(s+1)$-particle correlation functions is given in appendix 2.1.

2.4 Tests of primitive approximations to triplet correlation function

2.4.1 Superposition approximation

To make any further progress with the theory it is essential to relate the triplet function $g^{(3)}$ to the pair function $g(r)$. This can be done, by now, at various levels of sophistication.

Still, one of the useful approaches is to build from the so-called superposition approximation of Kirkwood (1935). By assuming, say, the measured form of $g(r)$, pair correlations can be treated exactly but $g^{(3)}$ is then built up as a product of pair terms†, namely

$$g^{(3)}(\boldsymbol{r}_1, \boldsymbol{r}_2, \boldsymbol{r}_3) = g(r_{12})g(r_{23})g(r_{31}) \qquad (2.22)$$

This is the basis of the Born–Green (1946) theory of structure which will be discussed in more detail later.

2.4.2 Experimental test

Our purpose at present is to confront the experiments referred to above on the pressure dependence of $S(k)$ with the predictions resulting from the superposi-

†The term superposition arises from a generalisation of the potential of mean force $U(r)$, in eqn (2.12), to describe $g^{(3)}$. The additivity of U_3 obviously leads to products of g's.

tion approximation. Inserting (2.22) in (2.21), leads to the relation

$$k_B T\left(\frac{\partial g(r)}{\partial p}\right)_T = g(r)\int d\mathbf{r}_3[g(r_{23})-1][g(r_{31})-1] \tag{2.23}$$

This has been tested by Egelstaff, Page and Heard (1971) using their own data on liquid rubidium near the triple point and the data of Mikolaj and Pings (1967) for liquid argon near the critical point (see *figure 2.1*).

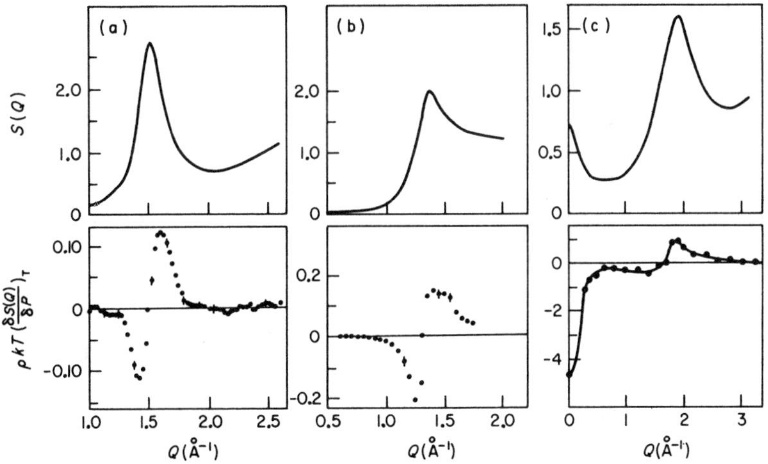

Figure 2.1 Experimental structure factors and isothermal pressure derivatives for: (*a*) rubidium at 333 K and 420 atm: (*b*) carbon tetrachloride at 296 K and 420 atm; (*c*) argon at 143 K and 52·5 atm (from Egelstaff, Page and Heard, 1971)

Our main interest until chapter 9 is with liquids near the triple point. Therefore, we simply comment here that for the critical point data on liquid argon (see *figure 2.2*) the superposition approximation is in reasonable agreement with the data except at small wave vectors where the experimental data for $(\partial S(k)/\partial p)_T$ show a large negative dip.† This can be interpreted as indicating that the triplet function is intrinsically longer ranged than the pair function, at the critical point.

But more important for our present purposes is the fact that the superposition approximation differs very significantly from the experimental data on liquid rubidium near the triple point (see *figure 2.3*). On the other hand, a modification of the Born–Green theory given in section 2.5.2, and known as the hypernetted-chain (HNC) approximation, gives very reasonable results, even though it fails to account for the shift in the main peak of the structure factor with the applica-

†This statement also applies to two other approximate theories, the Percus–Yevick and hypernetted-chain theories (see sections 2.5.2 and 2.6.3 below).

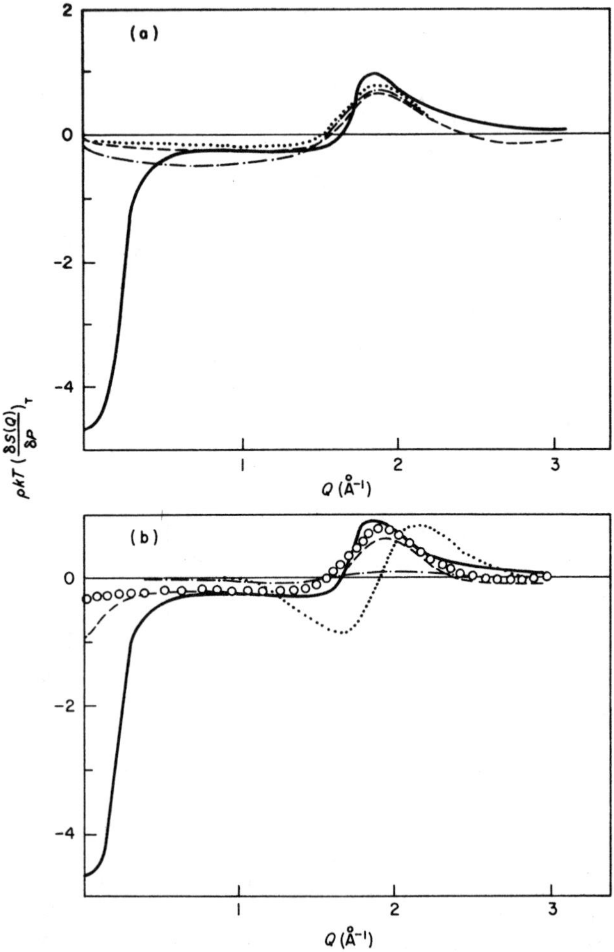

Figure 2.2 Comparison between experiment (full curves) and theories for the pressure dependence of the structure factor in liquid argon near the critical point (from Egelstaff, Page and Heard, 1971). The superposition, Abe and Percus–Yevick approximations are represented, respectively, by the broken curve in part (*a*), by the circles in part (*b*), and by the broken curve in part (*b*)

tion of pressure (see part (*b*) of *figure 2.3*). This shift arises through the rearrangement of the near-neighbour atoms and is most simply accounted for by the uniform compression of the liquid under application of pressure.

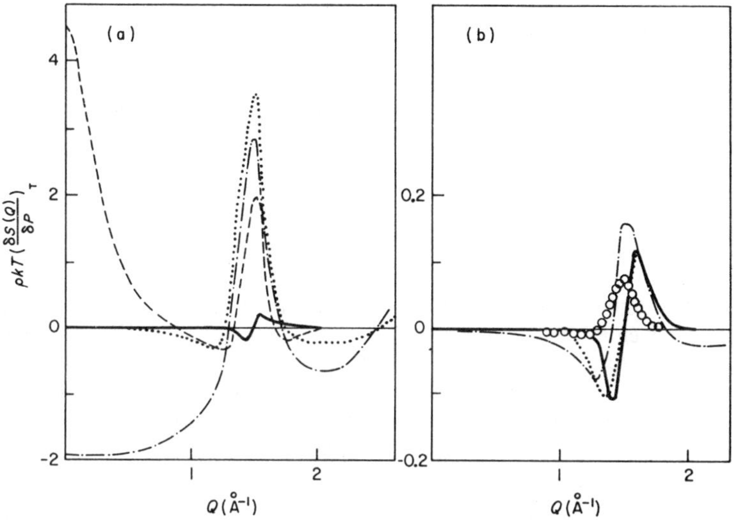

Figure 2.3 Comparison between experiment (full curves) and theories for the pressure dependence of the structure factor in liquid rubidium near the triple point (from Egelstaff, Page and Heard, 1971). The superposition approximation and the hypernetted-chain approximation are represented by the broken curve in part (*a*) and by the circles in part (*b*) respectively

2.4.3 Computer simulation test

To supplement the above discussion we record here briefly some results obtained by computer simulation on hard spheres and on liquid argon.

Alder (1964) has thereby shown that for configurations of three atoms in a fluid of hard spheres which are equilateral triangles, $g^{(3)} = [g(r)]^3$ to high accuracy.

This result has been confirmed for liquid argon at various temperatures by Krumhansl and Wang (1972), who demonstrate that the deviations are well within 10% over a limited range of r. For less symmetrical triplet configurations, they report up to 20% deviations from superposition at the lowest temperatures, such deviations correlating with the structure in $g(r)$ itself.

Two comments need making on the relation between these findings and those of Egelstaff, Page and Heard (1971):

(1) Though superposition is often numerically a good approximation to $g^{(3)}$, great care is obviously needed in deciding in what physical situations to use it.

(2) The pressure dependence of $g(r)$ is easily seen to depend only on the s term in the Legendre polynomial expansion (2.15), whereas the force equation depends on the p ($l = 1$) term. Thus, no direct deduction about the validity of the

Born–Green theory of structure follows from the pressure experiments, although obviously caution is indicated in the application to metals.

Nevertheless, the three simple theories of structure—Born–Green, Percus–Yevick and HNC—are still of sufficient interest to justify a fuller discussion, to which we now turn.

2.5 Approximate theories of structure

2.5.1 Born–Green theory

In utilising the superposition approximation (2.22) in the force equation (2.14), it should be noted that only the term $g(r_{23})$ depends on cos ϑ, and because of the argument given at the end of section 2.2 we can subtract a spherical part and therefore substitute $g(r_{12})g(r_{31})[g(r_{23})-1]$ in place of $g^{(3)}$ in (2.14), without changing the result. This latter substitution will be convenient, because the total correlation function, $h(r)=g(r)-1$, is a well-behaved function, tending to zero at infinity. Thus we find, after a simple calculation from (2.14),

$$\frac{\mathrm{d}}{\mathrm{d}s}\left[\ln g(s)+\frac{\phi(s)}{k_{\mathrm{B}}T}\right]=-\frac{\rho}{k_{\mathrm{B}}T}\int \mathrm{d}\boldsymbol{r}_3 h(r_{23})g(t)\,\frac{t_z}{t}\,\phi'(t) \tag{2.24}$$

where t_z is the resolved part of t on s. Integration over s yields

$$\ln g(s)+\frac{\phi(s)}{k_{\mathrm{B}}T}=\frac{\rho}{k_{\mathrm{B}}T}\int_s^\infty \mathrm{d}x\int \mathrm{d}\boldsymbol{r}_3 h(r_{23})g(t)\,\frac{t_z}{t}\,\phi'(t) \tag{2.25}$$

By adopting now, as integration variables, the vector $\boldsymbol{r}_{32}\equiv\boldsymbol{r}$ and the length t, and noticing that a change dx in the length x at constant $\boldsymbol{r}$ is equivalent to a change $t\,\mathrm{d}t/t_z$ in the length t, we can write this equation in the form (see, for instance Rushbrooke, 1960)

$$\ln g(s)+\frac{\phi(s)}{k_{\mathrm{B}}T}=\rho\int E(|\boldsymbol{r}-\boldsymbol{s}|)h(r)\,\mathrm{d}\boldsymbol{r} \tag{2.26}$$

where

$$E(t)=\frac{1}{k_{\mathrm{B}}T}\int_t^\infty g(x)\phi'(x)\,\mathrm{d}x \tag{2.27}$$

This is the Born–Green (1946) equation, which gives us an explicit integral equation connecting the structure $g(r)$ and the pair potential $\phi(r)$.

We shall consider some properties of the solution of this equation below, but for the moment let us comment on two properties of $E(t)$ which will be useful later:

(1) At sufficiently large t, we can evidently replace $g(x)$ by unity in (2.27) when

we immediately obtain

$$E(t) \sim -\frac{\phi(t)}{k_B T} \tag{2.28}$$

(2) We can calculate the integral of $E(t)$ over the volume of the fluid, in terms of the fluid pressure p. Defining the Fourier transform $\tilde{E}(k)$ of $E(t)$ by

$$\tilde{E}(k) = \rho \int \mathrm{d}\boldsymbol{t} \exp(-i\boldsymbol{k} \cdot \boldsymbol{t})\, E(t) \tag{2.29}$$

then evidently

$$\tilde{E}(0) = \rho \int \mathrm{d}\boldsymbol{t}\, E(t) \tag{2.30}$$

and substituting from (2.27) we may write

$$\tilde{E}(0) = \rho \int \mathrm{d}\boldsymbol{r} \int_0^\infty \mathrm{d}t H(r, t) g(t) \frac{\phi'(t)}{k_B T} \tag{2.31}$$

where

$$H(r, t) = \begin{cases} 1, & t > r \\ 0, & t < r \end{cases} \tag{2.32}$$

We can now interchange the order of integration and we find

$$\int H(r, t)\, \mathrm{d}\boldsymbol{r} = \int_0^t 4\pi r^2\, \mathrm{d}r = \frac{4\pi t^3}{3} \tag{2.33}$$

and hence

$$\tilde{E}(0) = \frac{4\pi\rho}{3k_B T} \int_0^\infty g(t)\phi'(t) t^3\, \mathrm{d}t \tag{2.34}$$

But we saw in section 1.2, eqn (1.15), that the fluid pressure involved this integral and we find (Gaskell, 1965; see also p. 25 and p. 31)

$$\tilde{E}(0) = 2\left[1 - \frac{p}{\rho k_B T}\right] \tag{2.35}$$

For a perfect gas, with $p = \rho k_B T$, we see that $\tilde{E}(0)$ vanishes, which is evidently correct from (2.31) when $\phi' = 0$.

Before discussing the use of these results in an explicit asymptotic solution of the Born–Green equation for van der Waals interactions, we shall discuss a further approximate equation of liquid state theory, given by Abe (1958) and many other workers.

2.5.2 Abe's approximate form of the Born–Green theory (hypernetted-chain theory)

To see how Abe's method results from the Born–Green equation we rewrite $E(r)$ from (2.27) in the form

$$E(r)=\frac{1}{k_B T}\int_r^\infty \mathrm{d}s\, g(s)\frac{\mathrm{d}}{\mathrm{d}s}[\phi(s)-U(s)]+\frac{1}{k_B T}\int_r^\infty \mathrm{d}s\, g(s)\frac{\mathrm{d}U}{\mathrm{d}s} \tag{2.36}$$

where $U(s)=-k_B T\ln g(s)$. Then, the last term can be integrated explicitly and is simply the total correlation function $h(r)$. Thus we find

$$E(r)=\frac{1}{k_B T}\int_r^\infty \mathrm{d}s\, g(s)\frac{\mathrm{d}}{\mathrm{d}s}[\phi(s)-U(s)]+h(r) \tag{2.37}$$

Abe's approximation is now obtained (see Gaskell, 1966) by putting $g(s)$ in the integral term in (2.37) as unity, that is, replacing it by its asymptotic value. Such a procedure would seem to lead to a less accurate theory than the Born–Green theory, although the possibility exists that this second approximation could counteract the superposition assumption of the Born–Green theory. We shall see that, in at least one respect, this latter circumstance seems to exist! Then, we replace the function $E(r)$ of (2.27) by $c(r)$, say, defined by

$$c(r)=\frac{1}{k_B T}[U(r)-\phi(r)]+h(r) \tag{2.38}$$

This leads to the second approximate equation of structure theory, namely,

$$\ln g(s)+\frac{\phi(s)}{k_B T}=\rho\int \mathrm{d}\boldsymbol{r}\, c(|\boldsymbol{r}-\boldsymbol{s}|)h(r) \tag{2.39}$$

This equation and the Born–Green theory are so similar in structure that we can apply essentially the same method of solution, usually numerical, to both. Before discussing these equations further, it will be convenient to deal with the physical significance of the quantity $c(r)$ above and to introduce the Percus–Yevick equation

2.6 Ornstein–Zernike direct correlation function and Percus–Yevick theory

Let us substitute for $U-\phi$ from (2.38) into the left-hand side of (2.39) when we find

$$h(r)=c(r)+\rho\int c(|\boldsymbol{r}-\boldsymbol{r}'|)h(\boldsymbol{r}')\,\mathrm{d}\boldsymbol{r}' \tag{2.40}$$

We see that this equation relates $c(r)$ to $h(r)$ for a given density and we take it as the fundamental definition of $c(r)$, independently of approximate theories.

Clearly, if we measure $S(k)$ by X-ray or neutron scattering and hence get $h(r)$, we can obtain $c(r)$ directly from experiment. The function $c(r)$, as defined by (2.40), was first introduced by Ornstein and Zernike in connection with critical fluctuations. We shall see in chapter 9 that it does indeed play a central role in any discussion of critical phenomena. Most usually, $c(r)$ is referred to nowadays as the *direct* correlation function, for reasons we shall briefly discuss at this point.

From (2.26) we see that the 'potential of mean force' U is split into a direct part ϕ and a convolution of E and h. Asymptotically, as we have seen, $E \backsim \phi$ and $h \sim U$, and if we make these replacements we see that (2.40) and the Born–Green equation (2.26) have then the same form. We are then, in defining c, splitting the total correlation function h into a direct part c and an indirect part. Strictly by analogy with the force equation, some three-body correlation function should be involved and we can expect $c(r)$ to have a simple physical significance, at most, asymptotically.

From the Abe theory, we then find

$$h - c = -\frac{U - \phi}{k_B T} \tag{2.41}$$

and hence, for large r,

$$c(r) \backsim -\frac{\phi(r)}{k_B T} \tag{2.42}$$

provided $h^2 < |c|$. This is true well away from the critical point. Since $c(r)$ is, in fact, defined in terms of $h(r)$, we can calculate it from the measured structure factor. Taking the Fourier transform of (2.40) we find

$$\tilde{h}(k) = \tilde{c}(k) + \tilde{h}(k)\tilde{c}(k) \tag{2.43}$$

or

$$\tilde{c}(k) = \frac{\tilde{h}(k)}{1 + \tilde{h}(k)} = \frac{S(k) - 1}{S(k)} \tag{2.44}$$

Thus the direct correlation function in $\boldsymbol{k}$ space is simply $1 - 1/S(k)$ but since $S(0)$ is typically 0.01–0.03 in liquid metals, the form of $c(k)$ is very different from $S(k)$, as can be seen by comparing *figure 1.1* for $S(k)$ with the form of $\tilde{c}(k)$ shown in *figure 2.4* for liquid thallium just above its melting point, as measured by neutron experiments.

Having discussed the Born-Green and the Abe (often called hypernetted-chain because of its connection with diagrammatic analysis) theories, and noting that they are most conveniently discussed in $\boldsymbol{r}$ space, we turn now to two further treatments which are more basically formulated in $\boldsymbol{k}$ space, although they will turn out to have a close relation to the Abe theory.

2.6.1 Density fluctuations and correlation functions

In order to reformulate the force equation (2.14) in $\boldsymbol{k}$ space, we introduce the

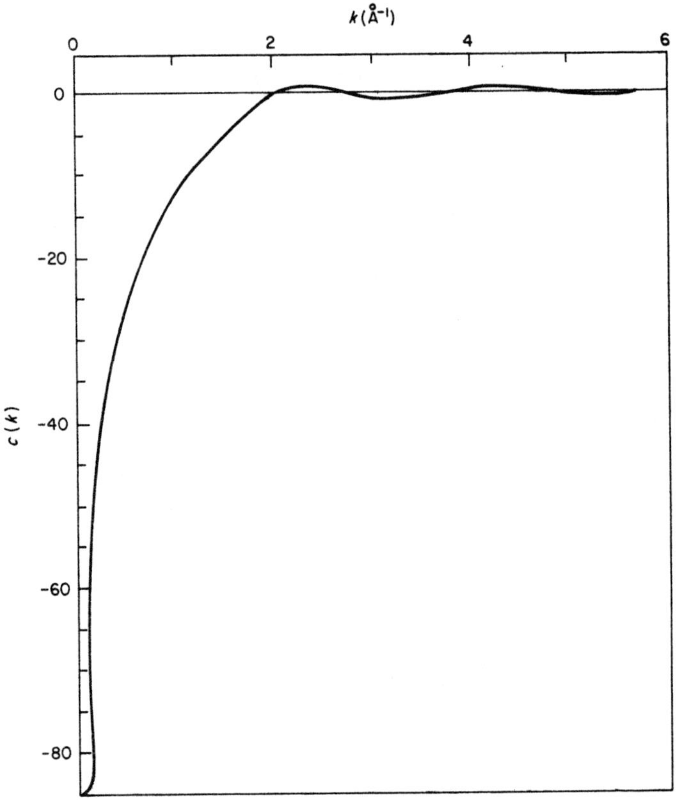

Figure 2.4 Direct correlation function $\tilde{c}(k)$ for liquid thallium at 600 K, as determined from the neutron-scattering data reported in *figure 1.1* (see Enderby and March, 1966)

Fourier components ρ_k of the density $\rho(\boldsymbol{r})$ defined in (1.16), which are

$$\rho_k = \sum_{i=1}^{N} \exp(i\boldsymbol{k} \cdot \boldsymbol{R}_i) \tag{2.45}$$

The variables ρ_k describe density fluctuations of given wave vector, and, since they involve the positions of all the atoms, they may be viewed as collective coordinates.

Let us now briefly consider how the two-particle distribution function $\rho^{(2)}(r) = \rho^2 g(r)$ may be expressed in terms of the density fluctuations. From (1.17) we can write, by use of the translational invariance of the fluid,

$$\rho g(r)=\frac{1}{N}\left\langle \sum_{i\neq j}\delta(\boldsymbol{r}-\boldsymbol{R}_i+\boldsymbol{R}_j)\right\rangle \tag{2.46}$$

We now show that the Fourier transform of this, the structure factor $S(k)$, is simply related to correlations between density fluctuations of given wave vector. To do this, we form

$$\rho_k\rho_{-k}=\sum_{i=1}^{N}\exp(i\boldsymbol{k}\cdot\boldsymbol{R}_i)\sum_{j=1}^{N}\exp(-i\boldsymbol{k}\cdot\boldsymbol{R}_j) \tag{2.47}$$

or

$$\rho_k\rho_{-k}-N=\sum_{i\neq j}\exp[i\boldsymbol{k}\cdot(\boldsymbol{R}_i-\boldsymbol{R}_j)] \tag{2.48}$$

Thus, by Fourier transform of (2.46), we obtain the result

$$S(k)=\frac{1}{N}\langle\rho_k\rho_{-k}\rangle \tag{2.49}$$

Likewise, we can obtain the higher order distribution functions in terms of the ρ_k. We now return to the exact force equation (2.14) and converting it into $\boldsymbol{k}$ space we find

$$S(k)=1+\frac{1}{Nk^2}\sum_{\boldsymbol{n}}\frac{\tilde{\phi}(n)}{k_BT}\langle\rho_{k+n}\rho_k\rho_n\rangle\,\boldsymbol{k}\cdot\boldsymbol{n} \tag{2.50}$$

where we have assumed, for the moment, that the pair potential $\phi(r)$ has a Fourier transform, given by

$$\tilde{\phi}(k)=\int \mathrm{d}\boldsymbol{r}\exp(i\boldsymbol{k}\cdot\boldsymbol{r})\,\phi(r) \tag{2.51}$$

This equation is exact for a potential that can be Fourier transformed. $\langle\rho_{k+n}\rho_k\rho_n\rangle$ involving three ρ_k's, comes from the three-body correlation function.

2.6.2 Random-phase approximation

As in the earlier theories based on the force equation, we must now approximate. Since we want to relate $S(k)$ and $\tilde{\phi}(k)$, we must somehow reduce $\langle\rho_{k+n}\rho_k\rho_n\rangle$ to $S(k)$ and, if we pick out from the sum the term $\boldsymbol{n}=-\boldsymbol{k}$, this is evidently related to $S(k)$. This is, in fact, the random-phase approximation which will be discussed more fully in chapter 7. The qualitative argument is that unless $\boldsymbol{k}$ and $\boldsymbol{n}$ bear this simple relation, destructive interference between the various oscillatory components will tend to annul the other terms. Actually, such an argument turns out to be a long wavelength approximation. We then find, collecting the terms in $S(k)$,

$$S(k)=\left[1+\frac{\rho\tilde{\phi}(k)}{k_BT}\right]^{-1} \tag{2.52}$$

Comparing this with the result (2.44), rewritten as

$$S(k)=[1-\tilde{c}(k)]^{-1} \tag{2.53}$$

we see that the direct correlation function is assumed in the random-phase approximation to have its asymptotic form

$$\tilde{c}(k)=-\rho\tilde{\phi}(k)/k_{B}T \tag{2.54}$$

or in $\boldsymbol{r}$ space

$$c(r)=-\phi(r)/k_{B}T \tag{2.55}$$

This is the same result as the Abe approximation yields for large r.

Actually, without going through the $\boldsymbol{k}$ space analysis, this same form arises from replacing $g^{(3)}$ inside the integration over $\boldsymbol{r}_3$ in the force equation by $h(r_{23})$. However, its basic theoretical justification for small $\boldsymbol{k}$ comes from the random-phase approximation.

2.6.3 Effective interatomic potential and the Percus–Yevick theory

The crippling limitation of the above approximation is that it assumes an interaction with a Fourier transform. For liquids, in general, the interaction has almost a hard core, and we cannot Fourier transform it. The Percus–Yevick (1958) method is an attempt to produce an effective potential which will replace the Fourier components $\tilde{\phi}(k)$ above.

The idea behind it is to use the $\rho_{\boldsymbol{k}}$ as collective coordinates. For them to afford an accurate approximation we must be able to express the Hamiltonian in terms of them and the corresponding momenta, and then to treat them as if they were 'almost' independent. There are $3N$ coordinates, $\boldsymbol{R}_1 \ldots \boldsymbol{R}_N$, in the original Hamiltonian and while, in a finite system with, say, periodic boundary conditions imposed over a large cube of side L, $\boldsymbol{k}$ has discrete though dense values, all such discrete $\boldsymbol{k}$ are allowed in enumerating the $\rho_{\boldsymbol{k}}$. Actually, we would transform naturally to centre of mass coordinates,

$$X=\frac{1}{N}\sum_{i=1}^{N}\boldsymbol{R}_i \tag{2.56}$$

and $(3N-3)$ $\rho_{\boldsymbol{k}}$†.

Suppose we consider the potential energy. Then, in (1.10), we found it in terms of $\phi(r)$ and $g(r)$ and by Fourier transform we get

$$\text{Potential energy}=\tfrac{1}{2}\sum_{\text{all}\,\boldsymbol{k}}\tilde{\phi}(k)\langle[\rho_{\boldsymbol{k}}\rho_{-\boldsymbol{k}}-N]\rangle \tag{2.57}$$

We can, at this stage, ask whether we can choose $\tilde{\phi}_{\text{eff}}(k)$ to get the 'best' possible approximation. There is ambiguity in this statement, but we might determine the

†Some difficulties arise from redundant variables as in, for example, collective coordinates theories of an electron gas (see Bohm and Pines, 1953). We need not elaborate this further for our present purposes.

effective potential

$$\phi_{\text{eff}}(\boldsymbol{r}) = \sum_{(3N-3)\boldsymbol{k}\text{'s}} \tilde{\phi}_{\text{eff}}(k) \exp(-i\boldsymbol{k}\cdot\boldsymbol{r}) \tag{2.58}$$

by requiring that the mean square difference

$$\left\langle \left\{ \phi(\boldsymbol{R}_i - \boldsymbol{R}_j) - \sum_{(3N-3)\boldsymbol{k}\text{'s}} \tilde{\phi}_{\text{eff}}(k) \exp[-i\boldsymbol{k}\cdot(\boldsymbol{R}_i - \boldsymbol{R}_j)] \right\}^2 \right\rangle \tag{2.59}$$

be a minimum. A first-order approximation to this indicates that the choice is

$$\tilde{\phi}_{\text{eff}}(k) = FT[\phi(r)g(r)] \tag{2.60}$$

where, by the right-hand side, we mean the Fourier transform of the product of $\phi(r)$ and $g(r)$. We see already that we have avoided the very strong repulsive potential inside the core by weighting the interaction with the probability of the occurrence of the pair of atoms i, j at separation $\boldsymbol{r} = \boldsymbol{R}_i - \boldsymbol{R}_j$.

Actually, since we wish to calculate structure, it is better to minimise averages of $\rho_{\boldsymbol{k}}\rho_{-\boldsymbol{k}}$ with respect to the exact distribution function (involving ϕ) and the approximation to it (involving ϕ_{eff}). We then find, after numerous approximations,

$$\tilde{\phi}_{\text{eff}}(k) = k_B T \, FT[g(r)\{\exp[\phi(r)/k_B T] - 1\}] \tag{2.61}$$

which clearly reduces to the earlier choice (2.60) if we take $\phi(r)/k_B T$ to be small. This is the choice of the $\tilde{\phi}_{\text{eff}}(k)$'s generally referred to as *the* Percus–Yevick approximation. Thus we have, from (2.54),

$$c(k) = -\rho FT[g(r)\{\exp[\phi(r)/k_B T] - 1\}] \tag{2.62}$$

and if we assume (doubtfully, because of the use of the random-phase approximation) that this is true for all k, then we may write finally,

$$c(r) = g(r)[1 - \exp[\phi(r)/k_B T]] \tag{2.63}$$

Clearly again†

$$c(r) \sim -\phi(r)/k_B T \tag{2.64}$$

for large r, and there is a direct equivalence with the Abe approximation if U and ϕ are small.

If we examine the asymptotic form of the Born–Green equation for the specific case of van der Waals interaction, then (2.64) is regained, but with a different multiplying constant (Gaskell, 1965). This is obviously a serious defect of the

†Sufficiently far from the critical point, the available evidence points to the fact that this is the *exact* asymptotic form. However, the authors know of rigorous proofs only in very special cases (e.g. Lebowitz and Percus, 1963 and other references given there).

Born–Green equation. Using the result (2.64) with $\phi(r) = -Ar^{-6}$, it is readily shown that the structure factor $S(k)$ has the small k form for van der Waals fluids (Enderby, Gaskell and March, 1965):

$$S(k) = S(0) + a_2k^2 + a_3k^3 + \cdots : \quad a_3 = \frac{\pi^2 \rho A [S(0)]^2}{12k_BT} \tag{2.65}$$

2.6.4 Direct correlation function for fluid argon

We are now in a position to construct an approximate $c(r)$ for fluid argon. To do so simply, we shall divide the pair potential $\phi(r)$ into two parts, which we shall call ϕ_{sr} and ϕ_{lr}. The short-range part we shall define as the part inside the atomic diameter σ, as shown in *figure 2.5*. If we take a Lennard–Jones (6–12) potential, to be quite specific, that is

$$\phi(r) = \frac{D}{r^{12}} - \frac{A}{r^6} \tag{2.66}$$

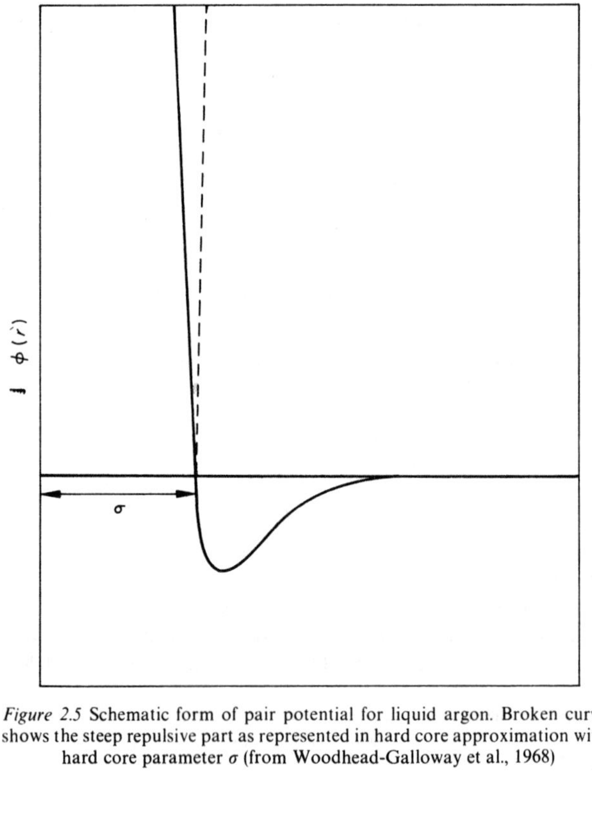

Figure 2.5 Schematic form of pair potential for liquid argon. Broken curve shows the steep repulsive part as represented in hard core approximation with hard core parameter σ (from Woodhead-Galloway et al., 1968)

then we define σ by $\phi(\sigma)=0$, and then $D=A\sigma^6$. However, for $r<\sigma$ we assume a rigid hard core, as shown by the dotted line in the figure. This is *not* essential, but we can now get an *exact* solution of the Percus–Yevick equation for hard spheres (Wertheim, 1963; Thiele, 1963). While this is an important result, we shall not derive it here, but refer to the account in Frisch and Lebowitz (1964) and to our subsequent discussion in section 6.5.1.

It is, in fact, immediately clear from (2.63) that

$$c_{hs}(r)=0 \quad \text{for} \quad r>\sigma \tag{2.67}$$

It turns out that $c_{hs}(r)$ is a polynomial inside σ and if we use σ as the unit of length and write $x=r/\sigma$ then

$$\begin{aligned} c_{hs}(x) &= A_0+A_1x+A_3x^3, \quad && x<1 \\ &= 0 \quad , && x>1 \end{aligned} \tag{2.68}$$

where A_0, A_1 and A_3 are functions of the packing fraction η given by

$$\eta=\tfrac{1}{6}\pi\rho\sigma^3 \tag{2.69}$$

Substituting in the Percus–Yevick equation, it can be shown that A_0, A_1 and A_3 are defined by

$$\begin{aligned} A_0 &= -(1+2\eta)^2(1-\eta)^{-4} \\ A_1 &= 6\eta(1+\tfrac{1}{2}\eta)^2(1-\eta)^{-4} \\ A_3 &= -\tfrac{1}{2}\eta(1+2\eta)^2(1-\eta)^{-4} \end{aligned} \tag{2.70}$$

$c(r)$ has then the form shown schematically in *figure 2.6* while, if we calculate the structure factor $S(k)$, we find, for a chosen value of η, the upper curve in *figure 2.7*. Actually, it seems likely that both the Born–Green and the Abe approximation would give similar results, but so far the corresponding equations have not been solved exactly. However, the lower part of *figure 2.7* shows density expansions of these theories and, except at small k, the results are quite similar. The other theories do not give $c(r)=0$ for $r>\sigma$ and, in fact, it may be shown that the exact $c(r)$ for hard spheres does not vanish. But we expect it to be small and we now turn to discuss the modified form the above theories suggest for $c(r)$ in fluid argon.

The form of $\tilde{c}_{hs}(k)/\tilde{c}_{hs}(0)$ is shown in *figure 2.8*, along with experimental results for fluid argon at 84 K. The model obviously has the general features of the experimental data.

2.6.5 Equation of state of fluid argon

We have seen in the Percus–Yevick theory that c_{hs} is zero outside σ. On the other hand, if we use the result (2.64) for the long-range part, we have immediately the form shown in *figure 2.6*, the tail directly reflecting the Lennard–Jones potential. This general form of $c(r)$ has been confirmed from scattering data by a

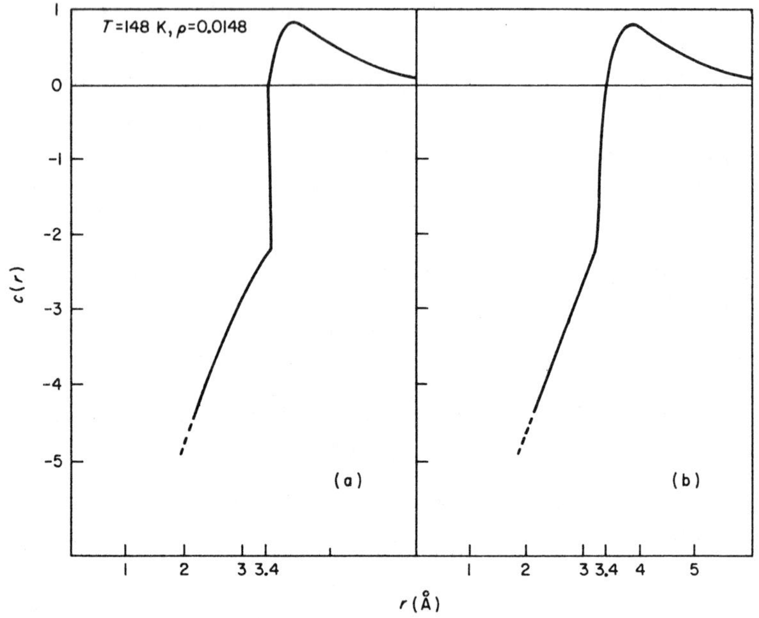

Figure 2.6 Direct correlation function $c(r)$ for a hard sphere fluid modified by an attractive potential tail (a) and after smoothing the singularities arising from the infinitely steep repulsive potential (from Woodhead-Galloway, Gaskell and March, 1968)

number of workers. We do not expect the step at σ, of course, in a real liquid, though we expect a rather steep rise which will lead to oscillations in $\tilde{c}(k)$ at large k.

One rather striking confirmation that this approach is quite appropriate to fluid argon comes from studying its equation of state at high density (i.e. near the triple point). Thus, Longuet-Higgins and Widom (1964) have shown that in this regime the equation of state is well represented by a modified van der Waals theory. In the van der Waals equation of state we have

$$p=\frac{\rho k_{\mathrm{B}}T}{1-b\rho}-a\rho^2 \tag{2.71}$$

where a and b are constants, independent of ρ and T. The first term is designed to take account of the 'finite size' of the molecules, i.e. it is the analogue of a hard sphere term, while the second term takes account of the attractive forces. In present terms, we want therefore to write

$$p=p_{\mathrm{hs}}(\rho, T)-a\rho^2 \tag{2.72}$$

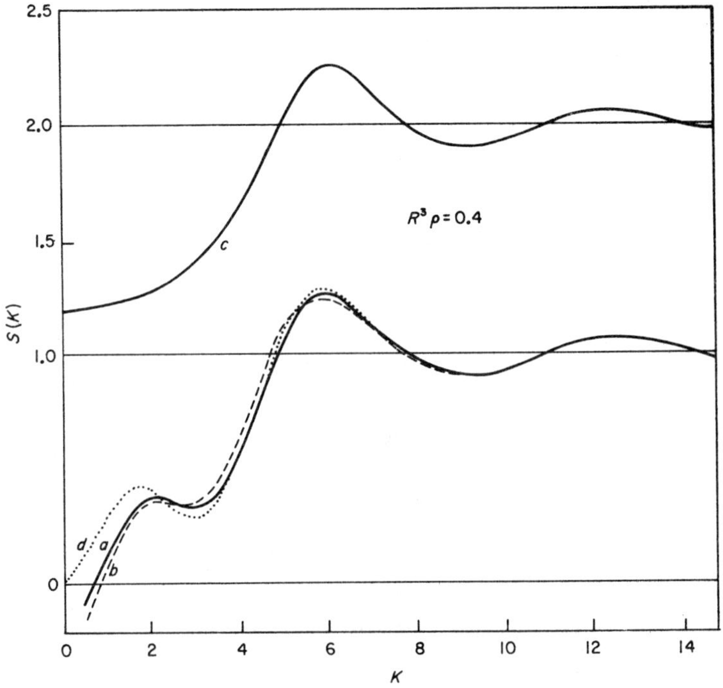

Figure 2.7 Static structure factor $S(k)$ for hard spheres with $\rho\sigma^3 = 0{\cdot}4$ (from Ashcroft and March, 1967). Upper curve—Percus–Yevick results: lower curves: results obtained by density expansion in (*a*) exact theory, (*b*) Born–Green theory, and (*d*) Abe theory

Longuet-Higgins and Widom show that this equation of state gives a quantitative account of the properties of fluid argon near the triple point. To show this, we reproduce in table 2.1 some results they obtained at the triple point, together with the corresponding experimental results.

Table 2.1 Properties of argon at triple point

	V_L/V_S	p	$\Delta S/Nk_B$	Cohesive energy
Theory	1.19	−5.9	1.64	−8.6
Experiment	1.11	−5.88	1.69	−8.53

The dimensionless quantities shown in table 2.1 are: first the ratio of the liquid and solid volumes at the triple point; second the pressure; third the entropy of fusion ΔS, in units of Nk_B; and fourth the cohesive energy of the liquid. Certain

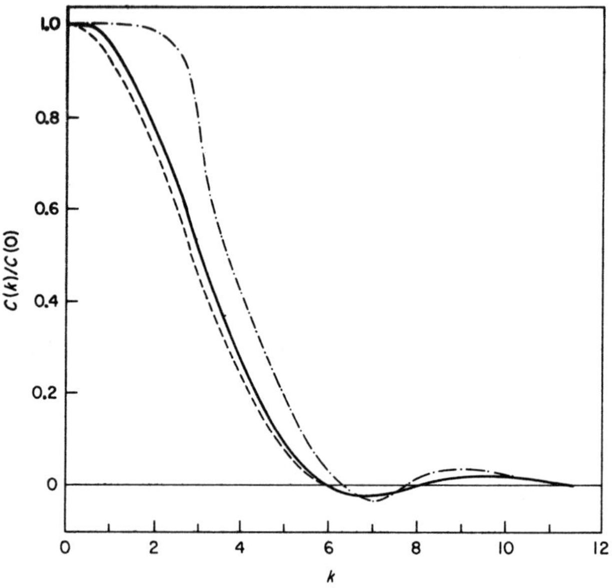

Figure 2.8 Direct correlation function in **k** space for fluid argon at 84 K. Dash-dot curve: experiment: other curves are slightly different ways of calculating hard sphere result (from Ashcroft and March, 1967)

second derivatives of the free energy are given less satisfactorily and, in particular, the configurational specific heat is zero. But the numbers above are given to illustrate that this is really a good equation of state for the high density fluid.

The link with our calculation of $c(r)$ may now be made by the fluctuation formula for $S(0)$, which relates $S(0)$ to the isothermal compressibility, eqn (1.27). From (2.55) we then have

$$\left(\frac{\partial p}{\partial \rho}\right)_T = \frac{k_B T}{S(0)} = k_B T[1 - \tilde{c}(0)] \tag{2.73}$$

Now we return to our results for $c(r)$, namely

$$c(r) \cong c_{hs}(r) + c_{lr}(r) \tag{2.74}$$

and

$$\tilde{c}(0) = \tilde{c}_{hs}(0) + \tilde{c}_{lr}(0) \tag{2.75}$$

Hence it follows that

$$\left(\frac{\partial p}{\partial \rho}\right)_T = \left(\frac{\partial p_{hs}}{\partial \rho}\right)_T - k_B T \tilde{c}_{lr}(0) = \left(\frac{\partial p_{hs}}{\partial \rho}\right)_T + \rho \tilde{\phi}_{lr}(0) \tag{2.76}$$

from (2.64). Differentiating (2.72) with respect to ρ at constant temperature we see immediately that these two expressions agree, provided

$$a = -\tfrac{1}{2}\tilde{\phi}_{lr}(0) \tag{2.77}$$

But this quantity, from the definition (2.64), is simply given by

$$a = -\tfrac{1}{2}\int_{\sigma}^{\infty} 4\pi r^2\left(\frac{A\sigma^6}{r^{12}} - \frac{A}{r^6}\right)\mathrm{d}r = \frac{4\pi}{9}A/\sigma^3 \tag{2.78}$$

This is in excellent agreement (within about 10%) of the empirical value of *a*. If we had used the Born–Green asymptotic form, we would have been quite wrong. Thus we conclude that this confirms the result (2.64) for argon and enables the equation of state to be calculated directly from the parameters in the force law (Woodhead-Galloway, Gaskell and March, 1968).

Actually Longuet-Higgins and Widom used the results of machine calculations for $p_{hs}(\rho, T)$. If we are content with slightly less accuracy, we could use the Percus–Yevick result. From (2.73) we find for the equation of state, using (2.68) and (2.70),

$$\frac{p}{\rho k_B T} = (1-\eta)^{-3}(1+\eta+\eta^2) \tag{2.79}$$

This result is not unique, but if we use the virial expression for the pressure we get a rather similar result and thermodynamic inconsistencies are not too serious in in this case.

If we use this argument in a metal, $\tilde{c}_{hs}(0)$ is made *more* negative by adding $\tilde{c}_{lr}(0)$, whereas in fluid argon it is made *less* negative. A crude calculation in a metal shows that c is derivable from the pseudo-potential which, as Ziman (1961) has discussed, tends to $\tfrac{2}{3}E_f$ as $k \to 0$. But $E_f \propto \rho^{2/3}$ from electron theory, and integrating this we get a term in the equation of state $\propto \rho^{5/3}$. This replaces the ρ^2 term in argon, the difference coming from the density dependence of the pair potential. Unfortunately, in a metal, σ defined by $\phi(\sigma) = 0$ for present purposes, is significantly density dependent. This means a more complex situation in a liquid metal than in fluid argon (see, for example, Watabe and Young, 1974). It would take us too far from our main theme to pursue matters specific to liquid metals further, at this point.

2.6.6 Softness corrections to hard-core potential

Considerable detailed refinements have been developed in recent work, taking full advantage of the computer simulation techniques. In particular, much effort has been devoted to the question of how best to correct the thermodynamic properties and the pair correlation function of a hard-core fluid for the finite steepness of the repulsive potential in real fluids. The most successful scheme, by comparison with Monte Carlo simulations (Hoover et al., 1970; Hansen, 1970), has been proposed by Andersen, Weeks and Chandler (1971). Defining the function

$$y(r)=g(r)\exp[\phi(r)/k_{\mathrm{B}}T] \tag{2.80}$$

where $\phi(r)$ is, as usual, the pair potential, taken to be purely repulsive, and similarly introducing the function $y_{\mathrm{d}}(r)$ for a hard-core fluid of core diameter d, these authors show that the relation

$$y(r)=y_{\mathrm{d}}(r) \tag{2.81}$$

is exact to first order in the range of the deviation between the two potentials provided that the hard-core diameter is chosen so as to satisfy the relation

$$\int \mathrm{d}\mathbf{r}\, y_{\mathrm{d}}(r)[\exp[-\beta\phi(r)]-\exp[-\beta\phi_{\mathrm{d}}(r)]]=0 \tag{2.82}$$

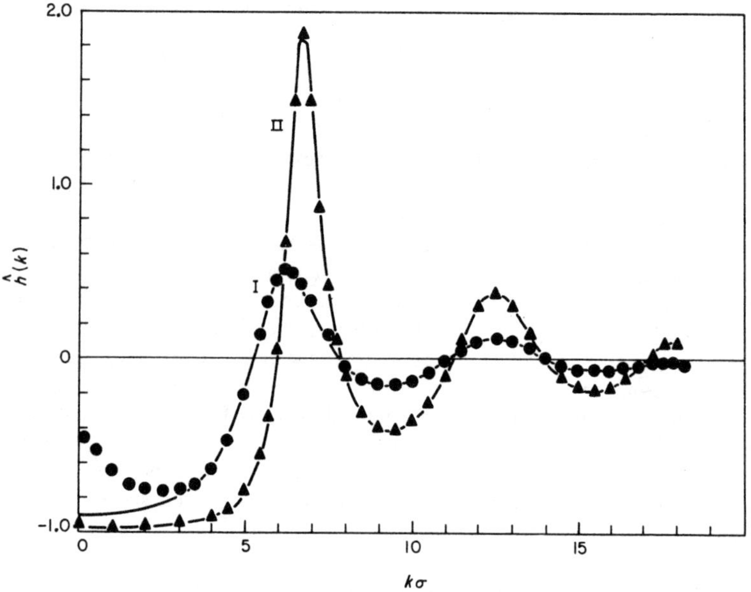

Figure 2.9 Fourier transform of the total correlation function, $S(k)-1$, for a fluid having the same repulsive forces as those of the Lennard–Jones potential, as calculated by the relations (2.81) and (2.82) from the Percus–Yevick hard-core fluid (from Chandler and Weeks, 1970). The two states refer to $\rho\sigma^3=0{\cdot}5426$ and $(\beta\varepsilon)^{-1}=1{\cdot}326$ (state I) and to $\rho\sigma^3=0{\cdot}844$ and $(\beta\varepsilon)^{-1}=0{\cdot}723$ (state II), σ and ε being the usual parameters of the Lennard–Jones potential. The circles and the triangles are molecular dynamics results by Verlet (1968) for the *full* Lennard–Jones potential, including an attractive tail

This simple approximation yields excellent agreement with the Monte Carlo data on the equation of state, at the expense of having a hard-core diameter which is dependent on temperature and density.

The results for $S(k)-1$ obtained by this approximation are compared with the molecular dynamics results of Verlet (1968) for a Lennard–Jones fluid in *figure 2.9* (from Chandler and Weeks, 1970). The agreement is especially remarkable because the theory has omitted the attractive part of the potential, but evidently the latter becomes noticeable only at relatively small wave vectors ($k\sigma \lesssim \pi$, where σ is the radial parameter in the Lennard–Jones potential) and that only if the density is not too high. The parallel results for $g(r)$ at high density are reported in *figure 2.10*, where they are also compared with the numerical solution of the Percus–Yevick theory for the *full* Lennard–Jones potential given by Mandel, Bearman and Bearman (1970). As noted by Chandler and Weeks, their results for $g(r)$ are much poorer at lower densities, where the errors in $S(k)$ at small k affect the form of $g(r)$ at all r.

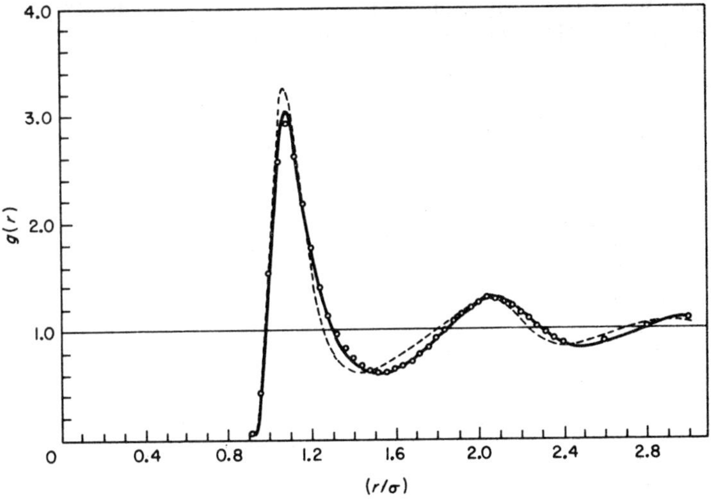

Figure 2.10 Pair correlation function $g(r)$ for a fluid having the same repulsive forces as those of the Lennard–Jones potential, as calculated by Chandler and Weeks (1970) for $\rho\sigma^3=0{\cdot}85$ and $(\beta\varepsilon)^{-1}=0{\cdot}88$. The results are compared with molecular dynamics results of Verlet (circles) and with the numerical Percus–Yevick solution (broken line) for the *full* Lennard–Jones potential

2.7 Summary and comments on approximate theories of structure

The following paragraphs summarise a few conclusions about the relative merits of the approximate theories:

(1) For short range forces, the Percus–Yevick method seems best. In partic-

ular, the equation of state for hard spheres which it predicts is in reasonable agreement with machine calculations.

(2) The random-phase approximation, while of value as we have seen for argon, is not directly applicable to fluids until the hard core is subtracted and treated separately. It is, on the other hand, a very useful approximation for fluids with strictly long-range forces, such as plasmas (see chapter 7).

(3) It appears that the correct asymptotic result for $c(r)$ is $c(r) \sim -\phi(r)/k_B T$ provided $h^2 < |c|$, the inequality defining the range of validity readily following from the Abe approximation. The condition is satisfied well away from the critical point.

One final comment seems called for. We began by asking how $g^{(3)}$ could be related to the two-body correlations. In retrospect, it seems that we should first have asked about the range of the forces! It seems clear that a relation between $g^{(3)}$ and g, which will be useful for long-range forces, is not likely to be valid for short-range interactions. This probably means that it is going to be hard to find a single equation for relating structure and forces which will successfully describe both liquid argon and liquid lead.

CHAPTER 3

LIQUID DYNAMICS AND TIME-DEPENDENT CORRELATION FUNCTIONS

3.1 Time-dependent density correlations

We shall introduce the van Hove correlation function in an intuitive way, and will later point out its intimate connection with neutron scattering (van Hove 1954*a*). We argue purely classically at first. Suppose that we define $G(r, t)$ as the average density of atoms at the point r at time t, if an atom was at the origin $\mathbf{r}=0$ at time $t=0$. Thus, $G(r, t)$ gives us the correlation in the positions of two atoms, which may or may not be different, at different times. This function can be expressed in the form

$$G(r, t)=\frac{1}{N}\left\langle \sum_{i,j=1}^{N} \delta[\mathbf{r}+\mathbf{R}_i(0)-\mathbf{R}_j(t)]\right\rangle \tag{3.1}$$

The average of the δ function involved in eqn (3.1) is obviously the probability that at time t the jth atom will be at $\mathbf{r}$ with respect to the position of the ith atom at time $t=0$. We then sum this probability over j and average over i. $G(r, t)$ is the space–time pair correlation function.

The quantum mechanical generalisation of eqn (3.1) is, in fact,

$$G_{\text{Quantum}}(r, t)=\frac{1}{N}\left\langle \sum_{i,j=1}^{N} \int \mathrm{d}\mathbf{r}'\delta[\mathbf{r}+\mathbf{R}_i(0)-\mathbf{r}']\delta[\mathbf{r}'-\mathbf{R}_j(t)]\right\rangle \tag{3.2}$$

where $\mathbf{R}_i(0)$ and $\mathbf{R}_j(t)$ are now Heisenberg operators which do not commute. If this failure to commute is ignored, it proves possible to integrate over $\mathbf{r}'$, and then eqn (3.1) is regained.

An alternative and sometimes useful form for $G(r, t)$ is obtained by introducing the density operator $\rho(\mathbf{r}, t)$ of the atoms at the point $\mathbf{r}$ at the time t (cf. eqn (1.16)):

$$\rho(\mathbf{r}, t)=\sum_{i=1}^{N} \delta[\mathbf{r}-\mathbf{R}_i(t)] \tag{3.3}$$

If we employ this form in the definition (3.2) and change the origin by substituting $\mathbf{r}''=\mathbf{r}'-\mathbf{r}$ we find

$$G(r, t)=\frac{1}{N}\left\langle \int \mathrm{d}\mathbf{r}''\rho(\mathbf{r}'', 0)\rho(\mathbf{r}''+\mathbf{r}, t)\right\rangle \tag{3.4}$$

Thus, we can interpret $G(r, t)$ as the space–time correlation of the density ρ.

If we now take explicit account of the homogeneity of the liquid, then the integrand in (3.4) is independent of $\boldsymbol{r}''$ and we find

$$G(\boldsymbol{r}-\boldsymbol{r}', t-t') = \frac{1}{\rho}\langle \rho(\boldsymbol{r}', t')\rho(\boldsymbol{r}, t)\rangle \tag{3.5}$$

At this point we follow van Hove, and take the diagonal term $i=j$ out of the sum over i and j in (3.2) when we obtain

$$G(r, t) = G_s(r, t) + G_d(r, t) \tag{3.6}$$

where

$$G_s(r, t) = \frac{1}{N}\left\langle \sum_{i=1}^{N} \int d\boldsymbol{r}'\delta[\boldsymbol{r}+\boldsymbol{R}_i(0)-\boldsymbol{r}']\delta[\boldsymbol{r}'-\boldsymbol{R}_i(t)]\right\rangle \tag{3.7}$$

and

$$G_d(r, t) = \frac{1}{N}\left\langle \sum_{i\neq j=1}^{N} \int d\boldsymbol{r}'\delta[\boldsymbol{r}+\boldsymbol{R}_i(0)-\boldsymbol{r}']\delta[\boldsymbol{r}'-\boldsymbol{R}_j(t)]\right\rangle \tag{3.8}$$

By this separation, we can interpret G_s as the correlation function which tells us the probability that a particle which was at the origin at time $t=0$, will be at $\boldsymbol{r}$ at time t. The part G_d obviously refers to the analogous conditional probability of finding a different atom at $\boldsymbol{r}$ at time t.

Let us now investigate the correlation functions at time $t=0$. Going back to (3.2) and noting that $\boldsymbol{R}_i(0)$ and $\boldsymbol{R}_j(0)$ commute, we can integrate over $\boldsymbol{r}'$ (cf. remarks after (3.2)) and we find

$$G(r, 0) = \frac{1}{N}\left\langle \sum_{i,j=1}^{N} \delta[\boldsymbol{r}+\boldsymbol{R}_i(0)-\boldsymbol{R}_j(0)]\right\rangle \tag{3.9}$$

Splitting this according to (3.6), we find almost immediately

$$G_s(r, 0) = \delta(\boldsymbol{r}) \tag{3.10}$$

$$G(r, 0) = \delta(\boldsymbol{r}) + \rho g(r) \tag{3.11}$$

where $g(r)$ is given by

$$\rho g(r) = \frac{1}{N}\sum_{i\neq j=1}^{N} \langle \delta(\boldsymbol{r}+\boldsymbol{R}_i-\boldsymbol{R}_j)\rangle \tag{3.12}$$

the usual radial distribution function we have discussed earlier.

In the limit of long times, we can assume there is no correlation between positions of particles. Thus, in (3.2) we can replace the average of the product of the δ-functions by the product of the averages:

$$\sum_{i,j=1}^{N} \langle \delta[\boldsymbol{r}+\boldsymbol{R}_i(0)-\boldsymbol{r}']\delta[\boldsymbol{r}'-\boldsymbol{R}_j(t)]\rangle \sim \left\langle \sum_{i=1}^{N} \delta[\boldsymbol{r}+\boldsymbol{R}_i(0)-\boldsymbol{r}']\right\rangle\left\langle \sum_{j=1}^{N} \delta[\boldsymbol{r}'-\boldsymbol{R}_j(t)]\right\rangle \tag{3.13}$$

Thus, for large t (or, in the absence of long-range order, for large r) we may write

$$G(r, t) \approx \frac{1}{N} \int \mathrm{d}\boldsymbol{r}' \, \rho(\boldsymbol{r} - \boldsymbol{r}')\rho(\boldsymbol{r}') \tag{3.14}$$

For systems with long-range order, $\rho(\boldsymbol{r})$ and $G(r, t)$ are periodic in space. For a fluid, $\rho = N/V$ where V is the volume, and therefore

$$G(r, \infty) \approx \rho \tag{3.15}$$

Similarly, for the self-correlation function we can show that for a homogeneous system

$$G_s(r, \infty) = \frac{1}{V} \tag{3.16}$$

which tends to zero as V tends to infinity. This is in marked contrast to the situation in which atoms are not free to move far from some 'lattice' sites. In this case, appropriate to solids when we neglect diffusion, $G_s(\boldsymbol{r}, \infty) \neq 0$.

For short times $G_s(r, t)$ approximates to a δ-function according to (3.10), while $G_d(r, t)$ is approximately the pair correlation function $g(r)$. As $t \to \infty$, $G_s(r, t) \to 0$ and $G_d(r, t) \to \rho$, and these forms are shown schematically in *figure 3.1*.

3.2 Intermediate scattering function and van Hove structure factor

Having defined the time-dependent correlation function $G(r, t)$, and having examined its general properties for short and for long times, we shall proceed to discuss two equivalent forms. Whereas $G(r, t)$ is a sum of two parts G_s and G_d, the latter part being the time-dependent generalisation of the radial distribution function $g(r)$ (actually $G_d \to \rho g(r)$ as $t \to 0$), the desired generalisation of the structure factor $S(k)$ is obtained by taking a double Fourier transform of $G(r, t)$. The resulting function $S(k, \omega)$, introduced by van Hove (1954*a*) and known as the dynamic structure factor, is directly related to inelastic scattering of neutrons from the fluid and gives the probability that there is a momentum transfer $\hbar\boldsymbol{k}$ and an energy transfer $\hbar\omega$ between the neutron and the fluid.

3.2.1 Intermediate scattering function

We begin by taking the space Fourier transform of $G(r, t)$, which defines the so-called intermediate scattering function $F(k, t)$,

$$F(k, t) = \int \mathrm{d}\boldsymbol{r} \exp(i\boldsymbol{k} \cdot \boldsymbol{r}) \, G(r, t) \tag{3.17}$$

with a corresponding relation between $G_s(r, t)$ and $F_s(k, t)$. Using (3.5), and taking account of the homogeneity of the liquid, we immediately find

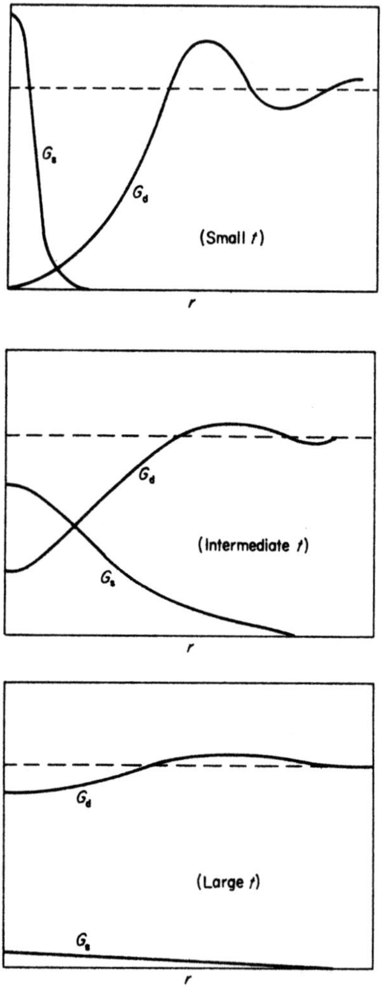

Figure 3.1 Schematic drawing of the spreading of the van Hove correlation functions with increasing time

$$\begin{aligned} F(k, t) &= \frac{1}{\rho} \int \mathrm{d}(\boldsymbol{r}-\boldsymbol{r}') \exp\,[i\boldsymbol{k}\cdot(\boldsymbol{r}-\boldsymbol{r}')]\langle \rho(\boldsymbol{r}', 0)\rho(\boldsymbol{r}, t)\rangle \\ &= \frac{1}{N} \iint \mathrm{d}\boldsymbol{r}\,\mathrm{d}\boldsymbol{r}' \exp\,[i\boldsymbol{k}\cdot(\boldsymbol{r}-\boldsymbol{r}')]\langle \rho(\boldsymbol{r}', 0)\rho(\boldsymbol{r}, t)\rangle = \frac{1}{N}\langle \rho_{-\boldsymbol{k}}(0)\rho_{\boldsymbol{k}}(t)\rangle \end{aligned} \qquad (3.18)$$

where

$$\rho_k(t) = \sum_{i=1}^{N} \exp\left[i\boldsymbol{k}\cdot\boldsymbol{R}_i(t)\right] \tag{3.19}$$

is the space Fourier transform of the density operator introduced in (3.3). As we see by comparison with (2.49), $F(k, t)$ is the time dependent generalisation of the structure factor $S(k)$, and describes time correlations between density fluctuations of equal and opposite wave vector.

The physical meaning of the intermediate scattering function is clear from (3.18). According to this equation, $F(k, t)$ expresses the probability amplitude that, having started with the fluid in its equilibrium state and having created a density fluctuation of wave vector $\boldsymbol{k}$ at time t and a density fluctuation of wave vector $-\boldsymbol{k}$ at time $t=0$, we find the system again in the same equilibrium state. We clearly must expect that, if such a density fluctuation is an exact eigenmode of the fluid, then $F(k, t)$ will vary with time as $\exp(-i\omega_k t)$ where ω_k is the frequency of the eigenmode. More generally, the decay of $F(k, t)$ in time will be determined by the lifetime of approximate eigenmodes of the system with wave vector $\boldsymbol{k}$. In practice, we shall usually meet situations where the density fluctuations are overdamped oscillations, except for special ranges of wave vector.

We also find from (3.7)

$$F_s(k, t) \equiv \int d\boldsymbol{r}\, \exp(i\boldsymbol{k}\cdot\boldsymbol{r})\, G_s(r, t) = \langle \exp\left[-i\boldsymbol{k}\cdot\boldsymbol{R}_1(0)\right] \exp\left[i\boldsymbol{k}\cdot\boldsymbol{R}_1(t)\right]\rangle \tag{3.20}$$

where $\boldsymbol{R}_1(t)$ denotes the position of any one particle in the fluid at time t.

3.2.2 Dynamic structure factor $S(k, \omega)$

We now define the dynamic structure factor $S(k, \omega)$ as

$$\begin{aligned} S(k, \omega) &= \int dt\, \exp(-i\omega t)\, F(k, t) \\ &= \iint d\boldsymbol{r}\, dt\, \exp\left[i(\boldsymbol{k}\cdot\boldsymbol{r} - \omega t)\right] G(r, t) \end{aligned} \tag{3.21}$$

with an analogous definition for $S_s(k, \omega)$ in terms of $F_s(k, t)$ and $G_s(r, t)$. If we introduce the frequency components of the density fluctuations,

$$\rho_k(\omega) = \int dt\, \exp(-i\omega t)\, \rho_k(t) \tag{3.22}$$

and take into account the invariance of the fluid at equilibrium under time translation, we find that $S(k, \omega)$ is determined by the correlations between the frequency components of the density fluctuations according to

$$\langle \rho_{-k}(\omega')\rho_k(\omega)\rangle = 2\pi S(k, \omega)\delta(\omega + \omega') \tag{3.23}$$

Thus, $S(k, \omega)$ gives the frequency spectrum of the correlations between density fluctuations of the given wave vector.

According to the argument developed in the preceding section, we expect that an approximate eigenmode of the fluid will appear in $S(k, \omega)$ as a more or less sharp peak at the frequency of the mode. Also, we note that

$$\int \frac{d\omega}{2\pi} S(k, \omega) = F(k, 0) = S(k) \tag{3.24}$$

the usual (static) structure factor.

3.2.3 Relation to neutron scattering

The process of inelastic scattering of a beam of incident particles (typically, neutrons) from a liquid serves as a microscopic probe of the dynamics of this many-particle system. More precisely, as first shown by van Hove (1954*a*), a measurement of coherent inelastic scattering yields $S(k, \omega)$ (see, for example, Lomer and Low, 1965; Marshall and Lovesey, 1973). Actually, this result is a special case of the fluctuation–dissipation theorem (see section 3.2.5 below) in statistical mechanics (Callen and Welton, 1951), according to which a measurement of the dissipation of energy by an external probe weakly coupled to a many-particle system is directly related to correlations between *equilibrium* fluctuations in the system.

To discuss the process of neutron inelastic scattering, we consider the neutron as coupled to the fluid through a Hamiltonian

$$H_1 = \sum_{i=1}^{N} \mathcal{V}_i(|\boldsymbol{R}_i - \boldsymbol{R}_n|) = \int d\boldsymbol{r} \sum_{i=1}^{N} \delta(\boldsymbol{r} - \boldsymbol{R}_i) \mathcal{V}_i(|\boldsymbol{r} - \boldsymbol{R}_n|) \tag{3.25}$$

where $\boldsymbol{R}_n$ is the position of the neutron and $\mathcal{V}_i(|\boldsymbol{R}_i - \boldsymbol{R}_n|)$ is the potential of interaction between the neutron and the nucleus of the *i*th atom. This interaction depends on the individual nucleus, as indicated by the suffix *i*, since even a pure liquid is usually a mixture of different isotopes. In practice, a Fermi δ-function potential (Fermi, 1936)

$$\mathcal{V}_i(|\boldsymbol{r} - \boldsymbol{R}_n|) = a_i \delta(\boldsymbol{r} - \boldsymbol{R}_n) \tag{3.26}$$

provides a good representation of the neutron-nucleus interaction, whose properties are thus fully described by just one nuclear parameter, the scattering length a_i.

If we assume the interaction (3.25) to be weak, we can use the golden rule formula of perturbation theory to write the probability of a process in which the neutron transfers momentum $\hbar\boldsymbol{k}$ and energy $\hbar\omega$ to the fluid as

$$P(\boldsymbol{k}, \omega) = \frac{2\pi}{\hbar} \sum_{m} w_m \sum_{m'} |M_{m',m}(\boldsymbol{k})|^2 \delta(E_{m'} - E_m - \hbar\omega) \tag{3.27}$$

where the transition matrix element $M_{m',m}(\boldsymbol{k})$ is given by

$$M_{m',m}(\boldsymbol{k})=\int d\boldsymbol{R}_n \exp(-i\boldsymbol{k}'_n\cdot\boldsymbol{R}_n)\langle\Psi_{m'}|H_1|\Psi_m\rangle\exp(i\boldsymbol{k}_n\cdot\boldsymbol{R}_n) \tag{3.28}$$

In this equation, the initial and the final state of the neutron are plane waves of wave vectors $\boldsymbol{k}_n$ and $\boldsymbol{k}'_n$, with $\boldsymbol{k}_n-\boldsymbol{k}'_n=\boldsymbol{k}$; Ψ_m and $\Psi_{m'}$ are the initial and the final state of the fluid, of energies E_m and $E_{m'}$; and we have supposed that the fluid is initially in its equilibrium state described by the distribution w_m. Use of (3.25) yields

$$M_{m',m}(\boldsymbol{k})=\sum_{i=1}^{N}\mathscr{V}_i(k)\langle\Psi_{m'}|\exp(i\boldsymbol{k}\cdot\boldsymbol{R}_i)|\Psi_m\rangle \tag{3.29}$$

where $\mathscr{V}_i(k)$ is the Fourier transform of the neutron-nucleus interaction, which, of course, reduces to a_i for the Fermi potential. By using the integral representation of the δ-function,

$$\delta(E_{m'}-E_m-\hbar\omega)=\frac{\hbar}{2\pi}\int dt\exp[it(E_{m'}-E_m-\hbar\omega)/\hbar] \tag{3.30}$$

and the usual definition of an operator in the Heisenberg representation, we can now write (3.27) in the form

$$\begin{aligned}P(\boldsymbol{k},\omega)&=\sum_{i,j=1}^{N}\mathscr{V}_i(k)\mathscr{V}_j^*(k)\int dt\exp(-i\omega t)\sum_{m,m'}w_m\langle\Psi_m|\exp[-i\boldsymbol{k}\cdot\boldsymbol{R}_i(0)]|\Psi_{m'}\rangle\\&\times\langle\Psi_{m'}|\exp[i\boldsymbol{k}\cdot\boldsymbol{R}_j(t)]|\Psi_m\rangle=\sum_{i,j=1}^{N}\mathscr{V}_i(k)\mathscr{V}_j^*(k)\int dt\exp(-i\omega t)\sum_m w_m\\&\times\langle\Psi_m|\exp[-i\boldsymbol{k}\cdot\boldsymbol{R}_i(0)]\exp[i\boldsymbol{k}\cdot\boldsymbol{R}_j(t)]|\Psi_m\rangle=\sum_{i,j=1}^{N}\mathscr{V}_i(k)\mathscr{V}_j^*(k)\\&\times\int dt\exp(-i\omega t)\langle\exp[-i\boldsymbol{k}\cdot\boldsymbol{R}_i(0)]\exp[i\boldsymbol{k}\cdot\boldsymbol{R}_j(t)]\rangle\end{aligned} \tag{3.31}$$

If the fluid contains only one type of nucleus, that is $\mathscr{V}_i(k)=\mathscr{V}(k)$, then the above expression becomes

$$\begin{aligned}P_{coh}(\boldsymbol{k},\omega)&=|\mathscr{V}(k)|^2\int dt\exp(-i\omega t)\sum_{i,j=1}^{N}\langle\exp[-i\boldsymbol{k}\cdot\boldsymbol{R}_i(0)\exp(i\boldsymbol{k}\cdot\boldsymbol{R}_j(t))\rangle\\&=N|\mathscr{V}(k)|^2\int dt\exp(-i\omega t)F(k,t)\\&=N|\mathscr{V}(k)|^2S(k,\omega)\end{aligned} \tag{3.32}$$

Thus, the differential scattering cross-section, which differs from $P(\boldsymbol{k},\omega)$ merely by a trivial kinetic factor, is proportional to the dynamic structure factor.

In a mixture of isotopes, instead, (3.32) gives only the coherent part of the scattering, with $\mathscr{V}(k)$ defined by

$$\mathscr{V}(k)=\frac{1}{N}\sum_{i=1}^{N} \mathscr{V}_i(k)\equiv\overline{\mathscr{V}(k)} \tag{3.33}$$

The remaining part, namely the incoherent scattering

$$P_{\text{incoh}}(\boldsymbol{k},\omega)=\sum_{i,j=1}^{N}\left[\mathscr{V}_i(k)\mathscr{V}_j^*(k)-|\overline{\mathscr{V}(k)}|^2\right]\int dt\,\exp(-i\omega t)\langle\exp[-i\boldsymbol{k}\cdot\boldsymbol{R}_i(0)]\exp[i\boldsymbol{k}\cdot\boldsymbol{R}_j(t)]\rangle \tag{3.34}$$

may be expressed through $S_s(k,\omega)$ if we assume that the expression in the square bracket becomes negligible for $i\neq j$. We then find

$$\begin{aligned}P_{\text{incoh}}(\boldsymbol{k},\omega)&=\sum_{i=1}^{N}\left[|\mathscr{V}_i(k)|^2-|\overline{\mathscr{V}(k)}|^2\right]\int dt\,\exp(-i\omega t)\langle\exp[-i\boldsymbol{k}\cdot\boldsymbol{R}_i(0)]\exp[i\boldsymbol{k}\cdot\boldsymbol{R}_i(t)]\rangle\\&=N\overline{|\delta\mathscr{V}(k)|^2}\int dt\,\exp(-i\omega t)\,F_s(k,t)\\&=N\overline{|\delta\mathscr{V}(k)|^2}S_s(k,\omega)\end{aligned} \tag{3.35}$$

with

$$\overline{|\delta\mathscr{V}(k)|^2}=\frac{1}{N}\sum_{i=1}^{N}|\mathscr{V}_i(k)|^2-|\overline{\mathscr{V}(k)}|^2=<|\mathscr{V}(k)|^2>-|\overline{\mathscr{V}(k)}|^2 \tag{3.36}$$

The neutron inelastic scattering from an isotopic mixture is discussed more systematically in appendix 6.6.

The above results actually describe inelastic scattering of any type of particles coupled to the fluid by an interaction of the type (3.25). Cold neutrons from a nuclear reactor provide the best microscopic probe for liquids because both their momenta and energies are in the range of the characteristic wave vectors and frequencies for the dynamics of liquids ($k\sim 1$ Å^{-1}, $\omega\sim 10^{13}$ sec^{-1}), and it is therefore feasible to perform an energy analysis of the radiation scattered within a given solid angle.† In an X-ray scattering experiment, with wave vectors of the order of 1 Å^{-1} the incoming photons have an energy which is larger than the energy transferred to atomic motions by several orders of magnitude. Thus, X-ray scattering experiments are used to measure $S(k)$, that is, the dynamic structure factor integrated over all energy transfers (see (3.24)). On the other hand, an energy analysis of inelastically scattered photons becomes feasible for photons in the infrared, but then the transferred wave vector is bound to be very small, and the experiment provides information only on long wavelength density fluctuations in the fluid. We shall briefly discuss such experiments in the next chapter, in connection with the hydrodynamic modes.

†For a recent review of neutron inelastic scattering from monatomic liquids, see Copley and Lovesey (1974).

3.2.4 Relation between $S(k, \omega)$ and self-function $S_s(k, \omega)$

Equation (3.18) for the intermediate scattering function can be written

$$F(k, t) = F_s(k, t) + \frac{1}{N} \sum_{i \neq j = 1}^{N} \langle \exp[-i\boldsymbol{k} \cdot \boldsymbol{R}_i(0)] \exp[i\boldsymbol{k} \cdot \boldsymbol{R}_j(t)] \rangle$$

$$= F_s(k, t) + \frac{1}{N} \sum_{i \neq j = 1}^{N} \langle \exp[-i\boldsymbol{k} \cdot [\boldsymbol{R}_i(0) - \boldsymbol{R}_j(0)]] \exp[-i\boldsymbol{k} \cdot \boldsymbol{R}_j(0)]$$

$$\times \exp[i\boldsymbol{k} \cdot \boldsymbol{R}_j(t)] \rangle \quad (3.37)$$

If we now approximate the thermal average in the above equation by the product of two thermal averages, we can write

$$F(k, t) \simeq F_s(k, t) + [F(k, 0) - 1] F_s(k, t)$$

$$= S(k) F_s(k, t)$$

or equivalently

$$S(k, \omega) = S(k) S_s(k, \omega) \quad (3.38)$$

which is the so-called convolution approximation proposed by Vineyard (1958). It is clearly equivalent to expressing the difference between $G(r, t)$ and $G_s(r, t)$ as a convolution of $G_s(r, t)$ and $g(r)$,

$$G(r, t) = G_s(r, t) + \rho \int d\boldsymbol{r}' g(r') G_s(\boldsymbol{r} - \boldsymbol{r}', t) \quad (3.39)$$

as is easily verified by Fourier transform. The approximation thus assumes that the motion of an atom which at time $t = 0$ is at a distance $\boldsymbol{r}'$ from the atom at the origin is not affected by the presence of the latter atom.

The convolution approximation provides a simple way of relating the coherent scattering to the incoherent scattering if the structure of the liquid is known. A number of its deficiencies have been pointed out over the years, one such deficiency being its inapplicability in the hydrodynamic regime, where it fails to predict the existence of the Brillouin doublet (see section 4.3). The approximation as originally proposed is also unsatisfactory in the finite frequency regime of interest in neutron scattering, as it violates the moment sum rules for the scattering function starting from the second moment (see section 3.3). This defect can, however, be partly remedied by simple modifications (Sköld, 1967; Rahman, 1967a) which satisfy up to the fourth moment sum rule and provide empirically useful relations between coherent and incoherent scattering functions through the static structure factor. For instance, the recipe proposed by Sköld (1967) satisfies the second moment sum rule by scaling the wave vector in $F_s(k, t)$ on the right-hand side of (3.38), which is written in the form

$$F(k, t) = S(k) F_s(k/\sqrt{S(k)}, t) \quad (3.40)$$

This relation has proved successful in the analysis of neutron scattering from liquid argon.

Actually, Vineyard's philosophy of expressing the dynamic structure factor $S(k, \omega)$ in terms of the self-function $S_s(k, \omega)$ and the static structure factor $S(k)$ has been shown by Gyorffy and March (1971) to be exact in the sense that $S(k, \omega)$ is a functional of $S_s(k, \omega)$, plus the static correlation functions. Some approximate forms of such a functional relation are available (see, for example, Hubbard and Beeby, 1969; Kerr, 1968; Singwi, Skold and Tosi, 1970) going beyond the Vineyard form corresponding to (3.38). Barker et al. (1973) discuss, from experiment, some features of such a functional relation.

3.2.5 Fluctuation–dissipation theorem

We conclude this section by giving the explicit relation between the dynamic structure factor $S(k, \omega)$ and the density response function $\chi(k, \omega)$. The latter is introduced by considering the fluid under a weak potential field $V_e(\boldsymbol{r}, t)$ coupled with the density fluctuations, which to linear terms produces a density change $\delta\rho(\boldsymbol{r}, t)$ given by

$$\delta\rho(\boldsymbol{r}, t) = \int\int d\boldsymbol{r}'\, dt'\, \chi(\boldsymbol{r}-\boldsymbol{r}', t-t')V_e(\boldsymbol{r}', t') \tag{3.41}$$

or, in Fourier transform,

$$\rho_{\boldsymbol{k}}(\omega) = \chi(\boldsymbol{k}, \omega)V_e(\boldsymbol{k}, \omega) \tag{3.42}$$

Explicit calculation by time-dependent perturbation theory yields

$$\chi(k, \omega) = \frac{1}{\hbar}\sum_{m,m'} w_m |\langle\Psi_{m'}|\rho_k^+|\Psi_m\rangle|^2 \left\{\frac{1}{\omega-\omega_{m'm}+i\eta} - \frac{1}{\omega+\omega_{m'm}+i\eta}\right\} \tag{3.43}$$

where $\omega_{m'm} = (E_{m'} - E_m)/\hbar$ and $\eta = 0^+$. If we write

$$\chi(k, \omega) = \chi'(k, \omega) + i\chi''(k, \omega) \tag{3.44}$$

and use the identity

$$\frac{1}{x \pm i\eta} = P\left(\frac{1}{x}\right) \mp i\pi\delta(x) \tag{3.45}$$

comparison with eqns (3.32) and (3.27) yields

$$\begin{aligned}\chi''(k, \omega) &= -\frac{\rho}{2\hbar}[S(\boldsymbol{k}, \omega) - S(-\boldsymbol{k}, -\omega)] \\ &= -\frac{\rho}{2\hbar}[1 - \exp(-\hbar\beta\omega)]S(k, \omega) \\ &\xrightarrow[(\hbar\to 0)]{} -\frac{\rho\omega}{2k_B T}S(k, \omega)\end{aligned} \tag{3.46}$$

This is the precise statement of the fluctuation–dissipation theorem in the present case. Of course, the imaginary part of the response function describes energy dissipation in the system, while its real part describes reversible polarisation processes.

Since $\delta\rho(\boldsymbol{r}, t)$ cannot depend on $V_e(\boldsymbol{r}', t')$ at times t' later than t, $\chi(\boldsymbol{r}, t)$ must vanish when $t<0$. This causality condition is ensured by the small imaginary part in the denominators of eqn (3.43). In turn, this mathematical statement implies the existence of a dispersion relation of the form

$$\chi(k, \omega)=\frac{1}{\pi}\int_{-\infty}^{\infty} \mathrm{d}\omega' \frac{\chi''(k, \omega')}{\omega'-\omega-i\eta} \tag{3.47}$$

namely, $\chi''(k, \omega)$ serves as the spectral density for $\chi(k, \omega)$. The above relation implies that $\chi'(k, \omega)$ and $\chi''(k, \omega)$ are related by Kramers–Kronig relations. For a classical fluid these relations yield in particular,

$$S(k)= -\frac{k_B T}{\rho}\chi(k, 0) \tag{3.48}$$

as is easily seen from (3.47) and (3.46).

3.3 Moments of van Hove correlation functions and short-time expansions

From the definitions given in section 3.2.2 we have for the second moment of $S_s(k, \omega)$:

$$\int_{-\infty}^{\infty} \frac{\mathrm{d}\omega}{2\pi}\omega^2 S_s(k, \omega)= -\left.\frac{\partial^2 F_s(k, t)}{\partial t^2}\right|_{t=0} \tag{3.49}$$

Similarly, for the fourth moment we may write

$$\int_{-\infty}^{\infty} \frac{\mathrm{d}\omega}{2\pi}\omega^4 S_s(k, \omega)= \left.\frac{\partial^4 F_s(k, t)}{\partial t^4}\right|_{t=0} \tag{3.50}$$

Thus, if the second and fourth time derivatives of $F_s(k, t)$ can be found from (3.20), we have the desired expressions for the moments.

To evaluate these derivatives, we find it convenient to develop the small time expansion of (3.20). Then, if we denote by $x_1(t)$ the component of $\boldsymbol{R}_1(t)$ along $\boldsymbol{k}$ and use

$$x_1(t)=x_1(0)+t\dot{x}_1(0)+\frac{1}{2!}t^2\ddot{x}_1(0)+\cdots \tag{3.51}$$

we have

$$F_s(k, t)=\langle \exp\,[ikt\dot{x}_1(0)]\,\exp\,[ikt^2\ddot{x}_1(0)/2!]\,\exp\,[ikt^3\dddot{x}_1(0)/3!]\cdots\rangle \tag{3.52}$$

The coefficient of t^2 in the short-time expansion of (3.52) is given by

$$\frac{ikt^2}{2!}\langle\ddot{x}_1(0)\rangle-\frac{k^2t^2}{2!}\langle\{\dot{x}_1(0)\}^2\rangle,$$

where the first term clearly averages to zero, while the second term gives the result

$$\left.\frac{\partial^2 F_s(k,t)}{\partial t^2}\right|_{t=0} = -k^2\langle\{\dot{x}_1(0)\}^2\rangle \tag{3.53}$$

But the thermal average of $\langle\{\dot{x}_1(0)\}^2\rangle$ is given by

$$\tfrac{1}{2}m\langle\{\dot{x}_1(0)\}^2\rangle = \tfrac{1}{2}k_B T \tag{3.54}$$

and combining (3.49), (3.53) and (3.54) we find

$$\int_{-\infty}^{\infty}\frac{d\omega}{2\pi}\,\omega^2 S_s(k,\omega) = k^2 k_B T/m \tag{3.55}$$

Actually, from the definition of $F(k, t)$ as

$$F(k,t) = \sum_i \langle \exp[+i\boldsymbol{k}\cdot\{\boldsymbol{R}_1(0)-\boldsymbol{R}_i(t)\}]\rangle$$

the distinct term $i \neq 1$ vanishes to order t^2, since the velocities of two atoms taken at the same instant of time are uncorrelated. Hence

$$\int_{-\infty}^{\infty}\frac{d\omega}{2\pi}\,\omega^2 S(k,\omega) = k^2 k_B T/m \tag{3.56}$$

for classical fluids. Clearly the behaviour of the fluid to order t^2 is thus described by a free particle model.

We show in appendix 3.1 that the fourth moments for classical fluids in terms of the radial distribution function $g(r)$ and the pair potential $\phi(r)$ are (de Gennes, 1959):

$$\int_{-\infty}^{\infty}\frac{d\omega}{2\pi}\,\omega^4 S_s(k,\omega) = 3k^4(k_B T)^2/m^2 + \frac{\rho k^2 k_B T}{3m^2}\int d\boldsymbol{r}\, g(r)\nabla^2\phi(r)$$

$$= 3k^4(k_B T)^2/m^2 + \frac{4\pi\rho k^2 k_B T}{3m^2}\int_0^{\infty} dr\, r^2 g(r)\left\{\phi''(r)+\frac{2}{r}\phi'(r)\right\} \tag{3.57}$$

and

$$\int_{-\infty}^{\infty}\frac{d\omega}{2\pi}\,\omega^4 S(k,\omega) = 3k^4(k_B T)^2/m^2 + \frac{\rho k^2 k_B T}{m^2}\int d\boldsymbol{r}\, g(r)(1-\cos kx)\frac{\partial^2\phi(r)}{\partial x^2} \tag{3.58}$$

Similarly, the higher moments can be expressed in terms of the pair potential and of higher static correlation functions.

3.3.1 Asymptotic form of $G(r, t)$ at large r and short times for van der Waals forces

Using these sum rules, plus the fact that

$$\int_{-\infty}^{\infty} \frac{d\omega}{2\pi} S(k, \omega) = S(k)$$

we can immediately expand the intermediate scattering function as

$$F(k, t) = S(k) - k^2 t^2 (k_B T)/2m + k^4 t^4 (k_B T)^2/8m^2$$
$$+ \rho k^2 t^4 k_B T/24m^2 \int d\boldsymbol{r}\, g(r)(1 - \cos kx) \frac{\partial^2 \phi}{\partial x^2} \tag{3.59}$$

We can employ Fourier transform arguments referred to in section 2.5 for van der Waals forces to show that $\int d\boldsymbol{r}\, g(r)(1 - \cos kx) \frac{\partial^2 \phi}{\partial x^2}$ has a term $-\pi^2 A k^5/12$ as the first odd power of k in its small-k expansion. Actually the presence of this odd power of k is signalled by examining the coefficient of k^6 in the above integral. This involves $\int d\boldsymbol{r}\, g(r) x^6 \frac{\partial^2 \phi}{\partial x^2}$ and at large r the integrand (replacing $d\boldsymbol{r}$ by $r^2\, dr$ for the dimensional argument) behaves as $r^8 \frac{\partial^2 \phi}{\partial r^2} \to$constant. Hence the integral diverges and we must have missed out a term in the expansion. Since

$$F(k, t) = \int d\boldsymbol{r} \exp(i\boldsymbol{k} \cdot \boldsymbol{r})\, G(r, t)$$

it is readily shown from the Lighthill theorem (1958) that the leading term in $G(r, t)$ at large r is precisely

$$G(r, t) \sim -\frac{70 A t^4}{m^2 r^{10}} (k_B T) \tag{3.60}$$

where, as usual, A is the strength of the van der Waals interactions. Should it prove desirable, the next term can also be calculated by this method for a Lennard–Jones (6–12) potential.

The non-analyticity arising from van der Waals interactions dominates the short-time dependence of $G(r, t)$ at large r.

3.3.2 Moments of van Hove function in quantal fluids

Finally, we shall briefly comment on the extension of the moment sum rules for $S(k, \omega)$ to the case of a quantal fluid. We note, first of all, that the moments (3.56) and (3.58) can be rewritten, in a classical fluid, as sum rules for the imaginary part of the density response function, using eqn (3.46) in the classical limit:

$$-\int_{-\infty}^{\infty} \frac{d\omega}{\pi} \omega \chi''(k, \omega) = \rho k^2/m \tag{3.56'}$$

and

$$-\int_{-\infty}^{\infty}\frac{d\omega}{\pi}\omega^3\chi''(k,\omega)=\frac{2\rho k^4 t}{m^2}+\frac{\rho^2k^2}{m^2}\int d\boldsymbol{r}\,g(r)(1-\cos kx)\frac{\partial^2\phi(r)}{\partial x^2} \quad (3.58')$$

where, in the first term on the right-hand side of (3.58′), we have replaced $\frac{3}{2}k_BT$ in favour of the mean kinetic energy per particle, denoted by t. In this form, the sum rules are in fact valid in general, that is independently of the classical approximation.

If one wishes, the above relations can be rewritten as sum rules on $S(k,\omega)$, using the general form of the relation (3.46) between $\chi''(k,\omega)$ and $S(k,\omega)$. It is preferable, however, to introduce the symmetrised correlation function, $\tilde{S}(k,\omega)$ by

$$\begin{aligned}\tilde{S}(k,\omega)&=\tfrac{1}{2}[S(\boldsymbol{k},\omega)+S(-\boldsymbol{k},-\omega)]\\&=\tfrac{1}{2}S(k,\omega)[1+\exp(-\hbar\beta\omega)]\\&=-\frac{\hbar}{\rho}\chi''(k,\omega)\coth(\tfrac{1}{2}\hbar\beta\omega)\end{aligned} \quad (3.61)$$

in terms of which the sum rules can be written as

$$\frac{1}{\hbar}\int_{-\infty}^{\infty}\frac{d\omega}{\pi}\omega\,\mathrm{tgh}(\tfrac{1}{2}\hbar\beta\omega)\tilde{S}(k,\omega)=k^2/m \quad (3.56'')$$

and

$$\frac{1}{\hbar}\int_{-\infty}^{\infty}\frac{d\omega}{\pi}\omega^3\,\mathrm{tgh}(\tfrac{1}{2}\hbar\beta\omega)S(k,\omega)=\frac{2k^4t}{m^2}+\frac{\rho k^2}{m^2}\int d\boldsymbol{r}\,g(r)(1-\cos kx)\frac{\partial^2\phi(r)}{\partial x^2} \quad (3.58'')$$

These are immediately seen to reduce to (3.56) and (3.58) in the classical limit, where, $\tilde{S}(k,\omega)=S(k,\omega)$ and $\mathrm{tgh}(\tfrac{1}{2}\hbar\beta\omega)=\tfrac{1}{2}\hbar\beta\omega$.

Similarly, the proper expression for the zero-moment sum rule, that is the static structure factor $S(k)=\frac{1}{N}\langle\rho_k\rho_{-k}\rangle$, is

$$\begin{aligned}S(k)&=\int_{-\infty}^{\infty}\frac{d\omega}{2\pi}\tilde{S}(k,\omega)\\&=-\frac{\hbar}{\rho}\int_{-\infty}^{\infty}\frac{d\omega}{2\pi}\chi''(k,\omega)\coth(\tfrac{1}{2}\hbar\beta\omega)\end{aligned} \quad (3.24')$$

The Ornstein–Zernike relation for $S(k)$ in the long-wavelength limit is in general no longer true, but it remains obviously true that the static density response of the fluid in the long wavelength limit is related to its compressibility,

$$\lim_{k\to 0}\chi(k,0)=-\rho^2K_T \quad (1.27')$$

In the classical limit, using (3.48), this reduces to the Ornstein–Zernike relation (1.27).

3.4 Simple models for $S_s(k, \omega)$ and $S(k, \omega)$

Before we begin to enquire into the general structure and properties of the dynamic structure factors $S_s(k, \omega)$ and $S(k, \omega)$, it will be instructive to evaluate these functions for a few simple models of the dynamics of fluids. The models actually approach the behaviour of real liquids under certain conditions of wave vector and frequency, as we shall see.

3.4.1 Free-particle model

The first simple model that we shall consider is that of a classical perfect gas. We expect that a real fluid will behave as an assembly of free particles in a scattering process at high momentum and energy transfer, or, in other words, that its dynamics will approach this model in the limit of extremely small distances and times.

From (3.20) we find that for a classical perfect gas, using $\boldsymbol{R}_1(t) = \boldsymbol{R}_1(0) + \boldsymbol{p}_1\, t/m$, the intermediate scattering self-function is

$$F_s(k, t) = \langle \exp(i\boldsymbol{k}\cdot\boldsymbol{p}_1 t/m)\rangle = \int \mathrm{d}\boldsymbol{p}_1 \exp[-p_1^2/(2mk_BT)] \exp(i\boldsymbol{k}\cdot\boldsymbol{p}_1\, t/m) \Big/ \int \mathrm{d}\boldsymbol{p}_1 \exp[-p_1^2/(2mk_BT)] = \exp[-k^2t^2k_BT/2m] \tag{3.62}$$

and correspondingly

$$\begin{aligned} S_s(k, \omega) &= \int_{-\infty}^{\infty} \mathrm{d}t \exp(-i\omega t) \exp(-k^2t^2k_BT/2m) \\ &= \left(\frac{2\pi m}{k_BTk^2}\right)^{1/2} \exp[-m\omega^2/(2k^2k_BT)] \end{aligned} \tag{3.63}$$

These are Gaussian functions with half-width determined by $k^2\langle v_x^2\rangle$, where $\langle v_x^2\rangle \equiv k_BT/m$ is the mean square thermal velocity along $\boldsymbol{k}$. Furthermore, since no correlations between positions or velocities of different particles exist in a classical perfect gas, we immediately find that in this model $G_d(r, t) = 0$ and $S(k, \omega) = S_s(k, \omega)$.

From the above result's for a classical perfect gas it is easily found that

$$\begin{aligned} \int_{-\infty}^{\infty} \frac{\mathrm{d}\omega}{2\pi} \omega^2 S_s(k, \omega) &= \int_{-\infty}^{\infty} \frac{\mathrm{d}\omega}{2\pi} \omega^2 S(k, \omega) \\ &= k^2k_BT/m \end{aligned} \tag{3.64}$$

in agreement with the exact results (3.55) and (3.56). Obviously, only the kinetic parts of the higher moments will be reproduced by this model.

3.4.2 Diffusion equation for $G_s(r, t)$

We can make contact with the macroscopic diffusional behaviour of the fluid by arguing that the self-correlation function $G_s(r, t)$ must be related to the solution of the diffusion equation, for it represents the meanderings of a particle initially at the origin at time $t=0$. Actually, this particle can exchange energy and momentum with its neighbours and we then fundamentally have only one conservation law for particle number. This is why the self-diffusion problem is much easier than viscosity and heat conduction.

The diffusion equation should be obeyed by $G_s(r, t)$ for times long compared with the collision time and for distances long compared with the mean free path. Bearing this in mind for later purposes, we write

$$D\nabla^2 G_s(r, t) = \frac{\partial G_s(r, t)}{\partial t} \tag{3.65}$$

As we have seen, we need a solution satisfying the condition

$$G_s(r, 0) = \delta(\mathbf{r}) \tag{3.66}$$

and such that, for the probabilistic interpretation of G_s

$$\int g(r, t)\,\mathrm{d}\mathbf{r} = 1, \text{ for all } t \tag{3.67}$$

Actually, (3.65) has the form of the Bloch equation for the density matrix (which is also useful for quantal fluids) and we obtain the desired solution from that equation in appendix 3.2. The result is

$$G_s(r, t) = \{4\pi D|t|\}^{-3/2} \exp\left(-\frac{r^2}{4D|t|}\right) \tag{3.68}$$

which is soon shown to satisfy (3.66) and (3.67).

From the results of appendix 3.2 we can immediately extract the intermediate scattering function $F_s(k, t)$ as

$$F_s(k, t) = \exp(-k^2 D|t|) \tag{3.69}$$

From phenomenological theory, we expect the diffusion coefficient to be related to some mean square distance over a characteristic time and so we next calculate the mean square displacement $\langle r^2 \rangle$ as

$$\langle r^2 \rangle = \int \mathrm{d}\mathbf{r}\, r^2 G_s(r, t) = 6Dt \tag{3.70}$$

from eqn (3.68). This result is valid for times long compared with collision times and we must expect that, for a $G_s(r, t)$ correctly calculated from a force law, $\langle r^2 \rangle$ will be proportional to t, the slope yielding the diffusion constant. Machine calculations have been made for a number of cases and *figure 3.2* shows typical results for two different potentials for liquid sodium. These

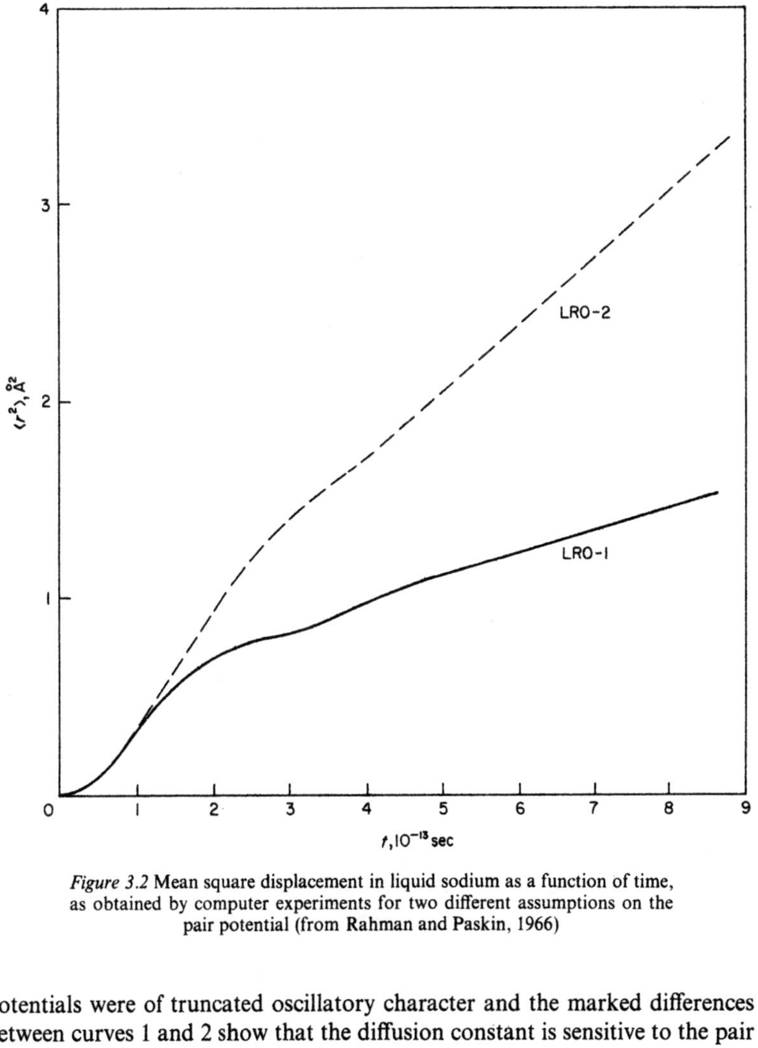

Figure 3.2 Mean square displacement in liquid sodium as a function of time, as obtained by computer experiments for two different assumptions on the pair potential (from Rahman and Paskin, 1966)

potentials were of truncated oscillatory character and the marked differences between curves 1 and 2 show that the diffusion constant is sensitive to the pair potential.

We turn now to obtain an important expression for D in terms of $S_s(k, \omega)$. To do so, we recall that

$$S_s(k, \omega) = \int \mathrm{d}t \exp(-i\omega t) F_s(k, t) \tag{3.71}$$

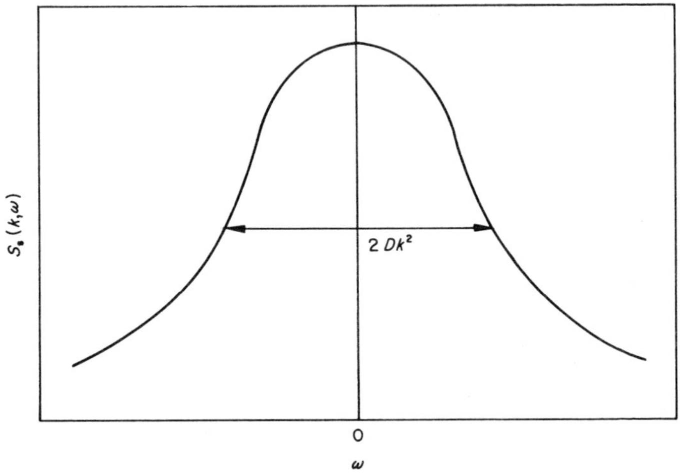

Figure 3.3 Schematic form of $S_s(k, \omega)$ for diffusion

and using (3.69) we find

$$S_s(k, \omega) = 2\int_0^\infty dt \cos \omega t \exp(-k^2Dt)$$

$$= \frac{2Dk^2}{\omega^2 + (Dk^2)^2} \tag{3.72}$$

The form of $S_s(k, \omega)$ as a function of ω is shown in *figure 3.3* and it may be seen that the total width at half maximum is determined by the diffusion constant D. Notice that $S_s(k, \omega)$ in a real liquid should change from the Lorentzian form (3.72) at small k and ω to the Gaussian form (3.63) at large k and ω.

We have already seen that the above theory can only work for long times or small ω and, in general, we expect macroscopic arguments to work in the long wavelength or small k limit. While, therefore, we cannot trust (3.72) in general, it is perfectly proper for the limiting case of small k and ω. If we wish to obtain a non-zero limit as $k \to 0$ we must, from (3.72), consider $S_s(k, \omega)/k^2$, and in the limit we have

$$\lim_{k \to 0} S_s(k, \omega)/k^2 = \frac{2D}{\omega^2} \tag{3.73}$$

In general, for a proper calculation of $S_s(k, \omega)$ the left-hand side of (3.73) will be a function of ω which will only become proportional to ω^{-2} as $\omega \to 0$. Thus, a *general* result for D may be written

$$D=\tfrac{1}{2}\lim_{\omega\to 0}\omega^2\lim_{k\to 0}S_s(k,\omega)/k^2 \qquad (3.74)$$

So far, no precise use of this result for any realistic interaction has been made, but it clearly gives interesting information about the structure of $S_s(k, \omega)$ near the origin of the (k, ω) plane, which any approximate theory should embody. The long-time behaviour of the self-motions in a liquid will be discussed in more detail in section 4.2.

3.4.3 Contribution of a collective mode to $S(k, \omega)$

As another important model for the calculation of the dynamic structure factor, we consider the case where the collective coordinate $\rho_k(t)$ describes a pure collective mode. This situation is closely realised at long wavelengths, both in quantal fluids (zero sound in liquid He^3 and phonon branch of the excitation spectrum in liquid He^4; see chapter 8) and in charged fluids (plasmon excitation; see chapter 7).

As a result of the standard transformation to normal coordinates in a problem of vibrations, the density fluctuation operator describing a pure collective mode can be written through boson creation and annihilation operators, a_k^+ and a_k, as (see for example Ziman, 1972)

$$\rho_k = N^{1/2}A_k(a_k+a_k^+) \qquad (3.75)$$

where A_k is an appropriate c-number giving the strength of the mode. The processes involved in the density fluctuation clearly are emission and absorption processes of one quantum. From (3.18) we find

$$\begin{aligned}F(k,t)&=|A_k|^2\sum_{\mathrm{m}} w_{\mathrm{m}}\langle\Psi_{\mathrm{m}}|(a_{-k}+a_k^+)\exp(iHt/\hbar)(a_k+a_{-k}^+)\exp(-iHt/\hbar)|\Psi_{\mathrm{m}}\rangle\\ &=|A_k|^2\sum_{\mathrm{m,m'}} w_{\mathrm{m}}\exp[i(E_{\mathrm{m'}}-E_{\mathrm{m}})t/\hbar]|\langle\Psi_{\mathrm{m}}|a_{-k}+a_k^+|\Psi_{\mathrm{m'}}\rangle|^2\\ &=|A_k|^2\sum_{\mathrm{m,m'}} w_{\mathrm{m}}\{\exp(i\omega_k t)|\langle\Psi_{\mathrm{m}}|a_{-k}|\Psi_{\mathrm{m'}}\rangle|^2+\exp(-i\omega_k t)|\langle\Psi_{\mathrm{m}}|a_k^+|\Psi_{\mathrm{m'}}\rangle|^2\}\\ &=|A_k|^2\{\exp(i\omega_k t)\langle a_{-k}a_k^+\rangle+\exp(-i\omega_k t)\langle a_k^+a_k\rangle\}\end{aligned} \qquad (3.76)$$

In the above derivation we have taken account of the selection rules for the harmonic oscillator, which ensure that the state $\Psi_{\mathrm{m'}}$ differs from the state Ψ_{m} by the addition or subtraction of a quantum of energy $\hbar\omega_k$, ω_k being the eigenfrequency of the mode. Since $\langle a_k^+a_k\rangle$ represents the mean number of quanta at equilibrium,

$$\langle a_k^+a_k\rangle \equiv n_k = [\exp(\hbar\omega_k/k_BT)-1]^{-1} \qquad (3.77)$$

we can finally write the intermediate scattering function as

$$F(k, t)=|A_k|^2\{(n_k+1)\exp(i\omega_k t)+n_k\exp(-i\omega_k t)\} \tag{3.78}$$

and the corresponding dynamic structure factor as

$$S(k, \omega)=2\pi|A_k|^2\{(n_k+1)\delta(\omega-\omega_k)+n_k\delta)\omega+\omega_k)\} \tag{3.79}$$

If we recall the interpretation of $S(k, \omega)$ as the probability of inelastic scattering of neutrons from the liquid, we can interpret the two terms in the brackets as representing, respectively, the emission and the absorption of a quantum by the neutron in a scattering process. The factor (n_k+1) in the emission term accounts, of course, for stimulated and spontaneous emission. We also note that (3.79) implies the relation

$$S(-\mathbf{k}, -\omega)=\exp(-\hbar\omega/k_B T)S(\mathbf{k}, \omega) \tag{3.80}$$

This is (cf. p. 44) a general property of the dynamic structure factor of a fluid, being just a consequence of the principle of detailed balancing, as can be proved by repeating the calculation of section 3.2.3 for the probability $P(-\mathbf{k}, -\omega)$. Of *course, in the classical limit* $(\hbar\omega \ll k_B T)$ we have $S(-\mathbf{k}, -\omega)=S(-\mathbf{k}, \omega)$.

Needless to say, all the simple models described above have a limited range of validity. To get a full microscopic theory of $S(k, \omega)$ which correctly includes the hydrodynamic (e.g. in $S_s(k, \omega)$ diffusion) regime remains a major goal of the theory. To do this from a given pair potential, it seems clear, will at least for a long while yet only be feasible by machine calculations (see for example Alder, 1973 for hard spheres). Therefore the philosophy pioneered by Vineyard (1958) of relating $S(k, \omega)$ to $S_s(k, \omega)$ and the static correlation functions still appears to be a worthwhile direction for further study.

CHAPTER 4

HYDRODYNAMIC FORMS OF CORRELATION FUNCTIONS AND GENERALISED HYDRODYNAMICS

The simple models evaluated in the preceding chapter drew attention to some of the basic features of time-dependent correlations in fluids. These are: (*a*) the meaning of their Fourier transforms as frequency spectra of dissipation by suitable external probes; (*b*) their relation to macroscopic fluid dynamics (i.e. hydrodynamics) and to transport coefficients in the limit $k \to 0$ and $\omega \to 0$; and (*c*) their short-time behaviour as expressible through moments of the frequency spectrum. The most convenient way to develop a theory of time-dependent correlations whose structure incorporates these properties, and thus allows us to draw conclusions on the dynamics of liquids, is through the response function and memory function approach (Mori, 1965*a*; see also Kubo, 1966). It will be instructive to introduce this formalism through a study of the simplest time-dependent correlation function, which is the velocity autocorrelation function for a particle in a classical fluid, before proceeding to the discussion of $S(k, \omega)$.

4.1 Velocity autocorrelation function

The velocity autocorrelation function is defined as

$$\phi(t) = \langle v(0) \cdot v(t) \rangle \tag{4.1}$$

where $v(t)$ is the velocity of a specified particle at time t. One reason for the importance of this function is that it is related to the mean square distance, $u(t)$ say, travelled by a particle over a time interval t. The relation, proved in appendix 4.1, is

$$u(t) = 2 \int_0^t d\tau (t - \tau) \phi(\tau) \tag{4.2}$$

$\phi(t)$ is especially suitable for treating the theory of the Brownian motion of a colloidal particle in a liquid, but it can also be used to describe the irregular zig-zag motion of a particle of a fluid under the impacts of the other particles. We then expect it to be related to the self-part of the dynamic structure factor, and it can in fact be shown (Rahman, Singwi and Sjolander 1962) that in a classical fluid

$$F_s(k, t) = \exp\left[-\tfrac{1}{6}k^2 u(t) + 0(k^4)\right] \tag{4.3}$$

the terms of order k^4 in (4.3) being determined by higher order velocity correlations. The behaviour of $\phi(t)$ in liquid argon, as obtained by Nijboer and Rahman (1966) by computer simulation, is reported in *figure 4.1*. The fact that $\phi(t)$ becomes negative seems to indicate that each particle on average recoils at short times after hitting its neighbours, but no oscillatory motion around a mean position is apparent in this fluid.

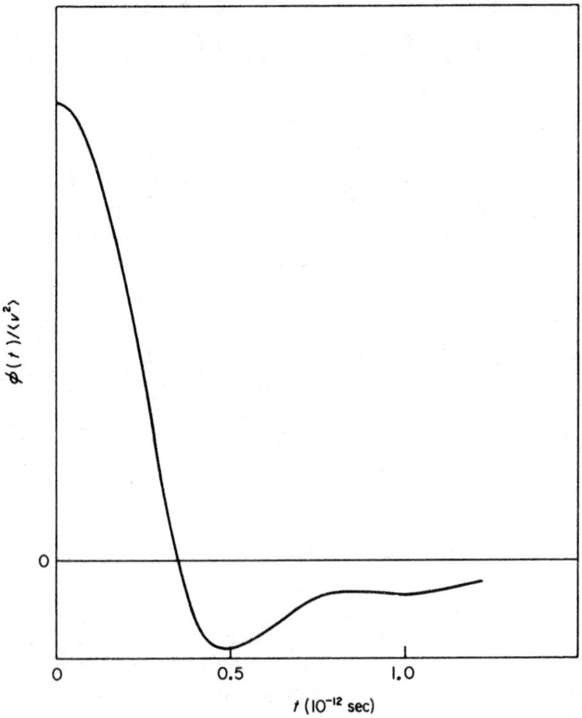

Figure 4.1 Velocity autocorrelation function in liquid argon near its freezing point (from Nijboer and Rahman, 1966)

Evidently from the above relation (4.2) and eqn (3.70) one obtains an expression relating the self-diffusion coefficient D to the time integral of $\phi(t)$,

$$D = \lim_{t \to \infty} u(t)/(6t) = \tfrac{1}{3} \int_0^{\infty} \mathrm{d}t\phi(t) \tag{4.4}$$

This provides, in the present instance, the connection of the correlation function to macroscopic dynamics.

We may now ask whether $\phi(t)$ is related to the dissipative response of the fluid to some external perturbation, just as $F_s(k, t)$ is related to incoherent neutron inelastic scattering. The answer to this question is affirmative and is implicit, in the macroscopic limit, in the well-known Einstein relation between the mobility μ and the self-diffusion coefficient,

$$\mu = D/k_B T \tag{4.5}$$

To prove this relation, it is sufficient to note that particles flow in a potential field $V(z)$ with a drift velocity

$$v_z = -\mu \frac{dV(z)}{dz} \tag{4.6}$$

so that the total particle current density is

$$i_z = -D \frac{d\rho(z)}{dz} - \rho(z)\mu \frac{dV(z)}{dz} \tag{4.7}$$

the first term being the diffusion current which is present even in the absence of external field provided that a density gradient $d\rho(z)/dz$ is maintained in the fluid. At equilibrium in the presence of the external field we must have $i_z = 0$ and $\rho(z) \propto \exp[-V(z)/k_B T]$, whence (4.5) follows. We now combine (4.4) and (4.5) to write

$$\mu = (3k_B T)^{-1} \int_0^\infty dt\phi(t) \tag{4.8}$$

This relation shows that the mobility, which describes the response of the particles of the fluid to an applied field, is determined by the velocity correlations in time, which are a property of the fluid in the absence of the field. Such a relation is evidently a deep consequence of the fact that both frictional loss under an applied field and velocity fluctuations at equilibrium derive from the same physical process, that is the collisions between the particles of the fluid. The extension of eqn (4.8) to time-dependent fields will be discussed below.

We now complete the discussion of the general properties of $\phi(t)$ by introducing its frequency spectrum, $f(\omega)$, through the Fourier analysis

$$\phi(t) = \int_{-\infty}^{\infty} \frac{d\omega}{2\pi} \exp(i\omega t) f(\omega) \tag{4.9}$$

The function $f(\omega)$ is real and even, as can be seen by inverting eqn (4.9) and using the invariance of $\phi(t)$ under time reversal, which is a consequence of its being a property of the fluid at equilibrium. From the definition of $\phi(t)$ we find

$$\langle v(\omega') \cdot v(\omega) \rangle = 2\pi f(\omega)\delta(\omega + \omega') \tag{4.10}$$

where $v(\omega)$ is the Fourier transform of $v(t)$. The behaviour of $f(\omega)$ in liquid argon is shown in the solid curve of *figure 4.2*, based on work of Nijboer and Rahman

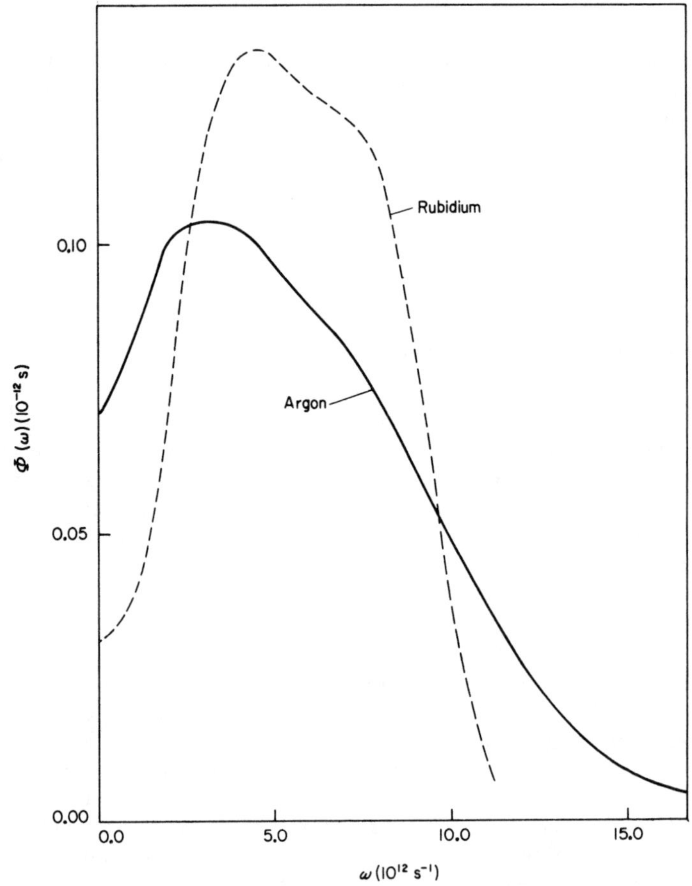

Figure 4.2 Frequency spectrum of the velocity autocorrelation function for liquid argon (solid curve—from Nijboer and Rahman, 1966) and liquid rubidium (broken curve—from Schommers, 1973) (from Copley and Lovesey, 1975)

(1966). The main notable features, which distinguish $f(\omega)$ from the frequency spectrum of an ideal harmonic crystal, are the absence of a high-frequency cut-off and its non-vanishing value at zero frequency. While the former feature is clearly due to the absence of lattice periodicity, the latter is related to self-diffusion, since from (4.4) we have

$$\lim_{\omega \to 0} f(\omega) = 6D \tag{4.11}$$

As we shall see, $f(\omega)$ is simply related to the frequency-dependent mobility in an applied field, and is therefore measurable in loss experiments.

Finally, we note that the moments of $f(\omega)$ give us the short-time expansion of $\phi(t)$, which from eqn (4.9) reads

$$\phi(t)=\sum_{n=0}^{\infty}\frac{a_n}{(2n)!}t^{2n} \tag{4.12}$$

with

$$a_n=(-1)^n\int_{-\infty}^{\infty}\frac{d\omega}{2\pi}\omega^{2n}f(\omega) \tag{4.13}$$

It is easily proved that

$$a_0=\langle v^2\rangle=3k_BT/m \tag{4.14}$$

and

$$a_1=-\langle \dot{v}^2\rangle \tag{4.15}$$

Thus, while a_0 reflects only the free-particle behaviour, a_1 is given by the mean square force acting on a particle in the liquid, and can thus be evaluated from the radial distribution function $g(r)$ (see appendix 3.1). The higher moments of $f(\omega)$ involve higher order static correlation functions, as well as the pair interaction.

4.1.1 Generalised Langevin equation

Having reviewed the main basic properties of the velocity autocorrelation function, we now consider the structure that a theory of it should take in order to incorporate these properties. The simplest theory of Brownian motion is based on the phenomenological Langevin equation (see for instance Wannier, 1966)

$$m\dot{v}(t)=-m\gamma v(t)+\boldsymbol{R}(t) \tag{4.16}$$

in which the force exerted by the particles of the fluid on the colloidal particle is split into a frictional term, systematically directed against the motion, and a random term $\boldsymbol{R}(t)$ which averages to zero and is not correlated with the velocity of the particle, namely

$$\langle \boldsymbol{R}(t)\rangle=0,\quad \langle v(t')\cdot\boldsymbol{R}(t)\rangle=0 \tag{4.17}$$

Equation (4.16) then gives

$$\frac{d}{dt}\langle v(0)\cdot v(t)\rangle=-\gamma\langle v(0)\cdot v(t)\rangle \tag{4.18}$$

whose solution is

$$\phi(t)=\langle v^2\rangle\exp\left[-\gamma|t|\right] \tag{4.19}$$

Equation (4.4) is satisfied provided that the frictional constant γ is related to the self-diffusion coefficient by

$$\gamma = \frac{\langle v^2 \rangle}{3D} = (m\mu)^{-1} \tag{4.20}$$

We note, however, that the exponential decay of $\phi(t)$ in time is certainly incorrect at short times, since it contradicts (4.12). Actually, the form (4.19) bears no resemblance to the behaviour of $\phi(t)$ for an argon atom in liquid argon, shown in *figure 4.1.*

The basic flaw of the Langevin equation lies in the assumption that the systematic force is determined only by the instantaneous value of the particle velocity. This implies that, at any given instant of time when the particle has velocity $v(t)$, the fluid should have been able to readjust itself instantaneously to the equilibrium state consistent with that given velocity, thus erasing its memory of the previous dynamics of the chosen particle. We should therefore extend the Langevin equation to read

$$m\dot{v}(t) = -m \int_0^t \mathrm{d}t' \gamma(t-t') v(t') + \boldsymbol{R}(t) \tag{4.21}$$

It should be stressed that this separation of the force into the sum of a systematic term, embodying memory of past dynamics through the 'memory function' $\gamma(t)$, and a random term, which still satisfies (4.17), is in fact exact since the form (4.21) of the equation of motion can be derived from the exact equation of motion (Mori, 1965*a*). Furthermore, in the presence of an external time-dependent force $\boldsymbol{F}(t)$ which is sufficiently weak to be treated by linear theory, the equation of motion simply becomes

$$m\dot{v}(t) = -m \int_0^t \mathrm{d}t' \gamma(t-t') v(t') + \boldsymbol{R}(t) + \boldsymbol{F}(t) \tag{4.22}$$

the memory function and the random force remaining unchanged in linear theory where the drift velocity produced by the applied force will be small.

Equation (4.22) is the correct form of the equation of motion to embody in the theory the basic relation between dissipation and correlations of equilibrium fluctuations. This is achieved through the memory function. Specifically, it is now easy to show that the velocity autocorrelation function in the absence of external force, as well as the response of the fluid to the external force, are fully determined by the memory function. In fact, from eqn (4.22) with $\boldsymbol{F}(t) = 0$, upon using $\langle v(0) \cdot \boldsymbol{R}(t) \rangle = 0$, we find

$$\int_0^\infty \mathrm{d}t \exp(i\omega t) \langle v(0) \cdot \dot{v}(t) \rangle = -\int_0^\infty \mathrm{d}t \exp(i\omega t) \int_0^t \mathrm{d}t' \gamma(t-t') \langle v(0) \cdot v(t') \rangle \tag{4.23}$$

which is immediately integrated to yield

$$\tilde{\phi}(\omega)=\frac{\langle v^2\rangle}{-i\omega+\tilde{\gamma}(\omega)} \tag{4.24}$$

Here, $\tilde{\phi}(\omega)$ and $\tilde{\gamma}(\omega)$ are the Laplace transforms of the velocity auto-correlation function and of the memory function,

$$\tilde{\phi}(\omega)=\int_0^\infty \mathrm{d}t \exp(i\omega t) \langle v(0)\cdot v(t)\rangle \tag{4.25}$$

and

$$\tilde{\gamma}(\omega)=\int_0^\infty \mathrm{d}t \exp(i\omega t)\, \gamma(t) \tag{4.26}$$

On the other hand, with an external force of the type $F(t)=F_0 \exp(-i\omega t)$, upon using $\langle R(t)\rangle=0$, equation (4.22) gives the drift velocity as

$$\langle v(t)\rangle=\frac{1}{m}\frac{1}{-i\omega+\tilde{\gamma}(\omega)} F_0 \exp(-i\omega t) \tag{4.27}$$

Thus, the drift velocity acquired by a particle in response to a real monochromatic force $F(t)=F_0 \cos \omega t$ can be written

$$\langle v(t)\rangle=\mathrm{Re}\,[\mu(\omega)F_0 \exp(-i\omega t)] \tag{4.28}$$

where we have introduced the frequency-dependent mobility $\mu(\omega)$ as

$$\mu(\omega)=\frac{1}{m}\frac{1}{-i\omega+\tilde{\gamma}(\omega)} \tag{4.29}$$

Contact with macroscopic dynamics is achieved by letting $\omega\rightarrow 0$ in either (4.24) or (4.29), when we recover (4.20) with $\gamma=\tilde{\gamma}(0)$.

If we now combine (4.24) and (4.29), we can write

$$\mu(\omega)=\frac{1}{m\langle v^2\rangle}\tilde{\phi}(\omega) \tag{4.30}$$

which is the statement of the fluctuation–dissipation theorem in the present instance. Thus, the velocity response of the fluid to an applied time-dependent force is determined by the time correlations of the particle velocity in the unperturbed fluid. In terms of the frequency spectrum introduced in (4.9) we can write

$$f(\omega)=m\langle v^2\rangle[\mu(\omega)+\mu(-\omega)]=2m\langle v^2\rangle\, \mathrm{Re}\, \mu(\omega) \tag{4.31}$$

having used, in the last step, the identity $\tilde{\gamma}(-\omega)=\tilde{\gamma}^*(\omega)$ according to (4.26). Thus the frequency spectrum of the velocity autocorrelation is measurable as the spectrum of energy loss by an applied field.

4.1.2 Memory function and continued fraction expansion

While the introduction of the memory function has allowed us to incorporate in a natural way into the theory the fluctuation–dissipation theorem and the relation to transport coefficients, the theory of velocity autocorrelations is now faced with the problem of evaluating the memory function. To this end, it is crucial to realise that the memory function is itself a time-dependent correlation function. Precisely, it is directly related to the autocorrelations in time of the random force $\boldsymbol{R}(t)$, which has so far played no role in the development.

From eqn (4.21) we find for the Laplace transform of the time correlations of the random force

$$\tilde{\phi}_{\mathrm{R}}(\omega) \equiv \int_0^\infty \mathrm{d}t \exp(i\omega t) \langle \boldsymbol{R}(0) \cdot \boldsymbol{R}(t) \rangle$$

$$= m^2 \int_0^\infty \mathrm{d}t \exp(i\omega t) [\langle \dot{\boldsymbol{v}}(0) \cdot \dot{\boldsymbol{v}}(t) \rangle + \tilde{\gamma}(\omega) \langle \dot{\boldsymbol{v}}(0) \cdot \boldsymbol{v}(t) \rangle] \tag{4.32}$$

By integrating by parts the first term in the brackets and using the identities $\langle \dot{\boldsymbol{v}}(0) \cdot \boldsymbol{v}(0) \rangle = \frac{1}{2} \frac{\mathrm{d}}{\mathrm{d}t_0} \langle \boldsymbol{v}(t_0) \cdot \boldsymbol{v}(t_0) \rangle = 0$ and $\frac{\mathrm{d}}{\mathrm{d}t_0} \langle \boldsymbol{v}(t_0) \cdot \boldsymbol{v}(t_0 + t) \rangle = 0$, we find

$$\tilde{\phi}_{\mathrm{R}}(\omega) = m^2[-i\omega + \tilde{\gamma}(\omega)] \int_0^\infty \mathrm{d}t \exp(i\omega t) \langle \dot{\boldsymbol{v}}(0) \cdot \boldsymbol{v}(t) \rangle$$

$$= -m^2[-i\omega + \tilde{\gamma}(\omega)] \int_0^\infty \mathrm{d}t \exp(i\omega t) \langle \boldsymbol{v}(0) \cdot \dot{\boldsymbol{v}}(t) \rangle \tag{4.33}$$

which, upon integration by parts and use of (4.24), yields

$$\tilde{\phi}_{\mathrm{R}}(\omega) = m^2 \langle v^2 \rangle \tilde{\gamma}(\omega) \tag{4.34}$$

Thus, the frequency spectrum of the random force is determined by the generalised friction. This result was first obtained by Nyquist (1928), who showed that the random electromotive force across a resistor is determined by its impedance.

The Nyquist theorem (4.34) now suggests a simple representation for the frequency dependence of the transform $\tilde{\phi}(\omega)$ of the velocity autocorrelations (Mori, 1965*b*). We can, in fact, write the equation of motion for the random force $\boldsymbol{R}(t)$ in the form of a generalised Langevin equation, involving a new memory function, $\gamma_{\mathrm{R}}(t)$ say. We shall then find

$$\tilde{\phi}_{\mathrm{R}}(\omega) = \frac{\Delta_{\mathrm{R}} m^2 \langle v^2 \rangle}{-i\omega + \tilde{\gamma}_{\mathrm{R}}(\omega)} \tag{4.35}$$

where Δ_{R} is some constant, and consequently from (4.24) and (4.34)

$$\tilde{\phi}(\omega) = \cfrac{\langle v^2 \rangle}{-i\omega + \cfrac{\Delta_{\mathrm{R}}}{-i\omega + \tilde{\gamma}_{\mathrm{R}}(\omega)}} \tag{4.36}$$

Clearly, $\tilde{\gamma}_R(\omega)$ will itself enter a new Nyquist theorem, and if we can repeat this process for any desired number of times, we arrive at an expression for $\tilde{\phi}(\omega)$ in the form of a continued fraction involving a set of constants such as Δ_R above. The values of these constants can be determined with the help of the known moments of $f(\omega)$, using a theorem for Laplace transforms according to which the coefficients of the high-frequency expansion of $\tilde{\phi}(\omega)$ give the coefficients of the short-time expansion of $\phi(t)$, that is the moments of $f(\omega)$ (for further developments see Sears, 1969).

The obvious assumption underlying the above discussion is that a generalised Langevin equation of the simple form (4.21) may be written for any dynamical variable of the fluid. More generally, we shall instead find that several coupled variables must be considered simultaneously, as is the case, for instance, in the hydrodynamics of longitudinal phenomena in one-component liquids (see section 4.3) or for velocity autocorrelations in multicomponent fluids. In this situation the memory function becomes a matrix coupling the rate of change of any one of the dynamical variables to the values of all the others at previous times, and the main results derived above remain valid when restated in matrix notation (Mori, 1965*b*). We shall explicitly illustrate this procedure in the calculation of $S(k, \omega)$ in the hydrodynamic regime in section 4.3.

4.2 Some explicit results for the velocity autocorrelation function

Having discussed a formally exact framework for treating the velocity autocorrelation function, we shall in this section first give an approximate theory of it for hard spheres. Secondly, we shall discuss briefly the long-time behaviour of the velocity autocorrelation function in real liquids (i.e. *not* hard spheres) and the consequence for the spectral density or frequency spectrum.

4.2.1 Hard sphere form

In a fluid like argon some of the physical properties can be usefully discussed by using a hard-sphere model. For this case, adopting the procedure we used in deriving the moment theorems, let us make a short-time expansion of the velocity autocorrelation function, namely

$$\langle v(0)\cdot v(t)\rangle = \langle v^2\rangle + t\left\langle \frac{\mathrm{d}}{\mathrm{d}t}[v(0)\cdot v(t)]\right\rangle_{t=0} + \tfrac{1}{2}t^2\left\langle \frac{\mathrm{d}^2}{\mathrm{d}t^2}[v(0)\cdot v(t)]\right\rangle_{t=0} + \cdots \tag{4.37}$$

In most cases it can be shown that the odd powers of t have vanishing coefficients which, it turns out, is related to the fact that positions and velocities are uncorrelated in an equilibrium ensemble.

Unfortunately, when we take the limit of hard spheres, this is no longer true. The ensemble averaging has to be done *before* the time derivatives, because of the ill-defined properties of a single molecule when there is a pathological force law. The term in t remains, and to make progress we follow Longuet-Higgins and Pople (1956) and write

$$\langle v(0)\cdot v(t)\rangle \simeq \langle v^2\rangle \exp(-\gamma|t|) \tag{4.38}$$

where we stress that this is specific to hard spheres. (Actually, we pointed out in section 4.1.1 that this is precisely the form describing Brownian motion in the Langevin treatment. It is usually rejected as unrealistic for liquids, because of the long range of the Fourier transform of the velocity correlation function leading to a divergent second moment. This divergence, however, exists for hard spheres.)

From (4.20) we have immediately

$$D=\frac{k_B T}{m\gamma} \tag{4.39}$$

The decay time γ^{-1} of the velocity correlations may be calculated by a kinetic argument, if we write it in the alternative form

$$\gamma=-\lim_{\Delta t\to 0}\left\{\frac{\langle v\cdot \Delta v\rangle}{\langle v^2\rangle \Delta t}\right\} \tag{4.40}$$

where Δv is the vector increment in velocity of the molecule in time interval Δt. The argument is given in detail in appendix 4.2 and the final result for D is given by

$$D=\tfrac{1}{2}R\left(\frac{\pi k_B T}{m}\right)^{1/2}\left(\frac{p}{\rho k_B T}-1\right)^{-1} \tag{4.41}$$

where the pressure p has entered because, through the virial result (1.15), it is related to the value of $g(r)$ at the hard sphere diameter R. This result is probably valid over a wide density range. In particular, for low densities the bracket in (4.41) may be replaced by the second virial coefficient, which is, in fact, $\frac{16}{3}\pi R^3$ for hard spheres. Then D becomes

$$D=\frac{3}{32R^2\rho}\left(\frac{k_B T}{\pi m}\right)^{1/2} \tag{4.42}$$

which is the result obtained earlier by Chapman and Enskog (reported by Chapman and Cowling, 1939) for a low density system. More recent work by Ross and Schofield (1971) considers, by machine calculation, the effect on D of softening the hard sphere potential (actually by using an inverse twelfth power potential)

The frequency spectrum of the velocity autocorrelation function, introduced in (4.9), is anomalous for hard spheres because of the presence of an odd power of t in the short-time expansion. Substituting for the velocity correlation function assuming the Longuet-Higgins–Pople form (4.38) for hard spheres we find

$$f(\omega)=6(k_B T/m)\gamma/(\gamma^2+\omega^2) \tag{4.43}$$

The slow fall-off of $f(\omega)$ as ω^{-2} is a consequence of the non-analyticity of the velocity autocorrelation function for hard spheres at the origin of time. The value $f(0)$, on the other hand; from the above result for $f(\omega)$ is simply $6k_BT/m\gamma = 6D$, as it must be.

Actually, we can make contact with experiment via $f(\omega)$ and results obtained for liquid rubidium and liquid argon are shown in *figure 4.2*. The hard-sphere model is clearly inadequate, $f(\omega)$ being far too long-range, as remarked above, and not exhibiting the structure found experimentally. Machine calculations have shown that the shape for argon comes from a van der Waals tail on a hard-sphere potential, and it is interesting to note that the behaviour of the liquid metal is more markedly oscillator-like.

4.2.2 Non-analyticity of frequency spectra in classical liquids

We have seen that diffusion problems particularly are described very usefully via the incoherent neutron scattering function $S_s(k, \omega)$. The frequency spectrum $f(\omega)$, which is the Fourier transform of the velocity autocorrelation function, is related to $S_s(k, \omega)$ through

$$f(\omega) = \omega^2 \lim_{k \to 0} \frac{S_s(k\omega)}{k^2} \tag{4.44}$$

and this leads, for a Debye solid, to

$$f(\omega) \propto \omega^2, \qquad \omega < \omega_{\text{Debye}} = \omega_D \tag{4.45}$$

The diffusion constant D is simply $\frac{1}{6}f(0)$ and evidently $f(\omega)$ in a liquid describes not only vibrational modes but also diffusive modes in the low frequency part of the spectrum.

From hydrodynamics, Ernst, Hauge and van Leeuwen (1970; 1971) have shown that the velocity autocorrelation function has a long tail, decaying at long time as $t^{-3/2}$. This was first noticed by Alder and Wainwright (1970) in molecular dynamical calculations on hard spheres. This implies that the leading ω-dependent term in the expansion of $f(\omega)$, which is an even function of ω, is proportional to $\omega^{1/2}$. Thus, we assume an expansion in $\omega^{1/2}$ exists for small ω and we write (see Gaskell and March, 1970)

$$f(\omega) = 6D + d_1\omega^{1/2} + d_2\omega + d_3\omega^{3/2} + d_4\omega^2 + \cdots \tag{4.46}$$

This means that to account for the $t^{-3/2}$ decay of the velocity autocorrelation function, $f(\omega)$ has a cusp at $\omega = 0$ and an infinite slope there. *Figure 4.2* does not reveal this cusp as the long-time tail is not seen at the times shown in *figure 4.1*.

From the discussion of velocity correlation function we find (see Pomeau 1972; Levesque and Ashurst, 1974)

$$\frac{\langle v(0)\cdot v(t)\rangle}{\langle (v(0))^2\rangle}\simeq\frac{2}{3\rho}\left[4\pi\left(D+\frac{\eta}{\rho m}\right)t\right]^{-3/2} \tag{4.47}$$

and hence, using the explicit relation

$$f(\omega)=\frac{6k_BT}{m}\int_0^\infty\frac{\langle v(0)\cdot v(t)\rangle}{\langle (v(0))^2\rangle}\cos\omega t\,\mathrm{d}t \tag{4.48}$$

we obtain for the coefficient d_1

$$d_1=-(2\pi)^{1/2}\frac{4}{\rho}\left[4\pi\left(D+\frac{\eta}{\rho m}\right)\right]^{-3/2}\frac{k_BT}{m} \tag{4.49}$$

This establishes the non-analyticity of $f(\omega)$ at the origin, which is a quite general consequence of hydrodynamics for classical liquids. In fact, physically, the long-time tail comes about because the diffusing atom shares momentum with other atoms in a radius $r\sim(Dt)^{1/2}$. It is this which leads to a $t^{-3/2}$ tail, although it must be noted that the *precise* magnitude of the tail (cf eqn (4.49)) involves also the shear viscosity η, through coupling to collective transverse motions.

4.2.3 Experimental evidence and relation to the Debye frequency

The negatively infinite slope of $f(\omega)$ at $\omega=0$, predicted by the hydrodynamical argument, explains qualitatively the initial decrease in $f(\omega)$ from the value $6D$ observed experimentally by Egelstaff (1967) and also by Randolph (1964) for liquid Na. Experiments are currently in progress to attempt to test the above theory quantitatively.

Some additional comments are worth making, in connection with the expansion of $f(\omega)$. Thus:

(1) In the limit $D\to0$ and $\eta\to\infty$, which can be viewed as a trend to a glassy or a solid-like situation, d_1 above tends to zero and so does the first term $6D$ Whether d_2 and d_3 would then likewise tend to zero has yet to be established rigorously, but approximate arguments suggest that they behave quantitatively like d_1. If this is verified subsequently then a Debye-like term $d_4\omega^2$ remains in this limiting case.

(2) It is worth comparing results for $f(\omega)$ for various liquid metals at the melting temperature T_m. There (see Brown and March, 1968) we can write approximately

$$\left.\begin{aligned} Dm^{1/2}\rho^{1/3}/T_m{}^{1/2}&=\text{constant}\\ \eta/T_m{}^{1/2}m^{1/2}\rho^{2/3}&=\text{constant}\end{aligned}\right\} \tag{4.50}$$

and hence it follows that d_1 is proportional to $T_m{}^{1/4}/m^{1/4}\rho^{1/2}$.

We also note here that in the approximation in which we can use Lindemann's law, the first two terms scale with the Debye frequency ω_D in the manner proposed by Brown and March (1968), since

$$\omega_D{}^{1/2} = \text{const}\, T_m{}^{1/4} \rho^{1/6} / m^{1/4} \tag{4.51}$$

Finally, it is a rather remarkable result that in spite of the fact that $S_s(k, \omega)$, which in Fourier transform $G_s(r, t)$ represents the meanderings of a chosen particle in time, determines the behaviour of $f(\omega)$ entirely, through eqn (4.44), the behaviour of $f(\omega)$ around its non-analytical point $\omega = 0$ determines the shear viscosity. In particular, we can write, recalling that $f(\omega) = 0) = 6D$.

$$\lim_{\omega \to 0} \omega^{1/2} \frac{df}{d\omega} = -(8\pi)^{1/2} \left[4\pi \left(D + \frac{\eta}{\rho m} \right) \right]^{-3/2} \frac{k_B T}{m\rho} \tag{4.52}$$

This is clearly connected with the fact that there is a deep-lying relation between the self-function $S_s(k, \omega)$, the total scattering function and the stress–stress correlation function. (For the dependence of the long-time tails on dimensionality, see also Dorfman and Cohen, 1970; 1972.)

4.3 Hydrodynamic form of $S(k, \omega)$

The evaluation of $S(k, \omega)$ clearly requires the solution of the equation of motion for the density fluctuations in the liquid. In the macroscopic limit, namely for small k and ω, each particle is subject to many collisions as it travels over distances of the order of $2\pi/k$ and during times of the order of $2\pi/\omega$. Collisions act to bring about a state of local thermodynamic equilibrium, characterised by local (space- and time-dependent) densities of conserved quantities such as particle number and energy, or by the associated intensive variables such as pressure and temperature. This situation is described by the well-known laws of hydrodynamics, so that in the macroscopic limit we can have recourse to the hydrodynamic equations for the evaluation of correlations between density fluctuations.

In fact, linearised hydrodynamics suffices for the evaluation of $S(k, \omega)$, precisely for the same reason for which inelastic neutron scattering is given by $S(k, \omega)$ when the interaction of the fluid with the neutrons is supposed to be weak. The results of the treatment are of interest not only for the description of correlations that they afford in the hydrodynamic regime, but also because they can be extended, by the memory function formalism introduced in section 4.1, to give the structure of $S(k, \omega)$ for general wave vector and frequency (Kadanoff and Martin, 1963).

4.3.1 Hydrodynamic equations

Transport in a monatomic fluid entails the conservation of five quantities (number of particles, total momentum and total energy), as expressed by continuity equations for the particle density $\rho(\mathbf{r}, t)$, the momentum density $\mathbf{p}(\mathbf{r}, t)$, and the energy density $\varepsilon(\boldsymbol{r}, t)$:

$$\frac{\partial}{\partial t}\rho(\boldsymbol{r}, t)+\frac{1}{m}\nabla\cdot \boldsymbol{p}(\boldsymbol{r}, t)=0 \tag{4.53}$$

$$\frac{\partial}{\partial t}\boldsymbol{p}(\boldsymbol{r}, t)+\nabla\cdot \boldsymbol{\Pi}(\boldsymbol{r}, t)=0 \tag{5.54}$$

$$\frac{\partial}{\partial t}\varepsilon(\boldsymbol{r}, t)+\nabla\cdot \boldsymbol{J}(\boldsymbol{r}, t)=0 \tag{4.55}$$

Here, $\boldsymbol{\Pi}(\boldsymbol{r}, t)$ is the momentum-flux tensor and $\boldsymbol{J}(\boldsymbol{r}, t)$ is the energy current density. The state of local equilibrium in the hydrodynamic regime is characterised by the five local densities of conserved quantities, or by the associated intensive variables: the pressure $p(\boldsymbol{r}, t)$, the average local velocity $\boldsymbol{v}(\boldsymbol{r}, t)$, and the temperature $T(\boldsymbol{r}, t)$. For small deviations from complete equilibrium, and assuming that the fluid in complete equilibrium is homogeneous and at rest, we can write (see Landau and Lifshitz, 1959)

$$\langle \boldsymbol{J}(\boldsymbol{r}, t)\rangle=(\varepsilon+p)\boldsymbol{v}(\boldsymbol{r}, t)-\kappa\nabla T(\boldsymbol{r}, t) \tag{4.56}$$

where ε and p refer to the complete equilibrium state and κ is the coefficient of thermal conductivity. We can also write

$$\langle\Pi_{\alpha\beta}(\boldsymbol{r}, t)\rangle=\delta_{\alpha\beta}p(\boldsymbol{r}, t)-\eta\left[\frac{\partial v_\alpha(\boldsymbol{r}, t)}{\partial x_\beta}+\frac{\partial v_\beta(\boldsymbol{r}, t)}{\partial x_\alpha}\right]-\delta_{\alpha\beta}(\zeta-\tfrac{2}{3}\eta)\nabla\cdot \boldsymbol{v}(\boldsymbol{r}, t) \tag{4.57}$$

where η and ζ are the shear and bulk viscosities. The behaviour of the fluid is then completely specified by these equations in terms of the complete equilibrium state and of the transport coefficients κ, η and ζ.

Equation (4.55) can be rewritten

$$\left(\frac{\partial}{\partial t}-\frac{\eta}{m\rho}\nabla^2\right)\langle \boldsymbol{p}(\boldsymbol{r}, t)\rangle-\frac{\zeta+\frac{1}{3}\eta}{m\rho}\nabla(\nabla\cdot\langle \boldsymbol{p}(\boldsymbol{r}, t)\rangle)=-\nabla p(\boldsymbol{r}, t) \tag{4.58}$$

where we have used $\langle \boldsymbol{p}(\boldsymbol{r}, t)\rangle=m\rho\boldsymbol{v}(\boldsymbol{r}, t)$. Upon splitting the momentum density into a longitudinal (i.e. irrotational) part and a transverse (i.e. divergenceless) part, $\boldsymbol{p}=\boldsymbol{p}_{\mathrm{L}}+\boldsymbol{p}_{\mathrm{T}}$, we find that the transverse momentum density is not coupled with any of the other densities† and is determined by the equation

$$\left(\frac{\partial}{\partial t}-\frac{\eta}{m\rho}\nabla^2\right)\langle \boldsymbol{p}_{\mathrm{T}}(\boldsymbol{r}, t)\rangle=0 \tag{4.59}$$

We then take the divergence of (4.58) and use (4.53) to eliminate $\langle \boldsymbol{p}_{\mathrm{L}}(\boldsymbol{r}, t)\rangle$,

†This would not be valid in the presence of substantial angular correlations between the molecules. For some discussion of such correlations, see Buckingham (1967), and Egelstaff, Page and Powles, (1971).

finding the equation for the particle density in the form

$$m\left(\frac{\partial^2}{\partial t^2}-D_L\nabla^2\frac{\partial}{\partial t}\right)\langle\rho(\boldsymbol{r},t)\rangle=\nabla^2 p(\boldsymbol{r},t) \tag{4.60}$$

where $D_L=(\frac{4}{3}\eta+\zeta)/(m\rho)$. Finally, (4.55) can be rewritten as the heat diffusion equation,

$$\frac{\partial Q(\boldsymbol{r},t)}{\partial t}=\kappa\nabla^2 T(\boldsymbol{r},t) \tag{4.61}$$

where

$$Q(\boldsymbol{r},t)=\langle\varepsilon(\boldsymbol{r},t)\rangle-\frac{\varepsilon+p}{\rho}\langle\rho(\boldsymbol{r},t)\rangle \tag{4.62}$$

is the local heat density, according to the thermodynamic identity $TdS/V=d(\varepsilon V)/V+pdV/V$.

Since the local temperature depends on the local density, and the local pressure depends on the local heat density, eqns (4.60) and (4.61) are in fact two coupled equations. To solve them simultaneously we express $p(\boldsymbol{r},t)$ and $T(\boldsymbol{r},t)$ through $\langle\rho(\boldsymbol{r},t)\rangle$ and $Q(\boldsymbol{r},t)$ by means of the thermodynamic relations

$$p(\boldsymbol{r},t)=\left(\frac{\partial p}{\partial\rho}\right)_S\langle\rho(\boldsymbol{r},t)\rangle+\frac{V}{T}\left(\frac{\partial p}{\partial s}\right)_\rho Q(\boldsymbol{r},t) \tag{4.63}$$

and

$$T(\boldsymbol{r},t)=\left(\frac{\partial T}{\partial\rho}\right)_S\langle\rho(\boldsymbol{r},t)\rangle+\frac{V}{T}\left(\frac{\partial T}{\partial s}\right)_\rho Q(\boldsymbol{r},t) \tag{4.64}$$

The equations are then solved by taking their Laplace–Fourier transforms, to find the Laplace–Fourier transforms $\tilde{\rho}(\boldsymbol{k},\omega)$ and $\tilde{Q}(\boldsymbol{k},\omega)$ in terms of the initial density fluctuation $\langle\rho(\boldsymbol{r},0)\rangle$ and heat fluctuation $Q(\boldsymbol{r},0)$, or in terms of the initial pressure $P(\boldsymbol{r},0)$ and temperature $T(\boldsymbol{r},0)$ imposed in the fluid. Here,

$$\tilde{\rho}(\boldsymbol{k},\omega)=\int d\boldsymbol{r}\int_0^\infty dt\exp\left[-i(\boldsymbol{k}\cdot\boldsymbol{r}-\omega t)\right]\langle\rho(\boldsymbol{r},t)\rangle \tag{4.65}$$

with an analogous definition for $\tilde{Q}(\boldsymbol{k},\omega)$.

4.3.2 Hydrodynamic form of $S(k,\omega)$ at low temperatures

Let us consider first the simpler situation obtaining at low temperatures, where $\frac{1}{T}\left(\frac{\partial p}{\partial S}\right)_\rho\rightarrow 0$ and $\left(\frac{\partial T}{\partial\rho}\right)_S\rightarrow 0$ so that the two equations (4.60) and (4.61) become decoupled. We find

$$\tilde{\rho}(k,\omega)=\frac{i\omega-D_L k^2}{\omega^2-c_T{}^2k^2+i\omega k^2 D_L}\frac{p(k,0)}{mc_T{}^2} \tag{4.66}$$

and

$$\tilde{Q}(k,\omega)=\frac{\rho c_v}{-i\omega+k^2\kappa/(\rho c_v)}T(k,0) \tag{4.67}$$

where $c_T=\left(\frac{1}{m}\frac{dp}{d\rho}\right)^{1/2}$ is the (isothermal) sound velocity and c_v is the heat capacity per particle at constant volume. These equations are the analogues of (4.28), in that they relate the particle and heat density changes to the applied pressure and temperature changes which have produced them. By analogy with (4.31), we can now calculate the frequency spectrum of the density fluctuations of given wave vector, that is $S(k,\omega)$, as k_BT times the real part of the function relating $\tilde{\rho}(k,\omega)$ to $p(k,0)$, that is

$$S(k,\omega)=\frac{2k_BTD_L}{m}k^4\left[(\omega^2-c_T{}^2k^2)^2+(\omega k^2D_L)^2\right]^{-1} \tag{4.68}$$

It is easily verified that this expression, although valid only for small frequency and wave vector, satisfies the sum rule (3.24) as well as the second moment theorem (3.56) in the long wavelength limit,

$$\lim_{k\to 0}\int\frac{d\omega}{2\pi}S(k,\omega)=\frac{k_BT}{mc_T{}^2} \tag{4.69}$$

and

$$\lim_{k\to 0}\int\frac{d\omega}{2\pi}\omega^2S(k,\omega)=\frac{k_BT}{m}k^2 \tag{4.70}$$

The fourth moment of (4.68), instead, diverges, as a consequence of the failure of the hydrodynamic equations at high frequency.

The dynamic structure factor (4.68) has two peaks at frequencies $\pm c_Tk$, with half-width determined by D_L(Brillouin doublet). The doublet corresponds to sound waves propagating with a velocity determined by the isothermal compressibility and with an attenuation determined by the viscosity coefficient $(\frac{4}{3}\eta+\zeta)$. These results are a consequence of our neglect of the coupling between particle and heat density fluctuations, as we shall see in the following.

Finally, we note that by the same procedure followed above we can derive from (4.67) the frequency spectrum for heat density fluctuations at low temperature. This reads

$$S_Q(k,\omega)=2k_BT^2\kappa k^2[\omega^2+(k^2\kappa/\rho c_v)^2]^{-1} \tag{4.71}$$

which represents a simple relaxation-type peak centred at $\omega=0$ and with half-width determined by $\kappa/(\rho c_v)$. The additional factor of T in the numerator of (4.71) arises from the fact that the variable coupled with heat density fluctuations is $T(k,0)/T$.

4.3.3 Hydrodynamic form of $S(k, \omega)$ at arbitrary temperatures

The general solution of the two coupled equations (4.60) and (4.61) reads

$$\tilde{\rho}(\boldsymbol{k}, \omega) = \left\{ \frac{c_v}{c_p} \frac{i\omega - D'_L k^2 - D'_T \left(\frac{c_p}{c_v} - 1\right) k^2}{\omega^2 - c_S{}^2 k^2 + i\omega k^2 D'_L} + \left(1 - \frac{c_v}{c_p}\right) \frac{1}{-i\omega + k^2 D'_T} \right\} \frac{p(\boldsymbol{k}, 0)}{m c_T{}^2}$$

$$+ \left\{ \frac{1}{-i\omega + k^2 D'_T} - \frac{k^2 D'_T}{\omega^2 - c_S{}^2 k^2 + i\omega k^2 D'_L} \right\} \left(\frac{\partial \rho}{\partial T}\right)_p T(\boldsymbol{k}, 0) \quad (4.72)$$

and

$$\tilde{Q}(\boldsymbol{k}, \omega) = \frac{\rho c_p}{-i\omega + k^2 D'_T} T(\boldsymbol{k}, 0) + \left\{ \frac{1}{-i\omega + k^2 D'_T} - \frac{k^2 D'_T}{\omega^2 - c_S{}^2 k^2 + i\omega k^2 D'_L} \right\}$$

$$\frac{T}{V} \left(\frac{\partial S}{\partial p}\right)_T p(\boldsymbol{k}, 0) \quad (4.73)$$

In these equations $c_S = \left[\frac{1}{m}\left(\frac{\partial p}{\partial \rho}\right)_S\right]^{1/2} = \left(\frac{c_p}{c_v}\right)^{1/2} c_T$ is the adiabatic sound velocity, and c_p is the heat capacity per particle at constant pressure; furthermore, $D'_T = \kappa/(\rho c_p)$ and $D'_L = D_L + (c_p/c_v - 1) D'_T$.

The corresponding van Hove correlation function is

$$S(k, \omega) = \frac{2 k_B T}{m c_T{}^2} \left\{ \frac{c_v}{c_p} \frac{D'_L c_S{}^2 k^4 - k^2(\omega^2 - c_S{}^2 k^2) D'_T (c_p/c_v - 1)}{(\omega^2 - c_S{}^2 k^2)^2 + (\omega k^2 D'_L)^2} + \left(1 - \frac{c_v}{c_p}\right) \frac{k^2 D'_T}{\omega^2 + (D'_T k^2)^2} \right\} \quad (4.74)$$

and thus contains three peaks†. The Brillouin doublet peaks at frequencies $\pm c_S k$, with half-width determined by D'_L; namely, sound waves propagate at the adiabatic sound velocity and are damped by viscosity as well as by heat diffusion mechanisms. The coupling between particle and heat density fluctuations also manifests itself in the presence of the Rayleigh peak at $\omega \doteq 0$, with a half-width determined by heat diffusion. The strength of the Brillouin doublet is determined by the ratio c_v/c_p, while that of the Rayleigh peak is determined by $(1 - c_v/c_p)$. Thus, in proximity of a phase transition, where c_p becomes abnormally large, the Rayleigh peak dominates the scattering and correspondingly narrows as $D'_T = \kappa/(\rho c_p)$ decreases.

It is also apparent from the above equations that the frequency spectrum of the heat fluctuations has the same structure as at low temperatures, the only change in (4.71) being the replacement of c_v by c_p. On the other hand, the correlations between particle and heat density fluctuations have their own spectral function, which from (4.72) can be written as

†The observation of these by light scattering is referred to in section 4.3.5 later.

$$S_{\rho Q}(k,\omega)=2k_BT^2\left(\frac{\partial\rho}{\partial T}\right)_p D_T'k^2\left\{\frac{1}{\omega^2+(k^2D_T')^2}-\frac{\omega^2-c_S{}^2k^2}{(\omega^2-c_S{}^2k^2)^2+(\omega k^2D_L')^2}\right\} \quad (4.75)$$

The same result is obtained from (4.73) on account of the thermodynamic identity $\frac{1}{V}\left(\frac{\partial S}{\partial p}\right)_T=-\frac{1}{\rho}\left(\frac{\partial\rho}{\partial T}\right)_p$. This is an example of the general symmetry property of correlation functions, which is an extension of the well-known Onsager symmetry relations between kinetic coefficients.

4.3.4 Kubo formulae for transport coefficients

It is easily verified that the hydrodynamic form of $S(k,\omega)$ given in (4.74) again satisfies the sum rules (4.69) and (4.70) in the long-wavelength limit, while the fourth moment again diverges. We shall see in the following section how this deficiency of the hydrodynamic results can be remedied by the memory function formalism.

We want here to discuss, instead, the relation of the dynamic correlation functions with the usual transport coefficients, for which the hydrodynamic forms are especially suited. This leads in a straightforward manner to the Kubo formulae† for the transport coefficients. In particular, from (4.74) we find

$$\lim_{\omega\to 0}\omega^4\left[\lim_{k\to 0}\frac{1}{k^4}S(k,\omega)\right]=\frac{2k_BT}{m^2\rho}\left(\tfrac{4}{3}\eta+\zeta\right) \quad (4.76)$$

while from (4.71) we obtain for the thermal conductivity

$$2k_BT^2\kappa=\lim_{\omega\to 0}\omega^2\left[\lim_{k\to 0}\frac{1}{k^2}S_Q(k,\omega)\right] \quad (4.77)$$

By the use of the hydrodynamic equations (4.53)–(4.55), the above expressions can be rewritten in terms of time integrals of appropriate correlation functions.

We report here the results for a classical fluid, referring the reader to the paper of Luttinger (1964) for an exhaustive discussion. The shear viscosity can be written

$$\eta=(k_BT)^{-1}\lim_{s\to 0}\int_0^\infty dt\exp(-st)\int d\boldsymbol{r}\langle\tilde{\Pi}_{xy}(\boldsymbol{r},t)\tilde{\Pi}_{xy}(0,0)\rangle \quad (4.78)$$

where $\tilde{\Pi}$ denotes the fluctuation of the stress tensor from the equilibrium value. Similarly the thermal conductivity is given by

$$\kappa=(k_BT^2)^{-1}\lim_{s\to 0}\int_0^\infty dt\exp(-st)\int d\boldsymbol{r}\langle J_z{}^T(\boldsymbol{r},t)J_z{}^T(\mathbf{0},0)\rangle \quad (4.79)$$

†General expressions for transport coefficients in terms of correlation functions were first derived by M. S. Green (1952; 1954) and Mori (1958). Analogous formulae for electrical transport coefficients were first derived by Kubo (1957).

where

$$J_z^{\mathrm{T}}(\boldsymbol{r}, t) = J_z(\boldsymbol{r}, t) - (\varepsilon + P)p_z(\boldsymbol{r}, t)/m \tag{4.80}$$

is the 'thermal current density', and the longitudinal viscosity is given by

$$\tfrac{4}{3}\eta + \zeta = (k_{\mathrm{B}}T)^{-1} \lim_{s \to 0} \int_0^\infty \mathrm{d}t \exp(-st) \int \mathrm{d}\boldsymbol{r} \langle \tilde{\Pi}'_{zz}(\boldsymbol{r}, t)\tilde{\Pi}'_{zz}(0, 0)\rangle \tag{4.81}$$

where

$$\tilde{\Pi}'_{zz}(\boldsymbol{r}, t) = \Pi_{zz}(\boldsymbol{r}, t) - p(\boldsymbol{r}, t) \tag{4.82}$$

The deep significance of the Kubo formulae lies in the fact that they relate transport coefficients to correlation functions in the fluid at equilibrium. This is to be contrasted with older theories of transport, which have always proceeded by approximate calculations of the distribution function for the non-equilibrium situation. In actual practice, no exact calculation of $S(k, \omega)$ has proved possible so far for any realistic force law, and generally the methods of calculation available at present have to proceed via some kind of Boltzmann equation. Nevertheless, these results for transport coefficients are central in the fundamental theory.

4.3.5 Relation to light scattering

The light scattering from a liquid is essentially given by correlations in the dielectric function or refractive index. In practice these are usually related to the density–density correlations, and hence to the scattering function $S(k, \omega)$.

The spectral distribution of the scattered light has in it information, then, on the time dependence of the density fluctuations. Laser techniques have been responsible for considerable advances in this field. However, we shall not go into details of experimental results. Rather we shall focus on the predictions of the spectral distribution which follow from the time dependence of the density fluctuations predicted by the linearised hydrodynamic equations. The discussion below follows that of Mountain (1966)—see also Benedek (1966).

Brillouin (1922) seems the first to suggest that light is scattered by thermal sound waves in the fluid. Because of the adiabatic character of sound propagation, density fluctuations should be decomposed into pressure fluctuations at constant entropy, and entropy fluctuations at constant pressure. This was accomplished in the discussion of section 4.3.3.

Accepting the connection between $S(k, \omega)$ and intensity of scattered light (for the explicit relation see eqn (4.84) below) the conclusions following from the results of section 4.3.3 are as follows:

(1) The frequency of the light scattered by the fluctuations at constant entropy (thermal sound waves) is changed by an amount proportional to the velocity of the sound waves. Conservation of energy and momentum requires that the

proportionality constant is the magnitude of the change in the wave vector of the scattered light. Two lines are observed, because scattering can occur from waves travelling in opposite directions, but with the same velocity. These lines are broadened by dissipative processes: they are the Brillouin lines.

(2) The light scattering by the fluctuations at constant pressure is not changed in frequency, although broadening due to thermal dissipative processes will again occur.

The fine structure, then, of the scattered light consists of three lines. The ratio of the intensity of the central line, I_R say, to that of the two shifted lines, $2I_B$ say, is determined as

$$I_R/2I_B = (c_p - c_V)/c_V \tag{4.83}$$

Landau and Placzek (1934) seem to have been the first to observe that the widths of these lines are determined by the lifetimes of the density fluctuations as obtained from linearised hydrodynamics.

Komarov and Fisher (1963) show that the intensity I of scattered light from a fluid is given by

$$I(\boldsymbol{R}, \omega) = (\alpha^2 \Omega^4 N/2\pi c^4 R^2) I_0 \sin^2 \varphi \, S(\boldsymbol{k}, \omega) \tag{4.84}$$

The light is observed at $\boldsymbol{R}$, having been scattered at the origin. The angular frequency of the scattered light is Ω, the shift in angular frequency is ω, while $\boldsymbol{k}$ is the change in the wave vector of the scattered light from that of the incident light in the medium, $\boldsymbol{k}_1$:

$$k = 2k_1 \sin(\tfrac{1}{2}\vartheta) \tag{4.85}$$

ϑ being the angle of scattering. N is the number of (spherically symmetric) molecules of polarisability α in the scattering volume. The incident light is assumed to be in the form of plane-polarised monochromatic radiation. Finally, φ is the angle between $\boldsymbol{R}$ and the electric vector of the incident wave, while I_0 is the incident intensity.

Some further remarks on light scattering from fluids near the critical point will be made in chapter 9.

4.4 Memory function for $S(k, \omega)$ and generalised hydrodynamics

The hydrodynamic form of $S(k, \omega)$, derived in the preceding section, holds only in the macroscopic limit, as signalled by the divergence of its fourth moment. The extension of this form to the range of frequencies and wave vectors of interest in neutron inelastic scattering from a fluid can be formally obtained, however, by the memory function formalism introduced in section 4.1.

Let us consider first the simpler case of low temperatures, discussed in section 4.3.2, and compare the expression (4.66) for $\tilde{\rho}(k, \omega)/p(k, 0)$ with the expression (4.29) for the frequency-dependent mobility. We see that the memory function for $S(k, \omega)$ in this case can be directly introduced by generalising the transport coefficient D_L to become a (complex) function, $D_L(k, \omega)$. We also need, however, to extend the sound velocity c_T to become a function of wave vector, and we can achieve this most simply by requiring that the zero-moment theorem (4.69) be satisfied for all values of the wave vector, that is

$$S(k) = \int \frac{d\omega}{2\pi} S(k, \omega) = \frac{k_B T}{m c_T{}^2(k)} \tag{4.86}$$

Upon splitting $D_L(k, \omega)$ into its real and imaginary parts,

$$D_L(k, \omega) = D_L'(k, \omega) + iD_L''(k, \omega) \tag{4.87}$$

with

$$\lim_{\omega \to 0} \lim_{k \to 0} D_L'(k, \omega) = D_L \tag{4.88}$$

we then find the general structure of $S(k, \omega)$ in this case as

$$S(k, \omega) = \frac{2k_B T}{m} D_L'(k, \omega) k^4 \left\{ \left[\omega^2 - \frac{k_B T}{mS(k)} k^2 - \omega k^2 D_L''(k, \omega) \right]^2 + [\omega k^2 D_L'(k, \omega)]^2 \right\}^{-1} \tag{4.89}$$

According to the Nyquist theorem, $D_L(k, \omega)$ is itself a (causal) response function, and therefore its real and imaginary parts are related by a Kramers–Kronig relation,

$$D_L''(k, \omega) = P \int \frac{d\omega'}{\pi} \frac{D_L'(k, \omega')}{\omega - \omega'} \tag{4.90}$$

Furthermore, $D_L(k, \omega)$ admits a continued fraction expansion of the Mori type. For a more rigorous derivation of these results, the reader should refer to the article of Kadanoff and Martin (1963).

We shall discuss the application of this formalism to neutron inelastic scattering from liquid argon in the next chapter. Here we note that the structure (4.89) for $S(k, \omega)$ allows, at least in principle, the possibility that peaks corresponding to the Brillouin doublet in hydrodynamics may be present in $S(k, \omega)$ in the range of frequencies and wave vectors relevant to neutron scattering. More precisely, $S(k, \omega)$ may have peaks at frequencies $\omega_0(k)$ such that

$$\left[\omega^2 - \frac{k_B T}{mS(k)} k^2 - \omega k^2 D_L''(k, \omega) \right]_{\omega = \omega_0(k)} = 0 \tag{4.91}$$

provided, however, that $\omega k^2 D_L'(k, \omega)$ is small and slowly varying with frequency for $\omega \simeq \omega_0(k)$. The solutions of (4.91) would then define an approximate dispersion relation for resonances in the response of the fluid, which would become exact normal modes if $D_L'[k, \omega_0(k)]$ were to vanish. Indeed, if the condition specified above for $D'_L(k,\omega)$ is satisfied we can write

$$S(k,\omega)\big|_{\omega\simeq\omega_0(k)} \simeq \frac{2k_BT}{m} D'_L(k,\omega_0(k))k^4 \left\{\left[(\omega^2-\omega_0{}^2(k))\frac{\partial}{\partial\omega^2}\left\{\omega^2-\frac{k_BTk^2}{mS(k)} -\omega k^2 D''_L(k,\omega)\right\}\right]^2 + [\omega_0(k)k^2D'_L(k,\omega_0(k))]^2\right\}^{-1}$$

$$= Z(k)\frac{2k_BT}{m}\bar{D}_L(k)k^4\,\{[\omega^2-\omega_0{}^2(k)]^2+[\omega_0(k)k^2\bar{D}_L(k)]^2\}^{-1} \tag{4.92}$$

where

$$Z(k)=\left\{1-\left(\frac{\partial}{\partial\omega^2}[\omega k^2 D''_L(k,\omega)]\right)_{\omega=\omega_0(k)}\right\}^{-1} \tag{4.93}$$

and

$$\bar{D}_L(k)=Z(k)D'_L[k,\omega_0(k)] \tag{4.94}$$

Equation (4.92) has the same form as the dynamic structure factor for a harmonic oscillator of frequency $\omega_0(k)$, damping $k^2\bar{D}_L(k)$ and strength $Z(k)$.

In actual practice, the van Hove function $S(k,\omega)$ as measured by neutron scattering from liquid argon, when plotted as a function of frequency for a given wave vector $k \gtrsim 1$ Å^{-1} (Sköld et al., 1972), does not display any structure which may be interpreted as a generalised Brillouin doublet. The concept of collective motions in a liquid, which is suggested by the generalised hydrodynamic form (4.89) is nevertheless a fruitful one even in such cases, as we shall see in the next chapter.

Molecular-dynamics studies of a Lennard–Jones fluid simulating liquid argon (Levesque, Verlet and Kürkijarvi, 1973) actually show clear evidence of collective modes for $k \lesssim 0.3$ Å^{-1}. Such modes have subsequently been observed in liquid rubidium for $k \lesssim 1$ Å^{-1} both by neutron inelastic scattering (Copley and Rowe, 1974*a* and *b*) and molecular dynamics studies (Rahman, 1974*a* and *b*), the experimental results being reported in *figures 4.3* and *4.4*. The quantitatively

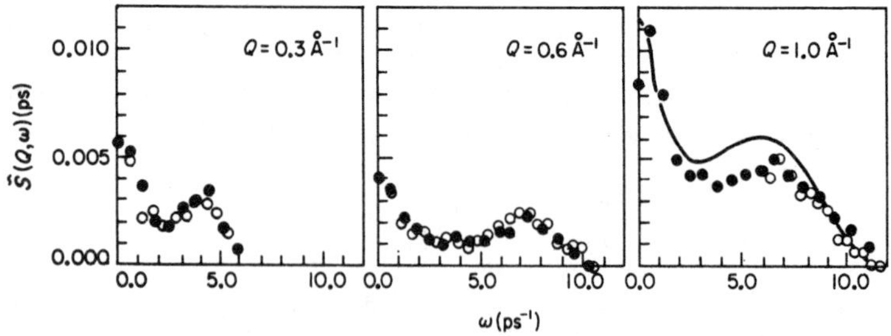

Figure 4.3 Scattering function of liquid rubidium for three values of the wave number below 1 Å^{-1} (from Copley and Rowe, 1974*a*). The side peak disappears at higher wave numbers

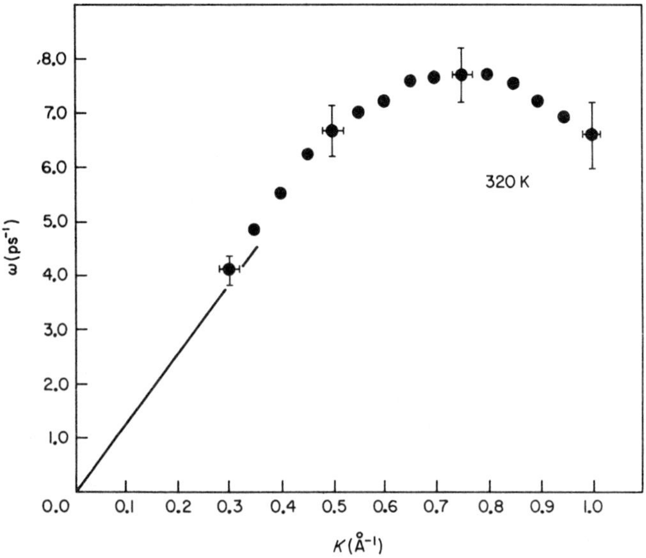

Figure 4.4 Position of side peak in scattering function of liquid rubidium, as a function of wave number (from Copley and Rowe, 1974*a*). The solid line is derived from the measured sound velocity

different behaviour of liquid rubidium and liquid argon may be interpreted (Copley and Lovesey, 1975) in terms of the elastic constant for longitudinal compressional waves (entering the fourth moment of $S(k, \omega)$) which is substantially larger in liquid metals. In this sense, and also as indicated in *figure 4.2*, a liquid metal appears to be considerably more rigid, and thus more solid-like, than a Lennard–Jones fluid.

Finally, in the case of arbitrary temperatures, we see from the hydrodynamic form of $S(k, \omega)$, given in (4.74), that we shall need two memory functions, related to the coefficients D'_L and D'_T for sound wave attenuation and thermal diffusivity. However, one expects (Chung and Yip, 1969) that generalised thermal diffusivity effects may not be too important at the comparatively high frequencies and wave vectors of interest in neutron scattering, though the expression (4.89) may not suffice for quantitative purposes below 1 Å^{-1}.

CHAPTER 5

MICROSCOPIC THEORIES OF THE VAN HOVE FUNCTION

The discussion of molecular dynamical calculations and neutron inelastic scattering for liquid metal rubidium at the end of the previous chapter suggests that, for such systems with relatively soft cores and long-range forces, an approach using essentially independent density fluctuations may be useful for some purposes. We immediately confirm that, if we note that the ratio of the specific heats γ for simple metals at their melting points is ~ 1.2 or 1.3, whereas for a hard-sphere like system such as argon $\gamma \sim 2.2$ at the triple point.

Therefore, before going on to microscopic theories of $S(k, \omega)$, we shall briefly consider the differences which arise between collective modes in crystals, liquids and gases, with a view to understanding these specific heat data.

5.1 Density fluctuations and phonons in fluids and solids

The well-founded theory of independent phonons in crystals leads to the standard result for the frequencies of longitudinal phonons as (see for instance Pines, 1963*b*):

$$\omega_k{}^2 = \frac{\rho}{m} \sum_{\boldsymbol{K}} \left[\left\{ \frac{\boldsymbol{k}}{k} \cdot (\boldsymbol{k} + \boldsymbol{K}) \right\}^2 \tilde{\phi}(\boldsymbol{k} + \boldsymbol{K}) - \left(\frac{\boldsymbol{k} \cdot \boldsymbol{K}}{k} \right)^2 \tilde{\phi}(k) \right] \tag{5.1}$$

Here $\tilde{\phi}(k)$ is the Fourier transform of the pair potential $\phi(r)$, given by

$$\phi(r) = \frac{1}{(2\pi)^3} \int \mathrm{d}\boldsymbol{k} \tilde{\phi}(k) \exp(-i\boldsymbol{k} \cdot \boldsymbol{r}) \tag{5.2}$$

and K denotes a reciprocal lattice vector. The total potential energy is

$$\Phi = \tfrac{1}{2} \sum_{ii'}{}' \phi(R_{ii'}) = \tfrac{1}{2} \sum_{ii'} \phi(R_{ii'}) - \tfrac{1}{2} N\phi(0) \tag{5.3}$$

where $\boldsymbol{R}_i$ is the position of atom i and $\boldsymbol{R}_{ii'} = \boldsymbol{R}_i - \boldsymbol{R}_{i'}$. The prime on the summation means that the term $i = i'$ is to be omitted.

If we make small displacements $\boldsymbol{r}_i$ from regular lattice sites $\boldsymbol{R}_i{}^0$ and, as in eqn

(5.1), confine ourselves to longitudinal phonons, then the normal coordinates q_k are defined by

$$\boldsymbol{r}_{\mathrm{i}} = \frac{i}{(Nm)^{1/2}} \sum_{k} \frac{\boldsymbol{k}}{k} q_k \exp(-i\boldsymbol{k}\cdot\boldsymbol{R}_{\mathrm{i}}^{0}) \tag{5.4}$$

In terms of the momenta $\boldsymbol{p}_k$ canonically conjugate to $\boldsymbol{q}_k$ which can be shown to take the form, in terms of the momenta $\boldsymbol{P}_{\mathrm{i}}$ of the atoms,

$$\boldsymbol{P}_{\mathrm{i}} = -i\left(\frac{m}{N}\right)^{1/2} \sum_{k} \frac{\boldsymbol{k}}{k} p_k \exp(i\boldsymbol{k}\cdot\boldsymbol{R}_{\mathrm{i}}^{0}) \tag{5.5}$$

and expanding the potential energy from eqns (5.3) and (5.4) we obtain the Hamiltonian H for the crystal as

$$H = T + V = \tfrac{1}{2}\rho N \sum_{\boldsymbol{K}} \tilde{\phi}(K) - \tfrac{1}{2} N\phi(0)$$

$$+\tfrac{1}{2}\sum_{k}\left[P_k P_{-k} + q_k q_{-k}\frac{n}{m}\sum_{\boldsymbol{K}}\left\{\left[\frac{k}{k}\cdot(k+K)\right]^2 \tilde{\phi}(k+K) - \left(\frac{k\cdot K}{k}\right)^2 \tilde{\phi}(K)\right\}\right] + \cdots \tag{5.6}$$

This is just the Hamiltonian for independent harmonic oscillators with frequencies given by eqn (5.1).

Whereas in the crystal we were interested in the ionic displacements $\boldsymbol{r}_{\mathrm{i}}$, or equivalently $\boldsymbol{q}_k$ above, in a fluid we are concerned with the density fluctuations

$$\rho_k = \sum_{\mathrm{i}} \exp(i\boldsymbol{k}\cdot\boldsymbol{R}_{\mathrm{i}}) = \rho^{\dagger}_{-k} \tag{5.7}$$

Evidently we must work with the momenta canonically conjugate to ρ_k in this case. In terms of these momenta P_k, it can be shown (see for example Bardasis, Falk and Simkin, 1965) that

$$T = \tfrac{1}{2}\sum_{k_1 k_2}\left(\frac{k_1}{k_1}\cdot\frac{k_2}{k_2}\right) P^{\dagger}_{k_1}\,\rho_{k_2-k_1}\,P_{k_2} \tag{5.8}$$

Expanding the potential energy, we again get a Hamiltonian representing uncoupled harmonic oscillators, this time with frequencies given by

$$\omega_k^{\,2} = \left(\frac{\hbar k^2}{2m}\right)^2 + \frac{k^2\rho}{m}\tilde{\phi}(k) \tag{5.9}$$

We shall content ourselves here with a brief discussion of the origin of the second term on the right-hand side of the expression (5.9) above. The classical expression for the second moment of $S(k, \omega)$ is given (see eqn (3.56)) by

$$\int \frac{\mathrm{d}\omega}{2\pi}\,\omega^2 S(k,\omega) = k_{\mathrm{B}}T\,k^2/m \tag{5.10}$$

If we assume that $S(k, \omega)$ is dominated by the collective mode $\omega(k)$ and write therefore

$$S(k, \omega) = 2\pi S(k)\delta[\omega - \omega(k)] \tag{5.11}$$

then we find

$$\omega^2(k) = \frac{k_B T k^2}{mS(k)} \tag{5.12}$$

Using the result that the direct correlation function $\tilde{c}(k) = 1 - \frac{1}{S(k)}$ is, in the random-phase approximation,

$$\tilde{c}(k) = -\rho\tilde{\phi}_k / k_B T \tag{5.13}$$

we get, if we can neglect the term unity relative to $1/S(k)$,

$$S(k) \sim k_B T / \rho\tilde{\phi}_k \tag{5.14}$$

and hence

$$\omega^2(k) = \frac{\rho k^2}{m}\tilde{\phi}_k \tag{5.15}$$

The conclusion is that such an approach can enable us to display the differences between sound waves in crystals and gases. The argument for gases is appropriate to liquids if it is corrected chiefly for the effects of backflow (Feynman and Cohen, 1956; see also chapter 8). To lowest order the above treatment yields ρ_k proportional to q_k. To include backflow, higher order corrections involving quadratic terms in the q_k's have to be included.

Having discussed such a unified approach to sound waves, we return to the discussion of the specific heats of liquids, starting from this collective coordinates approach.

5.2 Specific heat of liquids

5.2.1 Harmonic theory

Writing the Hamiltonian H as

$$H = H_0 + H_1 \tag{5.16}$$

where H_0 refers to independent oscillators, and H_1 is a coupling term which is assumed to be small, the logarithm of the phase integral can be expanded in powers of H_1 and averages taken, with the results

$$\ln\frac{Z}{Z_0} = \ln Z_1 + \ln Z_2 + 0(H_1{}^6) \tag{5.17}$$

where

$$\ln Z_1 = \tfrac{1}{2}\left\langle\left(\frac{H_1}{k_B T}\right)^2\right\rangle = \frac{k_B T}{4N^2}\sum_{\boldsymbol{k},\boldsymbol{k}'}[1 - \delta(\boldsymbol{k} - \boldsymbol{k}')]\left(\frac{\boldsymbol{k}\cdot\boldsymbol{k}'}{kk'}\right)^2\frac{1}{\tilde{\phi}_{\boldsymbol{k}-\boldsymbol{k}'}} \tag{5.18}$$

$$\ln Z_2 = \frac{1}{4!}\left[\left\langle\left(\frac{H_1}{k_B T}\right)^4\right\rangle - 3\left\langle\left(\frac{H_1}{k_B T}\right)^2\right\rangle^2\right] \tag{5.19}$$

Upper limits can be obtained by ignoring all restrictions on summations which are expressed in terms of delta functions and by replacing the scalar products of wave vectors by the products of their absolute values (Eisenschitz and Wilford, 1962). If, then, W is substituted for the $\tilde{\phi}_k$, one finds

$$\ln Z_1 \simeq \frac{2k_B T}{W} \ln Z_2 \simeq \frac{2}{N}\left(\frac{k_B T}{W}\right)^2 \tag{5.20}$$

If terms except $\ln Z_1$ are neglected, then the pressure p is readily obtained as

$$p = \frac{5}{2}\frac{Nk_B T}{V} + \tfrac{1}{2}\left[\frac{N}{V}\left(N\phi_0 - \sum_k \tilde{\phi}_k\right) + \frac{(k_B T)^2}{4VN^2}\sum_{k \neq k'}\left(\frac{\boldsymbol{k}\cdot\boldsymbol{k}'}{kk'}\right)^2 \frac{1}{\tilde{\phi}_{k-k'}}\right] \tag{5.21}$$

and the specific heat is given by

$$C_V = 3Nk_B\left[1 + \sum_{k \neq k'}\left(\frac{\boldsymbol{k}\cdot\boldsymbol{k}'}{kk'}\right)^2 \frac{k_B T}{6N^3\tilde{\phi}_{k-k'}}\right] \tag{5.22}$$

We shall come back to the evaluation of this expression. But we wish at this stage to discuss briefly the distribution function theory of specific heats.

5.2.2 Distribution function theory

As we discussed in connection with fluctuation theory in chapter 2, the specific heat at constant volume C_V can be written (see Schofield, 1966) in terms of higher order atomic distribution functions. The result is

$$\begin{aligned} C_V = \tfrac{3}{2}k_B + \frac{1}{k_B T^2}\Big[\tfrac{1}{2}\rho \int & \mathrm{d}\boldsymbol{r}\, g(r)\phi^2(r) + \rho^2 \int \mathrm{d}\boldsymbol{r}\,\mathrm{d}\boldsymbol{s}\, g_3(\boldsymbol{r}, \boldsymbol{s})\phi(r)\phi(s) \\ & + \tfrac{1}{4}\rho^2 \int \mathrm{d}\boldsymbol{r}\,\mathrm{d}\boldsymbol{s}\,\mathrm{d}\boldsymbol{t}\{g_4(\boldsymbol{r}, \boldsymbol{s}, \boldsymbol{t}) - g(r)g(|\boldsymbol{t}-\boldsymbol{s}|)\}\phi(r)\phi(|\boldsymbol{t}-\boldsymbol{s}|) \\ & - \left[\rho\int \mathrm{d}\boldsymbol{r}\, g(r)\phi(r) + \tfrac{1}{2}\rho^2 \int_\rho \mathrm{d}\boldsymbol{r}\,\mathrm{d}\boldsymbol{s}\,(g_3(\boldsymbol{r}, \boldsymbol{s}) - g(r)\}\phi(r)\right]^2 \Big/ S(0)\Big] \end{aligned} \tag{5.23}$$

A method can also be developed for calculating $C_p - C_V$ (Bratby, Gaskell and March, 1970), with the result

$$\begin{aligned} \frac{C_p - C_V}{S(0)} = 1 - \frac{2\pi\rho}{3k_B T}\int \mathrm{d}r\, r^3 g(r)\frac{\mathrm{d}\phi}{\mathrm{d}r} - \frac{\rho}{2k_B T S(0)} \\ \times\left[\int \mathrm{d}\boldsymbol{r} g(r)\phi(r) + \rho\int \mathrm{d}\boldsymbol{r}\,\mathrm{d}\boldsymbol{s}\,\{g_3(\boldsymbol{r}, \boldsymbol{s}) - g(r)g(s)\}\phi(r)\right] \end{aligned} \tag{5.24}$$

Clearly $C_p - C_V$ is a simpler quantity than either C_p or C_V separately, the four-particle correlation function g_4 appearing in C_p and C_V separately, but cancelling out in the difference. As we shall see below, all these formulae, used directly, seem to be *very* sensitive to the approximations employed to relate the higher order functions g_3 and g_4 to the pair function $g(r)$.

Because of this sensitivity to the approximations adopted for g_3 and g_4, it is worthwhile here to briefly record a result due to Schofield (1966), in the form of an inequality for the ratio of specific heats γ. This can be written in the form

$$\frac{\gamma}{S(0)} \leqslant \frac{5}{3} + \frac{2\pi\rho}{9k_B T} \int \mathrm{d}r\, r^2 g(r) \left[r^2 \frac{\mathrm{d}^2\phi}{\mathrm{d}r^2} - 2r \frac{\mathrm{d}\phi}{\mathrm{d}r} \right] \tag{5.25}$$

where as usual $S(0)$ is the long wavelength limit of the liquid structure factor.

5.2.3 Estimates of specific heats for simple liquids

5.2.3.1 Some estimates for liquid argon

The radial distribution function $g(r)$ for a Lennard–Jones potential $\phi(r)$ has been obtained by Rahman (1964). This was determined for a number density $\rho = 2.05 \times 10^{22}\mathrm{cm}^{-3}$ and $T = 94.4\mathrm{K}$. These results can be inserted into the inequality above to yield

$$\frac{\gamma}{S(0)} \leqslant 58 \tag{5.26}$$

The calculation of $S(k)$ for small k is difficult for the small system considered by Rahman, but since his method appears to be giving a $g(r)$ in very good agreement with experiment, it seems quite consistent to use the experimental value $S(0) = 0.06$ when we find

$$\gamma \leqslant 3.5 \pm 0.3 \tag{5.27}$$

More interesting than the numerical bound is the fact that the integrand in the inequality is quite short range. Of course, since we have *an* inequality, it is not possible to be quite certain, but there is a strong indication that the main contribution to γ is coming from a region inside 4 Å, the atomic diameter being 3.4 Å. Thus, for argon, the value of γ appears to be dominated by the short-range properties of $g(r)$ and $\phi(r)$. We discuss simple metals below: for Na the situation appears to be in contrast to argon—longer range effects come into play.

Secondly, for argon, let us briefly consider the sensitivity of $C_p - C_V$ to the form adopted for g_3 in eqn (5.24). It is straightforward to show that the superposition approximation is quite inadequate. On the other hand, if we make the alternative assumption that

$$g_3(\mathbf{r}, s) - g(r)g(s) \simeq g(r)h(|s - \mathbf{r}|) \tag{5.28}$$

inside the integral, then we find $C_p - C_V \sim 3k_B$ in good agreement with the measured value 2.8 k_B at the triple point of argon.

It is relevant here to mention an alternative inequality (Bratby, Gaskell and March, 1970) for γ to that of Schofield in (5.25), namely (see also Zemansky, 1951)

$$\gamma \leqslant 1+\tfrac{2}{3}S(0)\left\{\frac{P}{\rho k_B T}+\frac{1}{\rho k_B T}\left(\frac{\partial U}{\partial V}\right)_T\right\}^2 \tag{5.29}$$

Using the estimate of $\left(\frac{\partial U}{\partial V}\right)_T$ based on the decoupling (5.17), we find $\gamma \leqslant 3$, which is a slightly better bound than that given previously.

Finally, for argon, an attempt has been made to calculate C_V. Neglecting the term in g_4 one finds

$$\frac{\rho}{2(k_B T)^2}\int \mathrm{d}\boldsymbol{r}\, g(r)\phi^2(r)=5.9$$

$$\frac{\rho^2}{(k_B T)^2}\int \mathrm{d}\boldsymbol{r}\,\mathrm{d}\boldsymbol{s}\, g_3(\boldsymbol{r},\boldsymbol{s})\phi(r)\phi(s)=157 \tag{5.30}$$

and it is clear that *massive* cancellation must occur between the various terms in C_V. Therefore this formula is very sensitive to the approximations made for the higher order correlation functions. While it might be useful later as a test of more refined theories, it is not, as yet, a practical way to calculate C_V.

5.2.3.2 Liquid metals

To conclude this discussion we return to the harmonic theory which seems useful for simple liquid metals. The result (5.22) of Eisenschitz and Wilford (1962) can be rewritten (see Toombs, 1965; Bratby, Gaskell and March, 1970) using

$$S(q) \simeq \frac{1}{\beta N \tilde{\phi}(q)} \tag{5.31}$$

in terms of $S(q)$ and the experimentally determined structure factors can then be used to estimate the correction to the harmonic result. The important conclusion is that, since $S(q)$ is positive, the correction to the harmonic theory always increases C_V over its harmonic value $3k_B$.

An upper bound is immediately obtained by replacing $S(|k-k'|)$ by its maximum value within the range $0 \leqslant k \leqslant k_0$, k_0 being chosen to get the correct number of degrees of freedom, i.e. $k_0=(18\pi^2 N/V)^{1/3}$,

$$C_V \leqslant k_B\left[3+\tfrac{1}{2}S_{\max}\sum_{\boldsymbol{k},\boldsymbol{k}'}\frac{(\boldsymbol{k}\cdot\boldsymbol{k}')^2}{N^2k^2k'^2}\right]$$

$$\leqslant k_B[3+1.5\,S_{\max}] \tag{5.32}$$

For Na close to the melting point $N/V=0.0243$ atoms/Å^3 and $k_0=1.63$ Å^{-1}. Then $C_V \leqslant 3.6\ k_B$. For K, $C_V \leqslant 4$ by similar arguments, consistent with the measured value of $3.6\ k_B$.

The main conclusion is that, when a harmonic theory is useful, γ is near to 1 and $C_V > 3k_B$. These results are indeed true for quite a few of the simple liquid metals at the melting point. Having explored the relevance of independent density fluctuation theory to specific heats of simple metals, we now tackle the more basic problem of calculating $S(k, \omega)$ microscopically. This again means calculating $\langle\rho_k(0)\rho_{-k}(t)\rangle$ and taking the Fourier transform with respect to time.

5.3 Vlasov equation for $S(k, \omega)$

It is clear that to develop a fully microscopic theory of $S(k, \omega)$ from a pair potential we must make approximations. Instead of following this route, an alternative approach of great interest is to seek an approximate relation between $S(k, \omega)$ and the static structure $S(k)$ in a classical dense fluid. Such a relation is already apparent in the generalised hydrodynamic approach of section 4.4 (see eqn (4.89)), but, by developing detailed approximate theories, we shall be led to approximate expressions for the previously unknown functions entering eqn (4.89).

The time dependence of the correlation functions in a classical system must be determined by the Liouville equation for the phase space distribution function $f(\boldsymbol{r}_1, \ldots, \boldsymbol{r}_N; \boldsymbol{p}_1, \ldots, \boldsymbol{p}_N)$. This has the form

$$i\frac{\partial f}{\partial t} = Lf \tag{5.33}$$

where explicitly, in a classical fluid, L is given by

$$L = -i\sum_j \frac{\boldsymbol{P}_j}{m}\cdot\frac{\partial}{\partial \boldsymbol{r}_j} - i\sum_j \boldsymbol{F}_j\cdot\frac{\partial}{\partial \boldsymbol{P}_j} \tag{5.34}$$

where $\boldsymbol{F}_j$ is the total force on the jth molecule.

Now suppose for a moment that we could calculate the eigenfunctions Ψ_i of the Liouville operator L, with their eigenvalues λ_i. Then, in principle, we can construct a solution of (5.33) as

$$f = \sum_j c_j \exp(i\lambda_j t)\,\Psi_j \tag{5.35}$$

Thus the problem is reduced to the calculation of the eigenfunctions. This is obviously impossible for a general force law and we follow Zwanzig (1966) and employ the variational method. Thus, the requirement that $\langle\Psi^* L\Psi\rangle/\langle\Psi^*\Psi\rangle$ is stationary with respect to small variations in Ψ and Ψ^* leads directly to the eigenvalue equations for Ψ and Ψ^*.

We must then decide on the form of the trial function. Since, as we have noted, our basic object is to get the time-dependent correlation function in terms of $S(k)$ (or equivalently $g(r)$) we can only usefully allow $\Psi^*\Psi$ to contain two-body terms. Thus, Ψ itself must decompose into one-body terms to make progress and we write

$$\Psi = \sum_j \psi(\boldsymbol{r}_j, \boldsymbol{p}_j) = \sum_j \psi(j) \tag{5.36}$$

where we have written j for $(\boldsymbol{r}_j, \boldsymbol{p}_j)$.

Now it is very interesting that (5.36) already includes collective motions, for if it is applied to a crystal it yields the independent phonon description discussed above. The merit of this method is now that an Euler equation for $\psi(j)$ may be obtained in terms of the radial distribution function.

5.3.1 Euler equation for trial function

Because we are dealing with identical molecules, we can immediately reduce $\langle\Psi^*\Psi\rangle$ to the form

$$\langle\Psi^*\Psi\rangle = N\langle\psi^*(1)\psi(1)\rangle + N(N-1)\langle\psi^*(2)\psi(1)\rangle \tag{5.37}$$

Introducing the single- and two-particle distribution functions $f(1)$ and $f(1, 2)$ we have

$$\langle\Psi^*\Psi\rangle = N\int \mathrm{d}\boldsymbol{r}_1\,\mathrm{d}\boldsymbol{p}_1\,f(1)\psi^*(1)\psi(1) + N(N-1)\int \mathrm{d}\boldsymbol{r}_1\,\mathrm{d}\boldsymbol{p}_1\,\mathrm{d}\boldsymbol{r}_2\,\mathrm{d}\boldsymbol{p}_2\,f(1,2)\psi^*(2)\psi(1) \tag{5.38}$$

Similarly we have

$$\langle\Psi^*L\Psi\rangle = N\langle\Psi^*L\psi(1)\rangle \tag{5.39}$$

and L acting on $\psi(1)$ gives

$$L\psi(1) = -\frac{i}{m}\boldsymbol{p}_1\cdot\frac{\partial\psi(1)}{\partial\boldsymbol{r}_1} - i\boldsymbol{F}_1\cdot\frac{\partial\psi(1)}{\partial\boldsymbol{p}_1} \tag{5.40}$$

On integration by parts we find

$$\langle\Psi^*L\Psi\rangle = N\left\langle\psi^*(1)\left[-\frac{i}{m}\boldsymbol{p}_1\cdot\frac{\partial}{\partial\boldsymbol{r}_1} - i\boldsymbol{F}_1\cdot\frac{\partial}{\partial\boldsymbol{p}_1}\right]\psi(1)\right\rangle \tag{5.41}$$

We define the average force on a particular atom by

$$\langle\boldsymbol{F}_1, 1\rangle = \int \mathrm{d}(2)\ldots\mathrm{d}(N)f(1, 2, \ldots, N)\boldsymbol{F}_1/f(1) \tag{5.42}$$

and thus the quantity $\langle\Psi^*L\Psi\rangle$ becomes

$$\langle\Psi^*L\Psi\rangle = N\int \mathrm{d}(1)f(1)\psi^*(1)\left[-\frac{i}{m}\boldsymbol{p}_1\cdot\frac{\partial}{\partial\boldsymbol{r}_1} - i\langle\boldsymbol{F}_1, 1\rangle\cdot\frac{\partial}{\partial\boldsymbol{p}_1}\right]\psi(1) \tag{5.43}$$

Forming $\langle\Psi^*L\Psi\rangle/\langle\Psi^*\Psi\rangle$ and varying with respect to ψ^*, we find the eigenvalue equation for ψ as

$$\left[-\frac{i}{m}\boldsymbol{p}_1\cdot\frac{\partial}{\partial \boldsymbol{r}_1}-i\langle \boldsymbol{F}_1, 1\rangle\cdot\frac{\partial}{\partial \boldsymbol{p}_1}\right]\psi(1)=\lambda\left[\psi(1)+(N-1)\int \mathrm{d}(2)\frac{f(1,2)}{f(1)}\psi(2)\right] \tag{5.44}$$

This is the fundamental equation of the present method.

5.3.2 Solution for fluid and connection with Vlasov equation

We note that the one- and two-particle densities $\rho^{(1)}$ and $\rho^{(2)}$ discussed in chapter 1 are given in the canonical ensemble by

$$\begin{aligned}\rho^{(1)}(\boldsymbol{r}_1)&=f(1)/\phi_{\mathrm{B}}(\boldsymbol{p}_1)\\ \rho^{(2)}(\boldsymbol{r}_1, \boldsymbol{r}_2)&=f(1,2)/[\phi_{\mathrm{B}}(\boldsymbol{p}_1)\phi_{\mathrm{B}}(\boldsymbol{p}_2)]\end{aligned} \tag{5.45}$$

where ϕ_{B} is the Maxwell–Boltzmann distribution

$$\phi_{\mathrm{B}}=(2\pi m k_{\mathrm{B}}T)^{-3/2}\exp\left(-\frac{p^2}{2mk_{\mathrm{B}}T}\right) \tag{5.46}$$

In a fluid, the one-particle density $\rho^{(1)}$ is of course independent of position, $\rho^{(2)}\propto g(\boldsymbol{r}_1, \boldsymbol{r}_2)$ depends only on $|\boldsymbol{r}_1-\boldsymbol{r}_2|$ and $\langle \boldsymbol{F}_1, 1\rangle$ vanishes because of homogeneity. Hence the basic eigenvalue equation reduces to

$$-\frac{i}{m}\boldsymbol{p}_1\cdot\frac{\partial\psi(1)}{\partial \boldsymbol{r}_1}=\lambda\psi(1)+\lambda\rho\int \mathrm{d}\boldsymbol{p}_2\,\mathrm{d}\boldsymbol{r}_2\,\phi_{\mathrm{B}}(\boldsymbol{p}_2)g(|\boldsymbol{r}_1-\boldsymbol{r}_2|)\psi(\boldsymbol{r}_2, \boldsymbol{p}_2) \tag{5.47}$$

By Fourier transform, or noting that we can solve (5.47) using functions of the form

$$\psi(\boldsymbol{r}, \boldsymbol{p})=\exp(i\boldsymbol{k}\cdot\boldsymbol{r})\,\gamma_k(\boldsymbol{p}) \tag{5.48}$$

where $\boldsymbol{k}$ is an arbitrary vector, we find

$$\frac{\boldsymbol{k}\cdot\boldsymbol{p}_1}{m}\gamma_k(\boldsymbol{p}_1)=\lambda\gamma_k(\boldsymbol{p}_1)+\lambda\rho\int \mathrm{d}\boldsymbol{r}_2\,g(r_{12})\exp[i\boldsymbol{k}\cdot(\boldsymbol{r}_2-\boldsymbol{r}_1)]\int \mathrm{d}\boldsymbol{p}_2\,\phi_{\mathrm{B}}(\boldsymbol{p}_2)\gamma_k(\boldsymbol{p}_2) \tag{5.49}$$

To cast this into a form which can be interpreted physically, we transform (using velocity $\boldsymbol{v}$ rather than momentum $\boldsymbol{p}$ for convenience) from $\gamma_k(\boldsymbol{v})$ to $\chi_k(\boldsymbol{v})$ defined by

$$\gamma_k(\boldsymbol{v})=\frac{\chi_k(\boldsymbol{v})}{\phi_{\mathrm{B}}(\boldsymbol{v})}-\frac{\tilde{h}(k)}{1+\tilde{h}(k)}\int \mathrm{d}\boldsymbol{v}'\,\chi_k(\boldsymbol{v}') \tag{5.50}$$

where $g(r_{12})-1$ has been replaced by its Fourier transform $\tilde{h}(k)$. The quantity $\tilde{h}(k)/[1+\tilde{h}(k)]$ is simply the direct correlation function $\tilde{c}(k)$ and if we use the identity

$$-\boldsymbol{k}\cdot\boldsymbol{v}\phi_{\mathrm{B}}(\boldsymbol{v})=\frac{k_{\mathrm{B}}T\boldsymbol{k}}{m}\cdot\frac{\partial\phi_{\mathrm{B}}(\boldsymbol{v})}{\partial \boldsymbol{v}} \tag{5.51}$$

we find

$$\boldsymbol{k}\cdot v\chi_k)v)+\frac{k_BT}{m}\tilde{c}(k)\boldsymbol{k}\cdot\frac{\partial\phi_B(v)}{\partial v}\int dv'\chi_k(v')=\lambda\chi_k(v) \tag{5.52}$$

It should now be noted that this equation has exactly the form of the linearised Vlasov equation, widely used in plasma theory (see, for example, Montgomery and Tidman, 1964), with $-k_BT\tilde{c}(k)$ playing the part of an effective potential. This is an appealing result in the light of our earlier discussion of structure.

Actually, the Vlasov equation is a Boltzmann equation with, however, no collision term, but including instead a self-consistent treatment of the intermolecular (effective) potential.† Then, viewed as a kinetic equation, it can be used to calculate a distribution function in $\boldsymbol{r}$ and v which, when integrated over v will yield $G(r, t)$ or, by taking the Fourier transform, $S(k, \omega)$. The achievement of this calculation is that it determines $S(k, \omega)$ in terms of $S(k)$. Although the treatment is primitive we summarise the essential argument, due to Nelkin and Ranganathan (1967), in appendix 5.2, starting out from the Vlasov equation in customary Boltzmann equation form. We find that $S(k, \omega)$ is given by

$$S(k,\omega)=\frac{2\pi^{1/2}}{k(2k_BT/m)^{1/2}}\times$$

$$\frac{\exp(-x^2)}{\left[S^{-1}(k)+2\tilde{c}(k)x\exp(-x^2)\int_0^x\exp(t^2)\,dt\right]^2+[\tilde{c}(k)x\pi^{1/2}\exp(-x^2)]^2} \tag{5.53}$$

where $x=\frac{\omega}{k}(m/2k_BT)^{1/2}$. This expression, inserting $S(k)$ and $\tilde{c}(k)$ from experiment or, say, for hard spheres from the Percus–Yevick theory, has the correct general features as shown schematically in *figure 5.1*. The side peaks show the sound propagation and have finite breadth. In the range of wave vectors of interest in neutron inelastic scattering, however, eqn (5.53) grossly underestimates the spectral width (Singwi, Sköld and Tosi, 1968; 1970). We shall discuss subsequent developments of the mean-field theory of $S(k, \omega)$ immediately below.

5.5 Mean field theories of $S(k, \omega)$

We turn now to a discussion of more refined theories of the dynamic structure factor in a classical liquid. In the evaluation of such a microscopic dynamical property we cannot, of course, take too literally the concept of collective motions that we have developed in the preceding sections, It seems physically correct, however, also by analogy with the behaviour of quantal fluids (Pines, 1966), to suppose that collective, high-frequency oscillations in a classical fluid may

†The Boltzmann equation is derived in appendix 5.1, while the precise connection with the Vlasov equation is demonstrated in appendix 5.2.

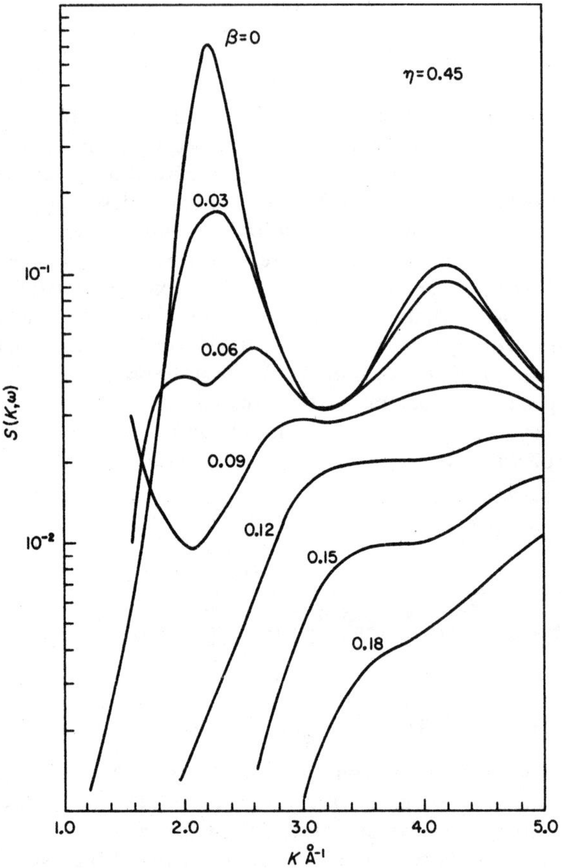

Figure 5.1 van Hove function versus wave number for several values of $\beta = \hbar\omega/k_B T$, as obtained from the Vlasov equation approach (from Nelkin and Ranganathan, 1967)

sustain themselves through their own mean field for times perhaps comparable with their period, with strong damping mechanisms arising mostly from disordered motions of the atoms. Mean field theories stemming from the Vlasov equation approach discussed in section 5.4 try to describe the dynamics of the liquid by a theoretical specification of these two components, adjusted to reproduce the first few moments of the spectrum. As noted in section 4.4, direct evidence for collective modes is now available for both a Lennard–Jones fluid and liquid rubidium, but only at rather small wave numbers.

It is especially convenient for the theory to focus attention on the spectrum of longitudinal current correlations, which by the continuity equation is given by the function $\frac{\omega^2}{k^2} S(k, \omega)$, as this spectrum in the liquid represents the closest analogue to the longitudinal phonon spectrum in the solid (Rahman, 1967*b*). At each wave vector this spectrum obviously has a peak at some frequency $\omega_m(k)$, because of the factor ω^2, but what is interesting is that with increasing wave vector $\omega_m(k)$ shows oscillations correlated with the liquid structure factor $S(k)$, before approaching a free-particle behaviour at very large wave vectors (see *figure 5.2*). The spectral width and the damping function $D_L'(k, 0)$ are also

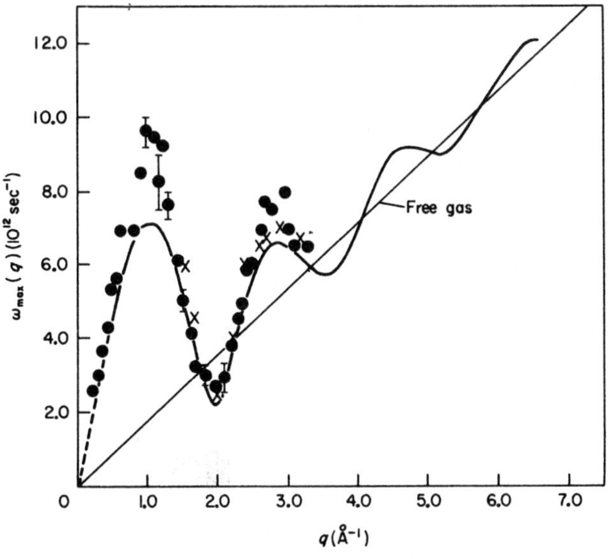

Figure 5.2 Position of the peak in the longitudinal-current correlation function in liquid argon near its freezing point (from Pathak and Singwi, 1970). The circles are from computer simulation of Rahman (1967*b*) and the crosses from neutron inelastic scattering of Sköld et al. (1972). The full curve is based on the theory discussed later in the text

correlated with the liquid structure (Rahman, 1967*b*; Tosi, Parrinello and March, 1974; see *figure 5.3*). Specifically, both $\omega_m(k)$ and the damping have a minimum at the wave vector of the first peak in $S(k)$: this wave vector corresponds in the crystal to the diameter of the first Brillouin zone, where the phonon frequency and damping vanish by lattice symmetry.

We shall here begin by discussing the theory of Hubbard and Beeby (1969), which attempts a rather detailed description of the collective and diffusive

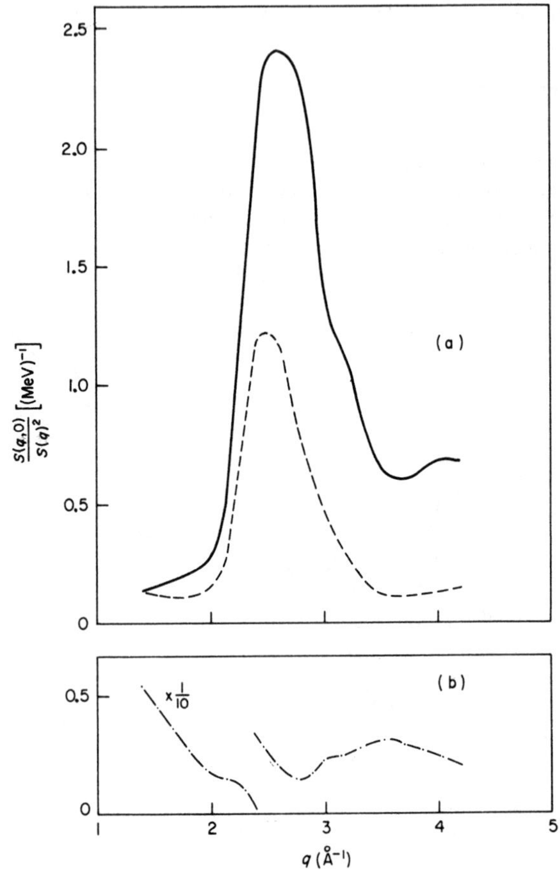

Figure 5.3 (*a*) Measured elastic scattering $S(k, 0)$ from liquid gallium at 295 K (broken curve) and static structure factor $S(k)$ (full curve) (from Barker et al., 1973). (*b*) Wave vector dependence of $S(k, 0)/[S(k)]^2$ which, as explained later in the text, measures the wave vector dependent viscosity in simple viscoelastic theory

motions in a liquid by analogy with the behaviour of amorphous solids, and shall then review the refinements of the Vlasov equation approach aimed at improving the description of the damping processes. In the next section we shall instead review the application of generalised hydrodynamics to the same problem. The latter approach, though starting from an apparently rather different standpoint, still basically leads to a mean-field type expression for $S(k, \omega)$. The various theories provide reasonably good descriptions of the spectrum of longitudinal current correlations in liquid argon over the range of wave vector and frequency of interest for neutron inelastic scattering ($k \sim 1$ Å^{-1}, $\omega \sim 10^{13}$ sec^{-1}), but it should be stressed that none of them can be extrapolated correctly into the hydrodynamic region ($k \sim 10^5$ cm^{-1}, $\omega \sim 10^8$ sec^{-1}).

5.4.1 Hubbard–Beeby theory

The theory of Hubbard and Beeby (1969) was designed to describe collective motions in liquids, starting from the standpoint of cold amorphous solids and then allowing diffusion to take place. The linear response of a disordered array of atoms is the starting point, and this is readily discussed within the framework of section 3.2.5. Making essentially an approximation equivalent to independent phonons, the response function $\chi(k, \omega)$ is obtained in the form appropriate to an isotropic system:

$$\chi(k, \omega) = \frac{\rho k^2}{m} \frac{1}{\omega^2 - \omega_k{}^2} \tag{5.54}$$

with

$$\omega_k{}^2 = \frac{\rho}{m} \int \mathrm{d}\boldsymbol{r}\, g(r) \frac{\partial^2 \phi}{\partial z^2} (1 - \cos kz) \tag{5.55}$$

For small k, the right-hand side of this equation is proportional to k^2 and thus $\omega_k = v_s k$, corresponding to a longitudinal sound velocity v_s. The van Hove function for independent phonons has the form

$$S(k, \omega) = \frac{\hbar k^2}{2m\omega_k} [(n_k + 1)\delta(\omega - \omega_k) + n_k \delta(\omega + \omega_k)] \tag{5.56}$$

where

$$n_k = [\exp(\hbar\beta\omega_k) - 1]^{-1} \tag{5.57}$$

The relation for the collective modes predicts non-zero frequencies for the transverse modes before going to the isotropic case, even for small k. This is reflecting the fact that an amorphous solid has rigidity and so we expect to find transverse modes. In contrast, the lack of rigidity in a liquid is a result of relaxation processes arising from the movement of the atoms in the liquid and is evidently not taken into account in the above treatment. We shall do this below. A consequence of the above form of the van Hove function $S(k, \omega)$ is that there will be a peak in the neutron scattering at the longitudinal collective mode frequency ω_k.

5.4.1.1 Effects of particle motion

The above discussion was based on the assumption that the atoms were fixed at a set of positions $\{\boldsymbol{r}_i\}$. Now we must take account of the atomic motions, and one must then consider not particular configurations $\{\boldsymbol{r}_i\}$ but particular histories $\{\boldsymbol{r}_i(t)\}$ of the system.

The equation of motion for a particular history $\{\boldsymbol{r}_i(t)\}$ can be derived in a straightforward manner, but the force constant now becomes time-dependent. Then a generalisation of the response function (5.54) is obtained, having the form

$$\chi(k,\omega) = -\frac{\rho k^2}{m}\frac{Q(k,\omega)}{1+\omega_k{}^2 Q(k,\omega)} \tag{5.58}$$

In obtaining this form the function $Q(k, \omega)$ involves, because of particle motion, the diffusive motion of an atom through the self-function $G_s(rt)$, the precise relation derived by Hubbard and Beeby being

$$Q(k,\omega) = \int_0^\infty \mathrm{d}t \int \mathrm{d}\boldsymbol{r}\, t\, G_s(rt) \exp(-i\boldsymbol{k}\cdot\boldsymbol{r} + i\omega t) \tag{5.59}$$

Rather than indicating in detail the approximations which led to these equations, (5.58) and 5.59), for which the original paper must be consulted, we shall stress below the similarity between the structure of this theory and that derived earlier from the Vlasov equation. To do this, we must form the van Hove scattering function, which is found from eqn (5.58) to be

$$\begin{aligned} S(k,\omega) &= \frac{2k^2}{m\beta\omega}\,\mathrm{Im}\left\{\frac{Q(k,\omega)}{1+\omega_k{}^2 Q(k,\omega)}\right\} \\ &= \frac{2k^2}{m\beta\omega}\frac{Q''(k,\omega)}{[1+\omega_k{}^2 Q'(k,\omega)]^2 + [\omega_k{}^2 Q''(k,\omega)]^2} \end{aligned} \tag{5.60}$$

where $Q(k, \omega) = Q'(k, \omega) + iQ''(k, \omega)$. If $Q''(k, \omega)$ is small (this *is* true for small k), then $S(k, \omega)$ is strongly peaked at the frequency determined by

$$1 + \omega_k{}^2 Q'(k,\omega) = 0 \tag{5.61}$$

This gives the collective mode frequency, which, however, at small k does not give the ordinary (isothermal or adiabatic) sound velocity but an 'instantaneous' sound velocity (Schofield, 1966) in that the calculation involves a high frequency approximation.

Hubbard and Beeby show that this approximation to $S(k, \omega)$ satisfies the second and fourth moment sum rules. Unfortunately it badly violates the zero moment sum rule, namely $S(k)$. This is because the approximate $S(k, \omega)$ is poor at low frequencies. The theory can, however, be extended (Pathak and Bansal, 1973) to account for these deficiencies.

5.4.1.2 Model for ω_k

Hubbard and Beeby have calculated $S(k, \omega)$ by making models for $\omega_k{}^2$ and for $Q(k, \omega)$, which, in turn, depend on $g(r)$ and $G_s(r, t)$. For $\omega_k{}^2$ they write

$$\omega_k{}^2 = \omega_E{}^2\left\{1 - 3\frac{\sin kr_0}{kr_0} - 6\frac{\cos kr_0}{(kr_0)^2} + 6\frac{\sin kr_0}{(kr_0)^3}\right\} \tag{5.62}$$

This amounts to evaluating (5.55) with $g(r)\dfrac{\partial^2\phi}{\partial z^2}\simeq A\delta(r-r_0)$, representing strong peaking of $g\phi''$ at some radius r_0. Their model for $Q(k, \omega)$ is more complicated and will not be given here. *Figure 5.4* shows $S(k, \omega)$ as a function of ω for several values of k. As Hubbard and Beeby stress, the theory is poor for small ω to the left of the dashed line.

5.4.2 Mean-field theory and single-particle motion

As noted in appendix 5.2, the Vlasov equation approach to the calculation of $S(k, \omega)$ assumes that the fluid can sustain collective oscillations through a mean field determined by the direct correlation function $\tilde{c}(k)$, the damping being determined by the disordered motions of the atoms treated as free particles. Successive developments of the approach have aimed at quantitative improvements of this picture, focusing especially on the damping processes which are seriously underestimated in the simple approach.

It is first of all interesting to compare the result of the Hubbard–Beeby theory given in eqn (5.60) with the Vlasov equation result reported in eqn (5.53). It is then apparent that the Hubbard–Beeby theory has basically the same form for the response function but (*a*) it fixes the mean field through the fourth moment theorem (compare eqn (5.55) with eqn (3.58)) rather than through the static structure factor $S(k)$; and (*b*) it relates the damping mechanisms to the true single-particle time evolution function $G_s(r, t)$ rather than to the free-particle time evolution function.

An approximate relation between the damping mechanisms and $G_s(r, t)$ also emerges from the theoretical development of Kerr (1968), and has been suggested independently by Singwi, Sköld and Tosi (1968; 1970). These alternative theories, however, still fix the mean field $\tilde{\phi}(k)$ through the structure factor $S(k)$, as in the simple Vlasov theory, and thus lead at small k to the ordinary (isothermal) sound velocity. The theory at this level reproduces exactly the zero and second moment of $S(k, \omega)$ but only approximately the fourth moment, and accounts qualitatively for the longitudinal current spectrum in liquid argon, while the Hubbard–Beeby theory still tends to overestimate the peak frequency and to underestimate the width of the spectrum (see *figures 5.5* and *5.6*).

A further refinement of the theory, which allows it to reproduce correctly the zero, second and fourth moment, has been proposed by Pathak and Singwi (1970). The basic ingredients of the theory still are a mean field $\tilde{\phi}(k)$ and a damping function $\tilde{\chi}(k, \omega)$, equivalent to $Q(k, \omega)$ in the Hubbard–Beeby theory. This function is taken in the form of a free-particle response function with finite lifetime effects included through a width function $\Gamma(k)$. Precisely, these authors write

$$\chi(k, \omega)=\frac{\tilde{\chi}(k, \omega)}{1-\tilde{\phi}(k)\tilde{\chi}(k, \omega)} \tag{5.63}$$

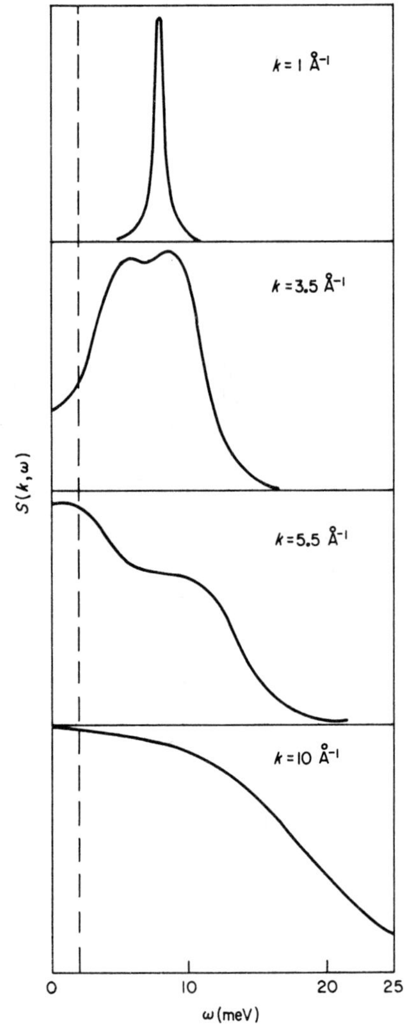

Figure 5.4 van Hove function as a function of ω for various values of the wave number, as given by the Hubbard–Beeby theory (from Hubbard and Beeby, 1969). The theory is poor for small ω to the left of the dashed line

with

$$\text{Im}\,\tilde{\chi}(k,\omega) = -\frac{\rho\pi^{1/2}}{k_B T} x \exp(-x^2), \qquad x^2 = \frac{m\omega^2}{2k_B T k^2 + \Gamma(k)} \tag{5.64}$$

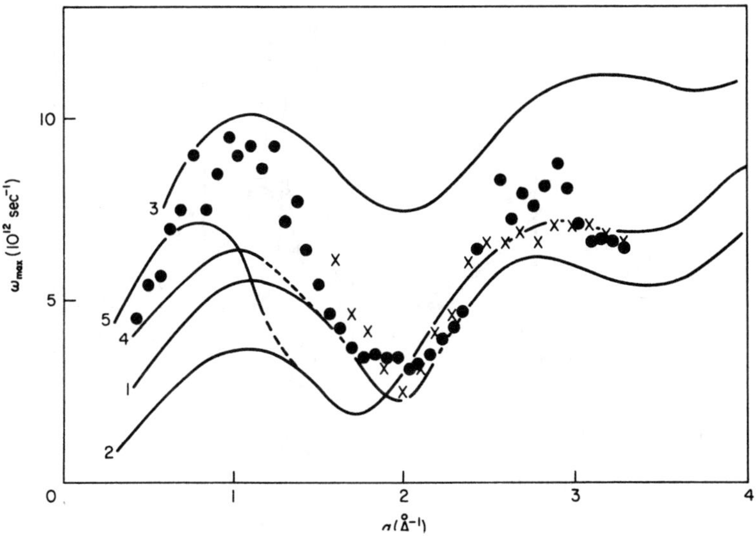

Figure 5.5 Peak frequency in the spectral function of longitudinal current correlations in liquid argon, as a function of wave number (from Singwi, Skold and Tosi, 1970). Circles, from molecular dynamics results of Rahman (1967*b*)); crosses, from neutron inelastic scattering experiments of Sköld and Larsson (1967). Theoretical curves: 1 and 2, simple Vlasov theories with slightly different choices of collective potential; 3, Vlasov theory with collective potential from fourth moment; 4 and 5, based on use of Kerr's approximation

The form (5.63) of the response function automatically satisfies the second moment theorem, and the unknown functions $\tilde{\phi}(k)$ and $\Gamma(k)$ are fixed so as to reproduce the zero and fourth moment. The agreement with the available data for liquid argon is satisfactory (see *figure 5.7*). In effect, as will be underlined by the discussion in the next section, it appears that the data on liquid argon, though very ample and detailed, do not sharply discriminate between alternative theories which fit the structure and the first few spectral moments. The theory can clearly be viewed as a sensible extrapolation of the free-particle behaviour into the (ω, k) region of interest in inelastic neutron scattering and is therefore expected to be especially useful at large k and ω.

5.5 Memory function approach to $S(k, \omega)$

A number of calculations of the dynamic structure factor for liquid argon have been based on the generalised hydrodynamics and memory function formalism that we have discussed in chapter 4 (Chung and Yip, 1969; Ackasu and Daniels, 1970; Murase, 1970; Lovesey, 1971). Alternative formulations of the application of this formalism involve either approximations on the functional form of

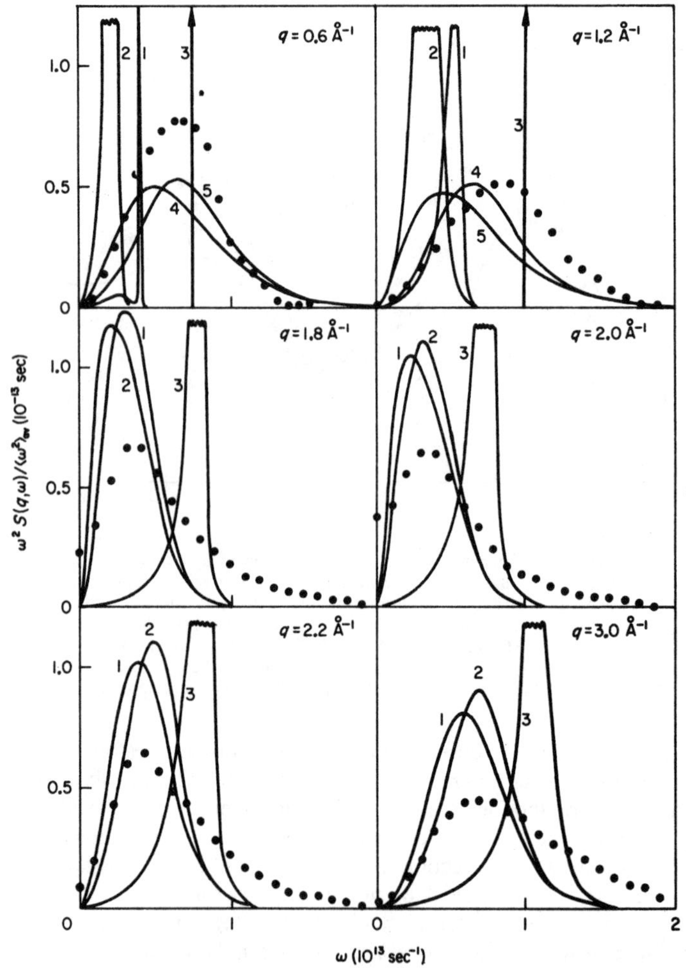

Figure. 5.6 Spectral function of longitudinal current correlations in liquid argon, for several values of the wave number (from Singwi, Sköld and Tosi, 1970). Symbols are as in *figure 5.5*

the generalised viscosity function $D_L(k, \omega)$ introduced in eqn (4.89), or a direct representation of the correlation functions through a continued fraction expansion of the Mori type. The same formalism has also been applied to the calculation of the transverse currents correlation spectra.

We briefly discuss here the calculation of Lovesey (1971) as an example of the

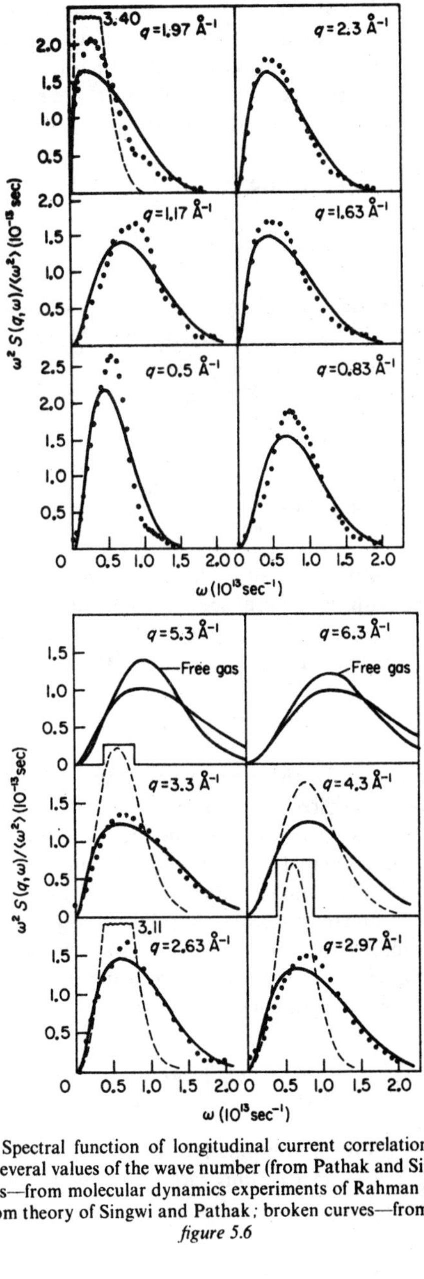

Figure 5.7 Spectral function of longitudinal current correlations in liquid argon, for several values of the wave number (from Pathak and Singwi, 1970). Solid circles—from molecular dynamics experiments of Rahman (1967*b*): full curves—from theory of Singwi and Pathak; broken curves—from theories of *figure 5.6*

continued fraction technique. The crucial approximation involved in the approach is the truncation of the expansion through the introduction of a suitable relaxation time, dependent on the wave vector. Thus, if we effect the truncation so as to enable the theory to reproduce the zero, second and fourth moments of the van Hove function, we can write

$$S(k,\omega)\propto \mathrm{Re}\,\cfrac{1}{\omega-\cfrac{a_1(k)}{\omega-\cfrac{a_2(k)}{\omega-i/\tau(k)}}} \tag{5.65}$$

where $a_1(k)$ and $a_2(k)$ are determined by the zero and fourth moment sum rules, and the relaxation time $\tau(k)$ is still to be determined. More precisely,

$$a_1(k)=\frac{k_B T k^2}{mS(k)} \tag{5.66}$$

and

$$a_1(k)+a_2(k)=\frac{m}{k_B T k^2}\int_{-\infty}^{\infty}\frac{d\omega}{2\pi}\,\omega^4 S(k,\omega) \tag{5.67}$$

the expression for the fourth moment in the above equation being given in eqn (3.58). By comparison with the generalised hydrodynamics expression reported in eqn (4.89) we see that eqn (5.65) is equivalent to an extension of the viscosity into the microscopic range by a simple relaxation assumption,

$$D_L(k,\omega)=\frac{a_2(k)/k^2}{i\omega+i/\tau(k)} \tag{5.68}$$

The same assumption underlies the calculation of $S(k,\omega)$ carried out by Ackasu and Daniels (1970) by using directly the generalised hydrodynamics expression. Alternatively, we may look upon the expression (5.65) as a mean field formula, in which the mean field is related to the liquid structure $S(k)$, as in the simple Vlasov approach, and the damping function is described by a simple relaxation process adjusted to the fourth moment sum rule.

We now note that information on the wave vector dependence of the relaxation time can be obtained from the elastic scattering function, $S(k,0)$. Indeed, from the generalised hydrodynamics expression of $S(k,\omega)$ reported in eqn (4.89), we find

$$D_L(k,0)=\frac{k_B T}{2m}\frac{S(k,0)}{[S(k)]^2} \tag{5.69}$$

The function $S(k,0)/[S(k)]^2$ as obtained from experimental data on liquid gallium is reported in *figure 5.3* to illustrate its minimum in correspondence with the first peak of $S(k)$.

For liquid argon, Lovesey chooses the wave vector dependence of the relaxation time from the known free-particle behaviour of the elastic scattering function $S(k, 0)$ at large wave vectors, which leads to $\tau(k) \propto [a_2(k)]^{-1/2}$. This prescription renders the theory totally inapplicable in the hydrodynamic region (as was the case for the mean field theories discussed in the preceding section) but still turns out to provide a reasonably accurate description of the longitudinal current correlation spectrum of liquid argon in the neutron region. The numerical results are, in fact, of comparable quality to those obtained by the mean-field approach and reported in *figure 5.7*. Very similar numerical results are obtained by the more careful determination of $\tau(k)$ carried out by Ackasu and Daniels, who estimate this function by an interpolation between its value

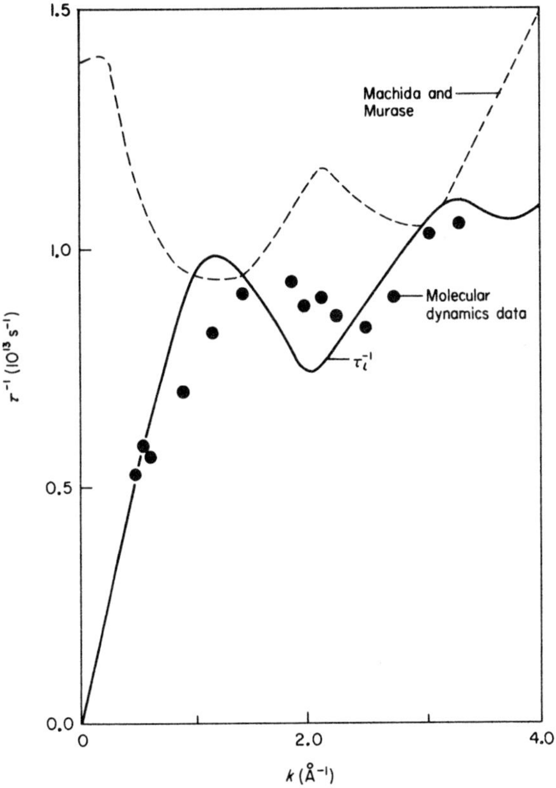

Figure 5.8 Inverse relaxation time as a function of wave number in liquid argon (from Copley and Lovesey, 1975). The dots are from a fit of molecular dynamics results by Ailawadi, Rahman and Zwanzig (1971), while the full line is calculated from the fourth moment of $S(k, \omega)$ according to Lovesey (1971)

for $k \to 0$, fixed by the longitudinal viscosity $(\frac{4}{3}\eta + \zeta)$, and its value in the opposite limit $k \to \infty$. In effect, one may fit the molecular dynamics results for liquid argon to extract $\tau(k)$ (Ailawadi, Rahman and Zwanzig, 1971) and the results are shown in *figure 5.8*.

It seems fair to say, in summary, that the main features of the current correlation spectrum in liquid argon can be understood on the basis of strongly damped collective motions, and that quantitative agreement with the available data can be obtained by approximate theories whose structure is suggested by the general theory of the preceding chapter and whose parameters are determined so as to fit the structure of the liquid and the first few moments of the van Hove function. The fundamental question still remains of how to construct an approximate theory of liquid dynamics that will correctly extrapolate to the known hydrodynamic limit. Forster and Martin (1970) have treated this problem in the case where the pair potential is weak, but little progress has been made for realistic pair potentials. The region of the (ω, k) plane involved in this extrapolation is unfortunately not easily accessible at present by experimental or simulation means.

CHAPTER 6
STRUCTURE AND DYNAMICS OF BINARY FLUIDS

We want now to turn from the discussion of pure one-component liquids to deal with mixtures. Although many of the arguments presented are easy to generalise, we shall present the treatment in terms of binary mixtures. The discussion, naturally enough, will start out with structure, at first static structure, but later we shall generalise this to dynamic properties.

6.1 Partial structure factors in a binary mixture

In a binary mixture, three pair correlation functions are evidently required for a complete description of static two-body correlation functions. Partial structure factors $a_{\alpha\beta}(k)$ for a binary system with a total of N atoms in a volume V are often defined by

$$a_{\alpha\beta}(k) = 1 + 4\pi\rho \int_0^\infty [g_{\alpha\beta}(r) - 1] \frac{\sin kr}{kr} r^2 \, dr \tag{6.1}$$

where $g_{\alpha\beta}(r)$ represent the pair distribution functions and $\rho = N/V$. This definition is usually adopted in dealing with thermodynamic properties of the mixture, but is not directly comparable with the structure factor in a one-component liquid. Structure factors giving directly the various contributions to the scattering intensity from the mixture are defined by

$$S_{\alpha\beta}(k) = \delta_{\alpha\beta} + 4\pi(\rho_\alpha\rho_\beta)^{1/2} \int_0^\infty [g_{\alpha\beta}(r) - 1] \frac{\sin kr}{kr} r^2 \, dr \tag{6.2}$$

where $\rho_\alpha = N_\alpha/V$ is the number of atoms of the αth component per unit volume.

It is sometimes convenient to build other structure factors as linear combinations of those defined above, to describe correlations between fluctuations of properties of the mixture which can be expressed as linear combinations of the partial density fluctuations. One such combination is the number N—concentration c correlation functions (Bhatia and Thornton, 1970)—defined by

$$g_{NN}(r) = c_1^2 g_{11}(r) + c_2^2 g_{22}(r) + 2c_1c_2 g_{12}(r) \tag{6.3}$$

$$g_{cc}(r) = c_1^2 c_2^2 [g_{11}(r) + g_{22}(r) - 2g_{12}(r)] \tag{6.4}$$

and

$$g_{Nc}(r)=c_1c_2[c_1g_{11}(r)-c_2g_{22}(r)+(c_2-c_1)g_{12}(r)] \tag{6.5}$$

where $c_\alpha=\rho_\alpha/\rho$ are the number concentrations. It is easily seen that these represent correlations between fluctuations in the total particle density (irrespective of the species) and between fluctuations in concentration, and cross-correlations between these fluctuations, respectively. The corresponding structure factors are

$$\begin{aligned} S_{NN}(k)&=1+\rho\int \mathrm{d}\boldsymbol{r}\exp(i\boldsymbol{k}\cdot\boldsymbol{r})\,[g_{NN}(r)-1] \\ &=c_1S_{11}(k)+c_2S_{22}(k)+2(c_1c_2)^{1/2}S_{12}(k) \end{aligned} \tag{6.6}$$

$$\begin{aligned} S_{cc}(k)&=c_1c_2+\rho\int \mathrm{d}\boldsymbol{r}\exp(i\boldsymbol{k}\cdot\boldsymbol{r})\,g_{cc}(r) \\ &=c_1c_2[c_2S_{11}(k)+c_1S_{22}(k)-2(c_1c_2)^{1/2}S_{12}(k)] \end{aligned} \tag{6.7}$$

and

$$\begin{aligned} S_{Nc}(k)&=\rho\int \mathrm{d}\boldsymbol{r}\exp(i\boldsymbol{k}\cdot\boldsymbol{r})\,g_{Nc}(r) \\ &=c_1c_2\left[S_{11}(k)-S_{22}(k)+\frac{c_2-c_1}{(c_1c_2)^{1/2}}S_{12}(k)\right] \end{aligned} \tag{6.8}$$

these being also defined directly as correlations between density and concentration fluctuations of the given wave vector, by relations similar to those that we give below in the dynamic case.

The X-ray diffraction intensity itself is a linear combination of the partial structure factors $S_{\alpha\beta}(k)$, with weighting factors given by products of the scattering amplitudes of the component species. Specifically,

$$I(k)\propto f_1{}^2S_{11}(k)+f_2{}^2S_{22}(k)+2f_1f_2S_{12}(k) \tag{6.9}$$

where f_α are the atomic scattering factors. An X-ray diffraction experiment is obviously insufficient to determine separately the three partial structure factors. Such a determination becomes possible (Enderby, North and Egelstaff, 1966) by neutron diffraction, if one takes advantage of the fact that different isotopes of the same element have sometimes very different scattering lengths. Mixtures built with the same concentrations c_α but different isotopes have the same structure in the classical limit, and three diffraction experiments on such mixtures allow a complete determination of the $S_{\alpha\beta}(k)$. The results of such an experiment on the Cu–Sn system are reported in *figure 6.1* (from Enderby, North and Egelstaff, 1966), while *figure 6.2* reports the number–concentration structure factors for the same system as constructed from the neutron scattering data by Bhatia and Thornton (1970). *Figure 6.3* reports data by Ruppersberg (1973) on the

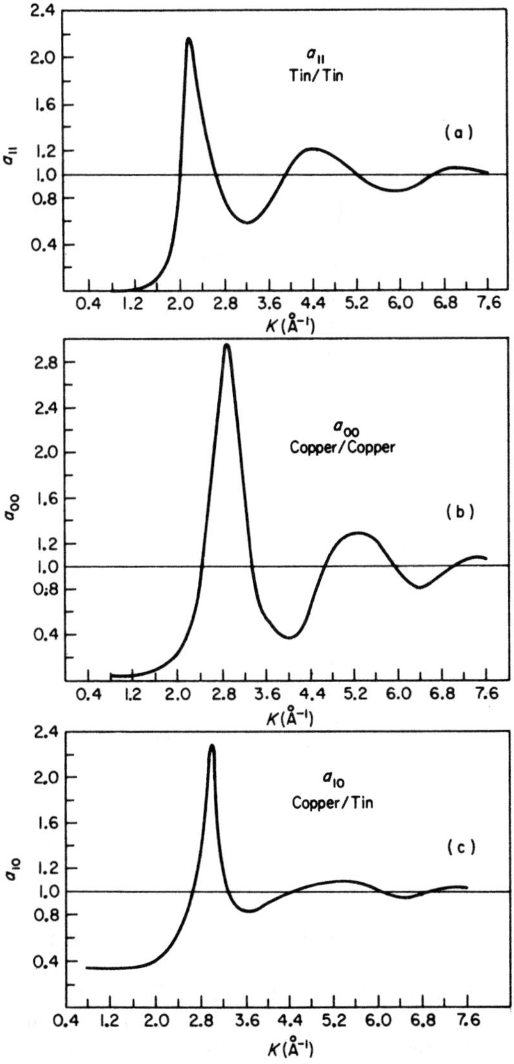

Figure 6.1 Partial structure factors in liquid Cu_6Sn_5 system, as obtained by neutron scattering through isotopic substitution (from Enderby, North and Egelstaff, 1966)

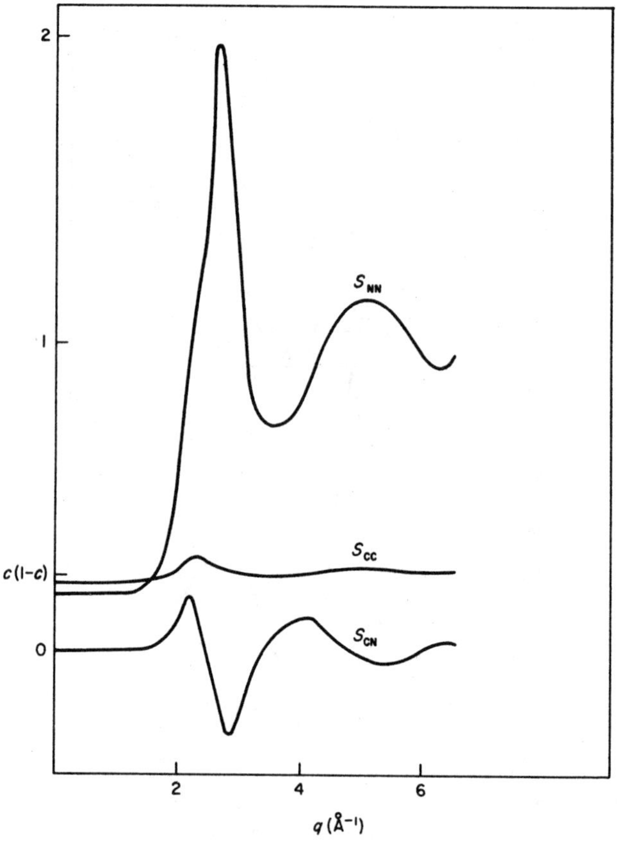

Figure 6.2 Number—concentration structure factors for the copper–tin alloy of *figure 6.1* (from Bhatia and Thornton, 1970)

scattering intensity from an Li–Pb alloy spanning the whole range of concentration.

The dynamic extension of the partial structure factors, $S_{\alpha\beta}(k, \omega)$, and of the N–c structure factors is immediate, with the relations (6.6)–(6.8) still being obviously valid. In treating dynamical properties it is, however, convenient to introduce mass density M—mass concentration X structure factors. If $\delta m_\alpha(\mathbf{r}, t) = m_\alpha \delta\rho_\alpha(\mathbf{r}, t)$ are the mass density fluctuations of each component, the total mass density and the mass concentration fluctuations are given by

$$\delta m(\mathbf{r}, t) = \delta m_1(\mathbf{r}, t) + \delta m_2(\mathbf{r}, t) \tag{6.10}$$

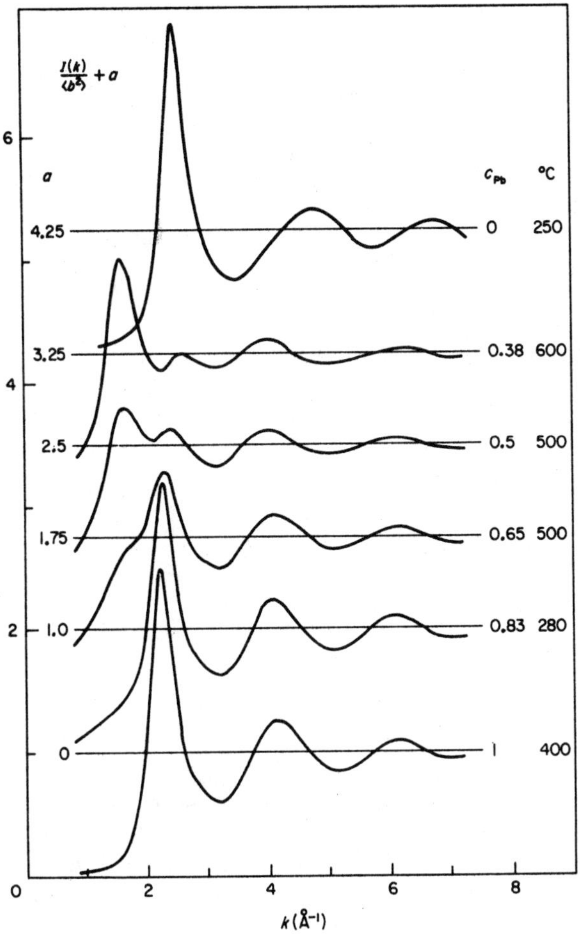

Figure 6.3 Neutron scattering intensities from Li–Pb alloys at various concentrations, indicated on the right of the figure (from Ruppersberg, 1973). The scattering lengths are such as to give a more pronounced effect from concentration fluctuations for c between 50% and 11% Pb. The pre-peaks at such concentrations are due to $S_{cc}(k)$

and

$$\delta x(\mathbf{r}, t) = \frac{1}{m^2} \left[c_2 m_2 \delta m_1(\mathbf{r}, t) - c_1 m_1 \delta m_2(\mathbf{r}, t) \right] \tag{6.11}$$

where $m = c_1 m_1 + c_2 m_2$. Denoting the Fourier transforms of δm and δx, respectively, by $M(\boldsymbol{k}, t)$ and $X(\boldsymbol{k}, t)$, then the M–X dynamical structure factors are given by

$$S_{\mathrm{MM}}(k, \omega)=\frac{1}{m}\int \mathrm{d}t \exp(-i\omega t)\,\langle M^{+}(\boldsymbol{k}, 0)M(\boldsymbol{k}, t)\rangle \tag{6.12}$$

$$S_{\mathrm{MX}}(k, \omega)=\frac{1}{2}\int \mathrm{d}t \exp(-i\omega t)\,\langle M^{+}(\boldsymbol{k}, 0)X(\boldsymbol{k}, t)+X^{+}(\boldsymbol{k}, 0)M(\boldsymbol{k}, t)\rangle \tag{6.13}$$

and

$$S_{\mathrm{XX}}(k, \omega)=m\int \mathrm{d}t \exp(-i\omega t)\,\langle X^{+}(\boldsymbol{k}, 0)X(\boldsymbol{k}, t)\rangle \tag{6.14}$$

Their relations to the partial structure factors are

$$S_{\mathrm{MM}}(k, \omega)=\frac{1}{m}[c_1 m_1{}^2 S_{11}(k, \omega)+c_2 m_2{}^2 S_{22}(k, \omega)+2(c_1c_2)^{1/2}m_1m_2S_{12}(k, \omega)] \tag{6.15}$$

$$S_{\mathrm{MX}}(k, \omega)=\frac{c_1c_2m_1m_2}{m^2}\left[m_1S_{11}(k, \omega)-m_2S_{22}(k, \omega)+\frac{c_2m_2-c_1m_1}{(c_1c_2)^{1/2}}S_{12}(k, \omega)\right] \tag{6.16}$$

and

$$\begin{aligned} S_{\mathrm{XX}}(k, \omega)&=\frac{c_1c_2(m_1m_2)^2}{m^3}[c_2S_{11}(k, \omega)+c_1S_{22}(k, \omega)-2(c_1c_2)^{1/2}S_{12}(k, \omega)] \\ &=\frac{(m_1m_2)^2}{m^3}\,S_{\mathrm{cc}}(k, \omega) \end{aligned} \tag{6.17}$$

6.2 Partial structure factors in the thermodynamic limit

We begin by discussing the partial structure factors of a binary mixture in the thermodynamic limit, $k \to 0$. Then, as Kirkwood and Buff (1951) have shown, the isothermal compressibility K_{T} of the mixture can be written in terms of the long wavelength limits of the partial structure factors defined in eqn (6.1), which we denote simply by a_{11}, a_{22} and a_{12}. If c is the concentration of 'impurity' 2, it is shown in appendix 6.1 that their result takes the form

$$\begin{aligned} \rho k_{\mathrm{B}}T K_{\mathrm{T}}&=\frac{[(1-c)a_{11}+c][ca_{22}+(1-c)]-c(1-c)(a_{12}-1)^2}{1+c(1-c)(a_{11}+a_{22}-2a_{12})} \\ &=\lim_{k\to 0}\frac{S_{11}(k)S_{22}(k)-S_{12}^2(k)}{S_{\mathrm{cc}}(k)/[c(1-c)]} \end{aligned} \tag{6.18}$$

Naturally when $c=0$ we recover from eqn (6.18) the usual formula for the compressibility of the solvent.

To study the compressibility in eqn (6.18) as a function of concentration, the

S's or equivalently the a's can be expanded in c. The slopes of the a's at $c=0$, as originally pointed out by Kirkwood and Buff (1951), can be expressed in terms of the triplet correlation functions at long wavelength in the pure solvent.

For arbitrary concentration, it is very helpful at $k=0$ to express the a's in terms of the concentration fluctuations $S_{cc}(k=0)$, a size factor and the compressibility of the alloy. Reference to appendix 6.1 (see also Bhatia and Thornton, 1970) permits one to write

$$a_{11}=\theta+\left[\frac{1}{(1-c)^2}-\frac{2\delta}{1-c}+\delta^2\right]S_{cc}-\frac{c}{1-c} \tag{6.19}$$

$$a_{22}=\theta+\left[\frac{1}{c^2}+\frac{2\delta}{c}+\delta^2\right]S_{cc}-\frac{1-c}{c} \tag{6.20}$$

$$a_{12}=\theta+\left[\delta^2-\frac{(2c-1)\delta}{c(1-c)}-\frac{1}{c(1-c)}\right]S_{cc}+1 \tag{6.21}$$

Explicitly θ, δ and S_{cc} are given by

$$\theta=\rho k_B T K_T \tag{6.22}$$

$$\delta=-\frac{1}{V}\left(\frac{\partial V}{\partial c}\right)_{T,p,N} \tag{6.23}$$

and

$$S_{cc}=N\langle(\Delta c)^2\rangle=Nk_B T\bigg/\left(\frac{\partial^2 G}{\partial c^2}\right)_{T,p,N} \tag{6.24}$$

where G is the Gibbs free energy.

6.3 Model of conformal solutions

To calculate the concentration fluctuations $S_{cc}(0)$, as well as δ and K_T, we must adopt a model. We shall adopt the approach of Longuet-Higgins (1951), which he refers to as the model of conformal solutions (Bhatia, Hargrove and March, 1973). The idea is to express the thermodynamic properties of the mixture in terms of those of a 'reference liquid'. We summarise in appendix 6.2 the essential equations of his treatment which indicate how the parameters used below for binary solutions relate to the properties of the 'reference liquid'. We will discuss briefly later how one might get information on the reference liquid from experiment. For our immediate purposes we shall content ourselves with estimating the parameters from experiment when the model of conformal solutions seems appropriate. Specialising to two components, the conformal solution has a Gibbs free energy given by

$$G=(1-c)G_1+cG_2+RT\{c\ln c+(1-c)\ln(1-c)\}+c(1-c)w \tag{6.25}$$

where G_1 and G_2 refer to the pure species 1 and 2 while $w=E_0 d_{12}$ depends on p

and T but is independent of concentration. The term in the braces gives the contribution from the entropy of mixing. The molar volume V then takes the form

$$V \equiv \left(\frac{\partial G}{\partial p}\right)_T = (1-c)V_1{}^0 + cV_2{}^0 + c(1-c)w' \ : \ w' = \left(\frac{\partial w}{\partial p}\right)_T \tag{6.26}$$

$V_1{}^0$ and $V_2{}^0$ evidently being the molar volumes of the pure species 1 and 2. Obviously expressions for θ and δ follow from eqn (6.26) by differentiating V with respect to p and c, respectively.

Furthermore the concentration fluctuations are obtained from the result

$$S_{cc} = \frac{c(1-c)}{1 - \dfrac{2w}{RT}c(1-c)} = Xc(1-c) \tag{6.27}$$

where this equation defines the quantity X. Equation (6.27) differs from regular solution theory only in that $w = w(p, T)$ (see appendix 6.2).

Using these results in eqns (6.19–6.21) and forming the quantity $2a_{12} - a_{11} - a_{22}$ we find immediately the result

$$2a_{12} - a_{11} - a_{22} = -\frac{2w}{RT} X \tag{6.28}$$

Thus, for a conformal solution, $2a_{12} - a_{11} - a_{22}$ is very simple and is plotted as a function of concentration for w positive ($w/RT = 1$) and w negative ($w/RT = -\frac{1}{2}$) in *figure 6.4*. It is clearly symmetrical around $c = \frac{1}{2}$.

We shall now briefly discuss the variation of a_{11}, a_{22} and a_{12} as functions of concentration for a Na–K alloy, as an example of the above model.

6.3.1 Numerical results for Na–K alloy

As the simplest prediction of the conformal solution model away from $c = 0$ we consider, first, eqn (6.28). Taking the experimental data presented by McAlister and Turner (1972) for Na–K, we find at $c = 0.5$

$$a_{11} \equiv a_{\text{NaNa}} = 2.9, \quad a_{22} = a_{\text{KK}} = 0.05, \quad a_{12} = a_{\text{NaK}} = -0.95$$

yielding $2a_{12} - a_{11} - a_{22} = -4{\cdot}85$. Using eqn (6.28) we then find

$$\frac{w}{RT} = 1.10 \tag{6.29}$$

in agreement with the experimental data on heat of mixing (Hultgren et al., 1963). At $c = 0.2$ eqn (6.28) then yields the value -3.4, whereas the experimental data leads to -3.3. The theory is symmetrical around $c = \frac{1}{2}$ and the experiments at $c = 0.8$ yield -3.8. There appears to be some asymmetry in the experi-

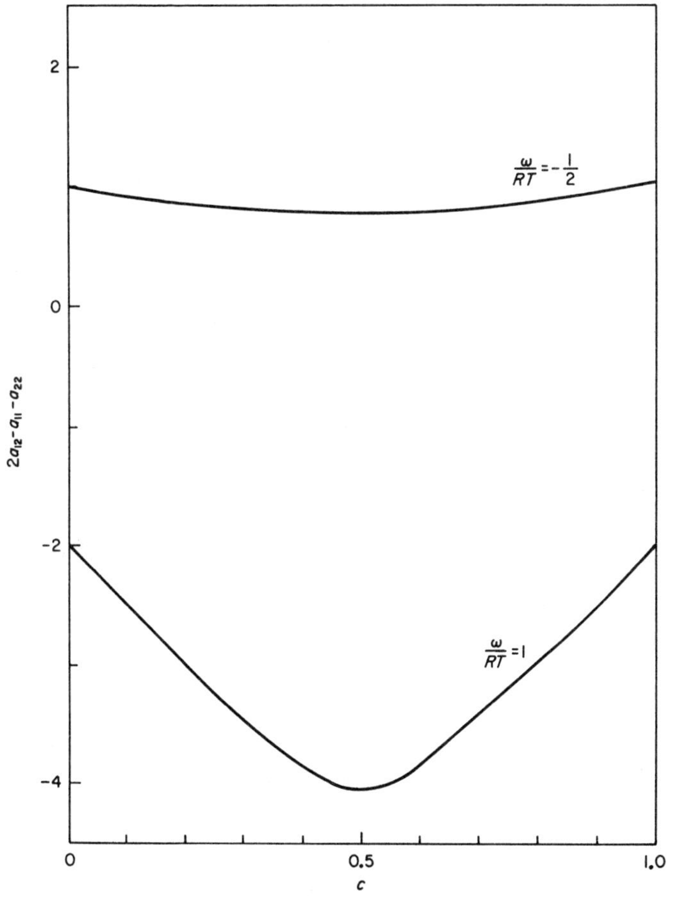

Figure 6.4 The structure factor $2a_{12}-a_{11}-a_{22}$ as a function of concentration in the thermodynamic limit, as evaluated by conformal solution theory for $w/RT=1$ and $w/RT=-1/2$ (from Bhatia, Hargrove and March, 1973)

ments, but the agreement between these concentration limits seems fairly satisfactory.

Adopting a value of δ obtained from McAlister and Turner, by taking $a_{12}^0-a_{11}^0$ at the Na-rich end, we find $\delta|_{c=0} \doteqdot -1$ and hence we obtain

$$a_{22}^0-a_{11}^0=2\delta+\frac{2w}{RT}\doteqdot 0.20 \tag{6.30}$$

yielding, since a_{11}^0 is small (~ 0.03) $a_{22}^0 \sim 0.2$.

Using measured pure metal properties the effect of varying w' and w'', the pressure derivatives of w, on the partial structure factors, has been investigated (Bhatia, Hargrove and March, 1973) and the results are plotted in *figure 6.5* for a selection of parameters α and β related to w'' and w' by

$$w'' = \alpha V_1^{(0)} K_T^{(1)}; \qquad w' = \beta V_1^{(0)} \tag{6.31}$$

The main features, for reasonable variations in α and β, are the maximum in a_{11} around a concentration $c \sim 0.75$ and a fairly flat region of a_{12} from $c \doteqdot 0.2$ to 0.5 passing through a node between 0.7 and 0.8. These features agree well with the thermodynamic analysis of McAlister and Turner. The greatest sensitivity occurs in a_{22} as can be seen from *figure 6.5* below $c = 0.3$, though of course a_{12} and a_{22} are connected at $c = 0$ through eqn (6.28), a_{11}^0 being very small. There is evidently no difficulty in finding parameters α and β for $c < 0.3$ to fit the general form of the thermodynamic data.

For all the parameters chosen, V is rather like a linear interpolation between $V_1^{(0)}$ and $V_2^{(0)}$. The best fit with experiment for V yields $\beta \doteqdot -0.1$, while from the V-K_T plot, by comparison with the experimental data of Abowitz and Gordon (1962) we find $\alpha \sim -0.5$. Thus, the partial structure factors corresponding to these parameters in *figure 6.5* should be considered as the most realistic.

The simplest prediction of conformal solution theory away from $c = 0$ is for the combination $2a_{12} - a_{11} - a_{22}$ given in eqn (6.28). When the concentration dependence follows this equation, as appears to be approximately the case of Na–K for $0.2 \leqslant c \leqslant 0.8$ the interchange energy w can be extracted. For this alloy at 100°C, $w/RT = 1.1$. The corresponding structure factors, although depending somewhat on w' and w'' have the main features required by the thermodynamic data of McAlister and Turner.

The conformal solution theory, however, does *not* appear to be appropriate to describe K–Hg, the other system considered by McAlister and Turner. It seems clear that this is because the difference in the forces between different species is too large.

6.3.2 Phase diagrams

As a further illustration of the use of conformal solution theory for liquid metal alloys, we shall discuss briefly how the phase diagram of Na–K can be calculated.

It has been known for a long time that the liquidus curve of an ideal solution is approximately described by the equation

$$\ln(1 - c_2) = (L_{10}/R)[T_1^{-1} - T^{-1}] \tag{6.32}$$

c_2 being the concentration of element 2 and L_{10} the latent heat at freezing temperature T_1 of pure liquid 1. To clarify the assumptions underlying (6.32) we shall summarise the thermodynamic equations along the liquidus (Bhatia and March, 1972). When no mixed crystals are present, we have, in terms of the chemical potentials (the subscript zero referring to a pure substance and super-

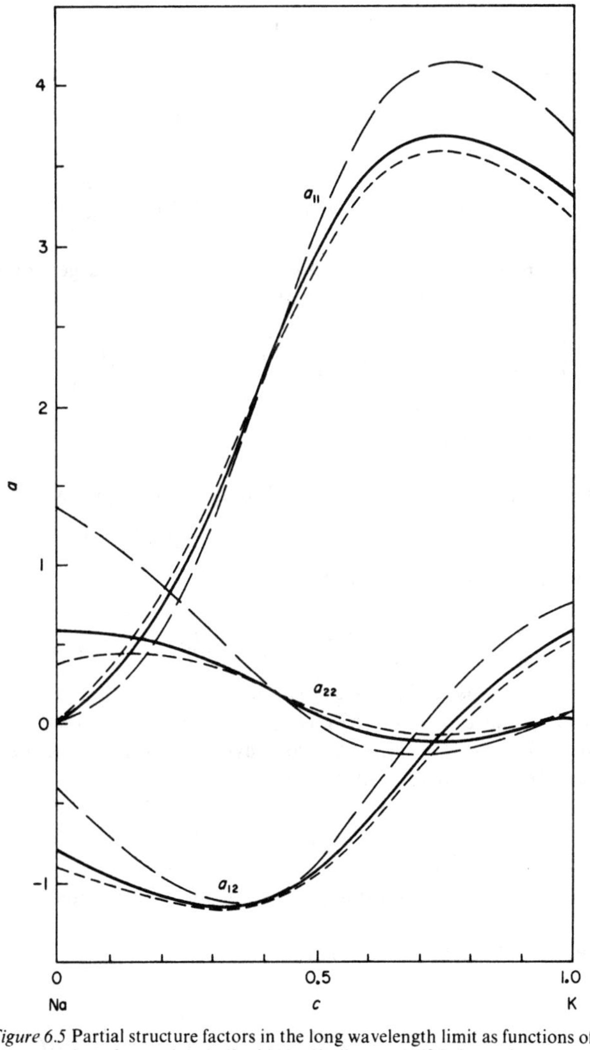

Figure 6.5 Partial structure factors in the long wavelength limit as functions of concentration of potassium in liquid Na–K alloys at 100°C, for various sets of the parameters α and β: $\alpha = -0.5$ and $\beta = -0.1$ (full curve): $\alpha = 0$ and $\beta = 0$ (short-dash curve): $\alpha = 0.5$ and $\beta = -0.5$ (long-dash curve) (from Bhatia, Hargrove and March, 1973)

script S denoting solid),

$$\mu_{10}^{S}(T)=\mu_1(T, c_2) \tag{6.33}$$

with the differential form

$$\begin{aligned}\frac{\Delta T}{\Delta c_2}&=-\left(\frac{\partial\mu_1}{\partial c_2}\right)_{p,T}\Big/\left[\left(\frac{\partial\mu_1}{\partial T}\right)_{c_2,p}-\left(\frac{\partial\mu_{10}^{S}(T)}{\partial T}\right)_{p}\right]\\&=\left(\frac{\partial\mu_1}{\partial c_2}\right)_{p,T}\Big/(L/T)=-c_2\left(\frac{\partial^2 G}{\partial c_2{}^2}\right)_{p,T}\Big/(L/T)=-\frac{RT^2c_2}{S_{cc}L}\end{aligned} \tag{6.34}$$

where $S_{cc}(0)$ represents the concentration fluctuations. L is a generalised concentration-dependent latent heat defined by

$$\begin{aligned}\frac{L}{T}&=\frac{L_{10}}{T_1}+\int_{T_1}^{T}\frac{\Delta c_{p10}}{T}\,\mathrm{d}T-\left(\frac{\partial}{\partial T}[RT\ln(\gamma_1(1-c_2))]\right)_{c_1,c_2,p}\\&\equiv F(T)-\left(\frac{\partial X}{\partial T}\right)_{p,c_2}\end{aligned} \tag{6.35}$$

γ_1 being the activity. The activity is related to μ_1 and to $S_{cc}(0)$ by

$$\mu_1=\mu_{10}(p,T)+RT\ln\gamma_1(1-c_2)\equiv\mu_{10}(p,T)+X \tag{6.36}$$

this equation serving to define X in (6.35) and

$$\frac{\partial}{\partial c_2}\ln\gamma_1=\frac{1}{1-c_2}-\frac{c_2}{S_{cc}(0)} \tag{6.37}$$

Since $S_{cc}(0)\to c_2$ as $c_2\to 0$ and from (6.35) $L\to L_{10}$ in this case, we see from (6.34) that $(\mathrm{d}T/\mathrm{d}c_2)_{\mathrm{liquidus}}$ at $c_2=0$ is $-RT_1{}^2/L_{10}$, an exact result, already contained in (6.32). In (6.35), $\Delta c_{p10}=c_{p10}^{L}-c_{p10}^{S}$ and evidently $F(T)$ is independent of concentration. It is then readily shown that

$$\int_{T_1}^{T}F(T)\,\mathrm{d}T=X(T, c_2)+\text{constant} \tag{6.38}$$

and by letting $T\to T_1$ the constant is easily shown to be zero. Hence

$$\begin{aligned}X(T, c_2)=RT\ln\gamma_1(1-c_2)&=\int_{T_1}^{T}F(T)\,\mathrm{d}T\\&=\int_{T_1}^{T}\mathrm{d}T'\frac{L_{10}}{T_1}+\int_{T_1}^{T}\mathrm{d}T'\int_{T_1}^{T'}\frac{\Delta c_{p10}(T'')}{T''}\mathrm{d}T''\end{aligned} \tag{6.39}$$

Thus, for an ideal solution with $\gamma_1=1$ and putting $c\equiv c_2$

$$1-c_{\mathrm{ideal}}=\exp\left[\frac{1}{RT}\int_{T_1}^{T}\mathrm{d}T'F(T')\right] \tag{6.40}$$

This will be referred to as the ideal liquidus curve. All the relevant properties of pure liquid 1 have been incorporated exactly, but the solution is taken as ideal.

In the true solution it then follows immediately that

$$\gamma_1 = \frac{1-c_{\text{ideal}}}{1-c(T)} \tag{6.41}$$

and from a knowledge of the properties of pure liquid 1 plus the experimentally determined liquidus curve the activity γ_1 can be extracted along the liquidus curve.

Knowledge of Δc_{p10} from the melting point T_1 to the eutectic temperature would require measurements on supercooled liquid and extrapolation will frequently be necessary. Therefore, to illustrate the above remarks, the model of conformal solution will again be adopted in which

$$RT \ln \gamma_1 \doteq wc^2 \tag{6.42}$$

where w is the interchange energy as before, and following earlier work we shall Taylor expand Δc_{p10} around T_1 and neglect higher terms than the first. Then we find

$$c(t)-1 = [c_{\text{ideal}} - 1] \exp(-Wc^2/t) \tag{6.43}$$

and

$$c_{\text{ideal}}(t) \doteq 1 - t^{\Delta} \exp[(l-\Delta)(1-t^{-1})] \tag{6.44}$$

$$t = T/T_1, \qquad \Delta = \Delta c_{p10}/R, \qquad l = L_{10}/RT_1, \quad W = \frac{w}{RT_1}$$

For Na–K, $l = 0.86$ at both ends of the phase diagram and $\Delta_{\text{Na}} = 0.06$, $\Delta_{\text{K}} = 0.16$. From the work on partial structure factors reported above, $w/RT_1{}^{\text{Na}} = 1.1$ and hence $W^{\text{K}} = 1.2$. In this way we find the liquidus curves in *figure 6.6*, shown together with measured data.

It should be cautioned that, in general, w is not necessarily independent of temperature. Further, the model of conformal solutions will not work when the forces between different species are markedly dissimilar.

6.3.3 Size effects in mixtures

Conformal solution theory is valid when (*a*) size differences are not too large, and (*b*) weak interaction theory is applicable.

In this section, we consider briefly how to modify the conformal (or regular) solution model to include size effects. As an example, we shall focus on Na–Cs, where there is a major size difference, the Cs ion being three times the size of Na. The work of Flory (1942) leads to a Gibbs free energy of mixing which we can write as

$$\Delta_m G = Nk_B T[c \ln \phi + (1-c)\ln(1-\phi)] + Ng(c)w \tag{6.45}$$

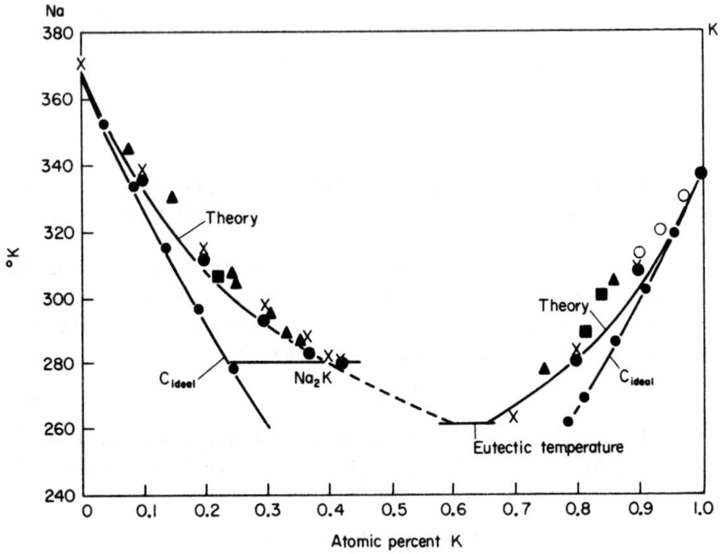

Figure 6.6 Liquidus curve in Na–K alloys (from Bhatia and March, 1972)

where the last term is the energy of mixing, the concentration dependence not being written out explicitly.

The modification introduced thereby is that ϕ is the concentration by volume of species 1. Explicitly, if v_1 and v_2 are the partial molar volumes, then ϕ is given by

$$\phi = \frac{cv_1}{cv_1 + (1-c)v_2} \tag{6.46}$$

Some arguments (see for example Guggenheim, 1945) can be given to derive such a Gibbs free energy of mixing. We shall not go into the details here as they rest on the use of so-called lattice models which are outside the scope of our discussion.

Rather we shall consider the application of Flory's model for Na–Cs, with $v_2/v_1 = 3$, as already remarked. The concentration fluctuations are readily calculated as

$$S_{cc}(0) = c(1-c)[1 + c(1-c)\delta^2 + c(1-c)g''(c)w/k_BT]^{-1} \tag{6.47}$$

[in terms of δ, $\phi = c + c(1-c)\delta$]. This formula can be used to calculate the activity, which has essentially been measured using a suitable electromotive cell by Ichikawa and Thompson (1974). The relation between activity and emf ε

is given by (the activity of species i being given by $\mu_i = \mu_i{}^0 + RT \ln a_i$)

$$\varepsilon = \frac{RT}{F} \ln a \tag{6.48}$$

F being the Faraday constant. For a fixed temperature, *figure 6.7* shows ε as a function of concentration, the measured points and the theoretical curve from

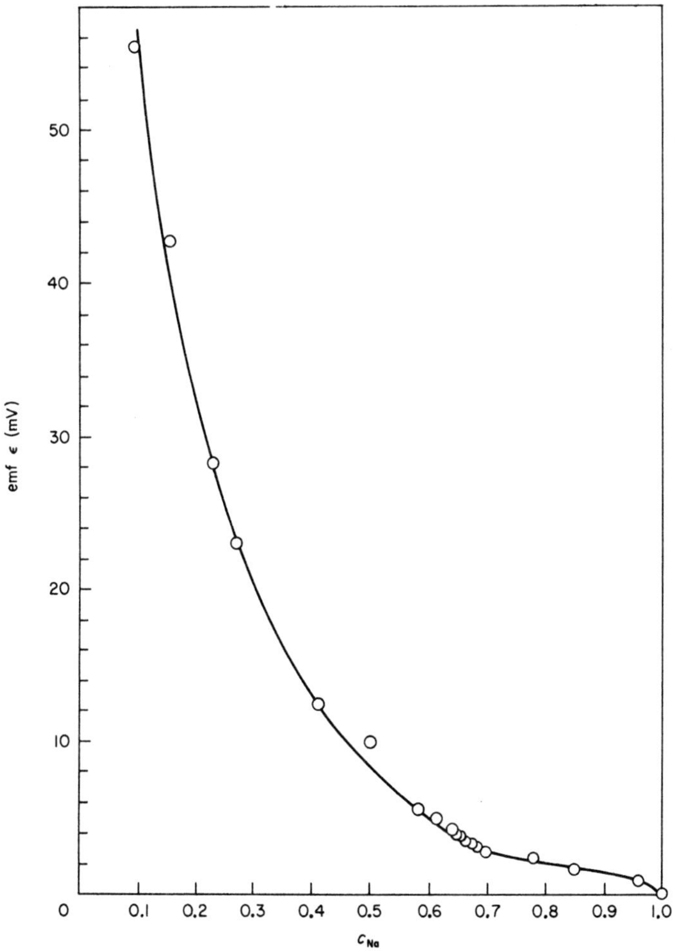

Figure 6.7 Emf (in mV) of the liquid Na–Cs alloy as a function of Na concentration. The points are experimental data of Ichikawa and Thompson (1974) and the curve is based on Flory's model, including size effects and interchange energy (from Bhatia and March, 1975)

Flory's model being compared. The parameter w/k_BT was found to be 1.14 to get a good fit.

With this parameter, $S_{cc}(0)$ is as shown in *figure 6.8*, a peak in the concentration fluctuations being found at $c_{Na} = 0.8$. Such a peak was deduced by Ichikawa and Thompson from their emf data. Its position depends on a balance between the size factor v_2/v_1 and the interchange energy w.

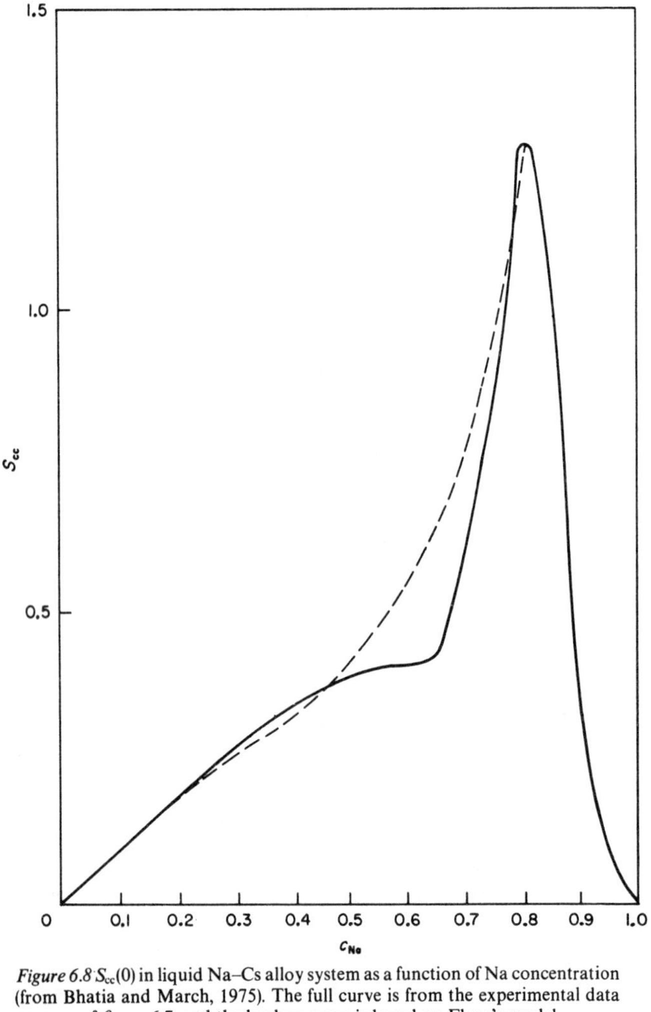

Figure 6.8 $S_{cc}(0)$ in liquid Na–Cs alloy system as a function of Na concentration (from Bhatia and March, 1975). The full curve is from the experimental data of *figure 6.7*, and the broken curve is based on Flory's model

It is interesting to note from eqn (6.34) that when $S_{cc}(0)$ is large the slope of the liquidus curve is small. This is immediately evident in the phase diagram of Na–Cs, as shown in *figure 6.9*. Indeed in quite a few metal alloys such flat portions of the liquidus curves are evident, presumably heralding pronounced peaks in the concentration fluctuations (see Bhatia and March, 1975 for details of these other alloy systems).

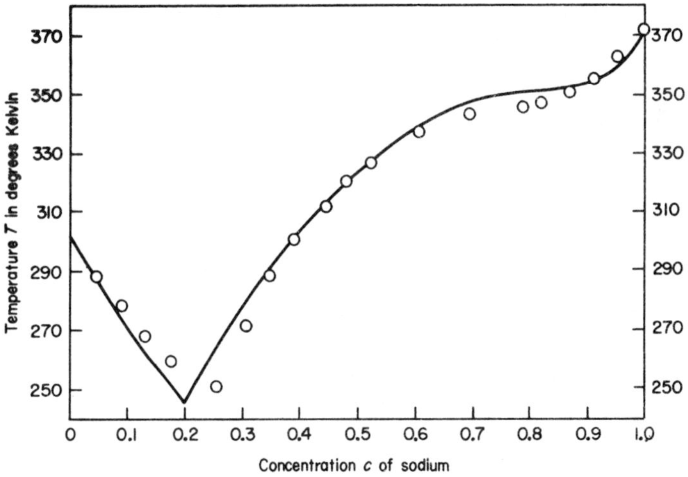

Figure 6.9 Liquidus curve of the Na–Cs system against Na concentration: symbols are as in *figure 6.7* (from Bhatia and March, 1975). The flat portion of the liquidus curve coincides with the concentration at which S_{cc} peaks in *figure 6.8*

6.4 Partial structure factors at finite wavelength in conformal solutions

Having discussed the application of conformal solution theory to the thermodynamic properties of mixtures, we shall now extend it to finite wave vectors, to relate the partial structure factors in the mixture to the correlation functions of the reference liquid. The q-dependent generalisation of the thermodynamic formalism introduced in section 6.2 is presented in appendix 6.3.

The specific assumption made in the conformal solution model (Longuet-Higgins, 1951) is that the pair potentials $\phi_{\alpha\beta}(r)$ between the different atoms in the mixture are related to the pair potential $\phi(r)$ in the reference liquid by

$$\phi_{\alpha\beta}(r) = A_{\alpha\beta}\phi(\lambda_{\alpha\beta}r) \tag{6.49}$$

In the lowest order theory it is further assumed that the A's and λ's deviate only slightly from unity, which means that the mixture has force laws very close to the reference liquid, and that $\lambda_{12} = \frac{1}{2}(\lambda_{11} + \lambda_{22})$. As we shall see, these assumptions allow one to express the thermodynamic properties of a binary mixture in terms of the product of the parameter $d_{12}=2A_{12}-A_{11}-A_{22}$ and of the quantity $E_0=\frac{1}{2}\rho\int \mathrm{d}\mathbf{r}\,\phi(r)g(r)$ or the reference liquid, the quantity w introduced in section 6.3 being simply $w=d_{12}E_0$.

The calculation of the pair functions in the mixture can be tackled by perturbative techniques, the perturbative parameters being $A_{\alpha\beta}-1$ and $\lambda_{\alpha\beta}-1$. A general discussion of perturbation theory as applied to correlation functions is given, because of its wider usefulness, in appendix 6.4; in the present instance the result is (Mo et al., 1974; Parrinello, Tosi and March, 1974b):

$$g_{\alpha\beta}(r)-g(r)=\sum_{\gamma}\frac{\Delta z_{\gamma}}{z}\left\{(\delta_{\alpha\gamma}+\delta_{\beta\gamma})g(r)+\rho_{\gamma}\int \mathrm{d}s\,[g_3(\mathbf{r},s)-g(r)]\right\}$$
$$-\frac{1}{k_{\mathrm{B}}T}\left\{\Delta\phi_{\alpha\beta}(r)g(r)+\sum_{\gamma}\rho_{\gamma}\int \mathrm{d}s\,[\Delta\phi_{\alpha\gamma}(s)+\Delta\phi_{\beta\gamma}(s)]g_3(\mathbf{r},s)\right.$$
$$\left.+\frac{1}{2}\sum_{\gamma,\gamma'}\rho_{\gamma}\rho_{\gamma'}\int \mathrm{d}s\,\mathrm{d}s'\,\Delta\phi_{\gamma\gamma'}(s')[g_4(\mathbf{r},s,s')-g(r)g(s')]\right\} \qquad (6.50)$$

Here, $g(r)$, $g_3(\mathbf{r},s)$ and $g_4(\mathbf{r},s,s')$ are the two-, three- and four-particle correlation functions in the reference liquid, and Δz_{γ} is the change in fugacity of the γth component in the mixture from the reference liquid. Furthermore,

$$\Delta\phi_{\alpha\beta}(r)=A_{\alpha\beta}\phi(\lambda_{\alpha\beta}r)-\phi(r)$$
$$\doteq(A_{\alpha\beta}-1)\phi(r)+(\lambda_{\alpha\beta}-1)P(r),\qquad P(r)=r\frac{\mathrm{d}\phi(r)}{\mathrm{d}r} \qquad (6.51)$$

We shall, in the following, indicate in a more intuitive way how the N–c correlation functions in the mixture are related to the correlation properties of the reference liquid.

6.4.1 Density–density correlations

It is first of all instructive to investigate how the pair function in a one-component liquid is modified under a rescaling of the units of energy and length in the pair potential,

$$\phi(r)\rightarrow\phi_{\mathrm{m}}(r)\equiv A\phi(\lambda r) \qquad (6.52)$$

It is easy to show that the chemical potential is modified to

$$\mu_{\mathrm{m}}(\beta,V)=A\mu(A\beta,\lambda^3V)+\frac{3}{\beta}\ln\lambda,\qquad \beta=\frac{1}{k_{\mathrm{B}}T} \qquad (6.53)$$

so that the fugacity becomes

$$z_{\mathrm{m}}(\beta,V)=\lambda^3A^{3/2}z(A\beta,\lambda^3V) \qquad (6.54)$$

The definition (2.3) of the pair function in the grand canonical ensemble then gives

$$\rho^2 g_{\mathrm{m}}(r) = \frac{1}{\Xi_{\mathrm{m}}} \sum_{N=2}^{\infty} \lambda^{3N} A^{3N/2} \frac{z^N(A\beta, \lambda^3 V)}{(N-2)!} \int \exp\left[-\beta A V_N(\lambda r_1, \ldots, \lambda r_N)\right] \mathrm{d}r_3 \ldots \mathrm{d}r_N \quad (6.55)$$

where

$$\Xi_{\mathrm{m}}(\beta, V) = \Xi(A\beta, \lambda^3 V, A^{3/2} z) \quad (6.56)$$

The transformation $\boldsymbol{r}_i \to \boldsymbol{r}'_i / V^{1/3}$ then yields

$$g_{\mathrm{m}}(r; \beta, V) = g(\lambda r; A\beta, \lambda^3 V, A^{3/2} z) \quad (6.57)$$

We can apply this result immediately to the density–density correlation function, $g_{NN}(r)$, in the mixture, writing

$$g_{NN}(r; \beta, V) = g(\lambda r; A\beta, \lambda^3 V, A^{3/2} z) \quad (6.58)$$

provided that we take the parameters A and λ to be given by

$$A = \sum_{\alpha,\beta} c_\alpha c_\beta A_{\alpha\beta} \quad (6.59)$$

and

$$\lambda = \sum_{\alpha,\beta} c_\alpha c_\beta \lambda_{\alpha\beta} \quad (6.60)$$

The result (6.58) is valid for arbitrary values of A and λ, not necessarily near to unity. When A and λ deviate only slightly from unity we can expand the above result in a Taylor series in $(A-1)$ and $(\lambda-1)$, finding to linear terms

$$g_{NN}(r) = g(r) + (A-1)\left[\beta\left(\frac{\partial g}{\partial \beta}\right)_\rho + \tfrac{3}{2} z \left(\frac{\partial g}{\partial z}\right)_\beta\right] + (\lambda - 1)\left[r\frac{\partial g}{\partial r} + 3V\left(\frac{\partial g}{\partial V}\right)_\beta\right] \quad (6.61)$$

Finally, by use of the identities

$$z\left(\frac{\partial g}{\partial z}\right)_\beta = \frac{\rho}{\beta}\left(\frac{\partial g}{\partial p}\right)_\beta \quad (6.62)$$

and

$$V\left(\frac{\partial g}{\partial V}\right)_\beta = -\rho\left(\frac{\partial p}{\partial \rho}\right)_\beta \cdot \left(\frac{\partial g}{\partial p}\right)_\beta \quad (6.63)$$

we can write

$$g_{NN}(r)-g(r)=(A-1)\beta\left(\frac{\partial g(r)}{\partial\beta}\right)_\rho+3\rho\left[\frac{A-1}{2\beta}-(\lambda-1)\left(\frac{\partial p}{\partial\rho}\right)_\beta\right]\left(\frac{\partial g(r)}{\partial p}\right)_\beta$$
$$+(\lambda-1)r\frac{\partial g}{\partial r} \tag{6.64}$$

This result agrees with a direct calculation based on eqn (6.50), when one uses the expressions for the various derivatives of the pair function in terms of the three- and four-particle functions that we have given in appendix 2.1, and the expression for the Gibbs free energy in conformal solution theory. It also agrees with the thermodynamic results of section 6.3, which lead to

$$\begin{aligned}S_{NN}(0)&=\rho k_BT\,K_{Tm}+\delta^2S_{cc}(0)\\&\doteqdot\rho k_BT\,K_{Tm}\end{aligned} \tag{6.65}$$

to first order in the small parameters. Here K_{Tm} is the isothermal compressibility of the 'modified' one-component liquid with pair potential $A\phi(\lambda r)$.

6.4.2 Concentration–concentration correlations

Since $g_{cc}(r)$ as defined in eqn (6.4) involves the differences $g_{11}(r)-g_{12}(r)$ and $g_{22}(r)-g_{12}(r)$, we expect strong cancellations to occur, as a change of the temperature and of the pressure will not affect these differences, at least to first order. In fact, only the first term in the second brace of eqn (6.50) survives, and one finds the very simple result

$$g_{cc}(r)=c_1{}^2c_2{}^2\beta d_{12}g(r)\phi(r) \tag{6.66}$$

Correspondingly,

$$S_{cc}(k)=c_1c_2[1+c_1c_2\beta d_{12}(\overline{\phi g})_k],\qquad(\overline{\phi g})_k=\rho\int d\boldsymbol{r}\exp(i\boldsymbol{k}\cdot\boldsymbol{r})\,\phi(r)g(r) \tag{6.67}$$

These equations suggest a very simple test of the applicability of conformal solution theory to a given mixture, since they make a simple explicit prediction on the dependence on concentration. Specifically, we expect the functions $g_{cc}(r)/(c_1{}^2c_2{}^2)$ and $\left[\dfrac{S_{cc}(k)}{c_1c_2}-1\right]/(c_1c_2)$ to be independent of concentration if the model applies.

For $k\to0$ the quantity $(\overline{\phi g})_k$ introduced in eqn (6.67) coincides with the quantity $2E_0$ entering the thermodynamic application of the model, defined as $E_0=\frac{1}{2}\rho\int d\boldsymbol{r}\,\phi(\boldsymbol{r})g(r)$, and thus we recover from eqn (6.67), to first order in d_{12}, the thermodynamic result (6.27) of section 6.3,

$$S_{cc}(0)=\frac{c_1c_2}{1-2\beta c_1c_2d_{12}E_0} \tag{6.68}$$

with $w=d_{12}E_0$.

Calculations using the relation of $S_{cc}(k)$ to the Fourier transform of the product $g\phi$, g and ϕ referring to the reference liquid, have been reported by Johnson et al. (1975). g and ϕ were chosen to reflect whether the mixture was insulating (like argon–krypton) or metallic (like sodium–potassium). Though the variety of potentials that appear possible in metals seems greater than for simple insulators, the characteristic forms for the Fourier transforms were not very different, the schematic form being shown in *figure 6.10*.

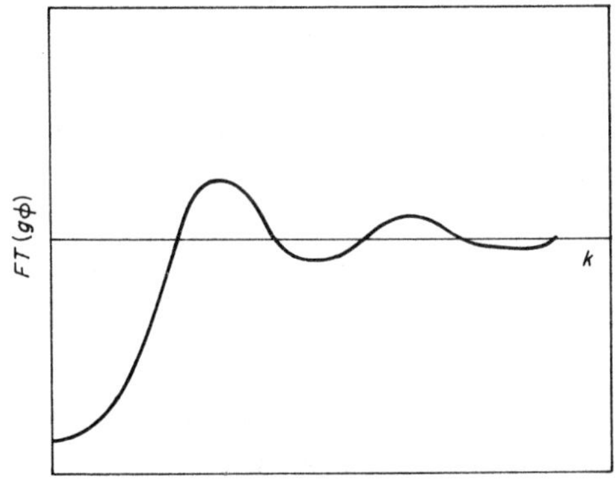

Figure 6.10 Schematic drawing of the Fourier transform of $\phi(r)g(r)$ versus wave number k

The form shown in the figure is reminiscent of the direct correlation function $\tilde{c}(k)$ of the reference liquid, although the connection is not better than qualitative (not only the amplitude is wrong but there is a substantial phase shift). This relation would be quantitative if RPA were applicable, for then (see eqn (2.54)) we would have

$$\phi_{eff}(k) \simeq FT(g\phi) \simeq -\tilde{c}(k)k_B T \qquad (6.69)$$

A more refined theory, such as Percus–Yevick, is needed, but even that is not fully quantitative for the cases examined by Johnson et al.

The most significant result of this work is that there is a direct relation between the position in k space of the first peak in $S_{cc}(k)$ and the peak in $S(k)$ of the reference liquid. This observation bears a relation to the pre-peaks found in the experiments of Ruppersberg (1973) on Li–Pb and shown in *figure 6.3*. Here, the pre-peaks are found not to vary in position in k-space over a substantial range of concentration. The first peak in $S(k)$ for both Li and Pb occurs at the same k value, and the peak in $S_{cc}(k)$ relates to this precisely in the manner predicted by Johnson et al. The same situation obtains for $S_{cc}(k)$ in Cu–Sn, calculated by Bhatia and Thornton (1970) from the experimental data of Enderby et al. (1966), provided we take $S_{NN}(k)$ to simulate the structure factor $S(k)$ of the 'reference' liquid. Probably Cu–Sn is *not* a conformal solution because of the valence difference of three (this is also the valence difference for Li–Pb discussed above).

Constructing $g_{Nc}(r)$, only the four-particle correlation term in (6.50) cancels and we shall therefore not quote the result here. Naturally, however, in the long wavelength limit we regain the thermodynamic results

$$\begin{aligned} S_{Nc}(0) &= -\delta S_{cc}(0) \\ &\doteqdot -c_1 c_2 \delta \end{aligned} \tag{6.70}$$

to first order in the small parameters.

A preliminary test of the applicability of conformal solution theory to the argon–krypton mixture has been carried out by Mo et al. (1974). They used directly eqn (6.50), choosing the reference liquid as the one-component liquid whose pair distribution coincides, at each concentration, with the corresponding $g_{NN}(r)$ and adopting the Kirkwood superposition approximation to estimate the terms involving three-body correlations. The test is very satisfactory except in dilute mixtures. As noted above, a simpler and very direct test can be achieved by examining the correlations between concentration fluctuations. Computer simulation calculations have shown that while small size differences affect the thermodynamic properties in an unimportant way, the conformal solution model gives too little structure in $S_{cc}(k)$.

In addition to liquid mixtures of rare gas atoms of not too different sizes, favourable candidates for a description as conformal solutions should be afforded by liquid mixtures of some organic molecules and by liquid metal alloys of equi-valent ions and small size differences. It should be noted, however, that while in organic mixtures the constraint of central pair forces may eventually have to be relaxed, for equi-valent metal alloys electronic effects will introduce an implicit concentration dependence of the pair potential whenever the volume per atom (and thus the Fermi momentum) varies appreciably with concentration. Only in some equi-valent alloys (such as the Mg–Cd and the In–Tl systems) is this effect truly negligible.

6.5 Partial structure factors in mixtures of hard spheres

In liquid mixtures where the components are not sufficiently similar for conformal solution theory to apply, one must appeal to other, different models for a description of their structure. One such model is the mean spherical model (MSM), whose assumptions are:

(1) The short range part of the interaction is an infinitely hard potential characterised by hard core diameters $\sigma_{\alpha\beta} = \frac{1}{2}(\sigma_{\alpha\alpha} + \sigma_{\beta\beta})$, namely

$$g_{\alpha\beta}(r) = 0 \quad \text{for} \quad r < \sigma_{\alpha\beta} \tag{6.71}$$

(2) Outside the hard cores the direct correlation functions are simply related to the pair potential by

$$c_{\alpha\beta}(r) = -\frac{1}{k_B T}\,\phi_{\alpha\beta}(r) \quad \text{for} \quad r > \sigma_{\alpha\beta} \tag{6.72}$$

The model is to be solved by taking advantage of the general relation between the direct and the radial correlation functions, which in a multi-component system reads (Pearson and Rushbrooke, 1957):

$$g_{\alpha\beta}(r) - 1 = c_{\alpha\beta}(r) + \sum_{\gamma} (\rho_\alpha \rho_\beta)^{1/2} \int \mathrm{d}\mathbf{r}'\, c_{\alpha\gamma}(\mathbf{r}-\mathbf{r}')[g_{\gamma\beta}(\mathbf{r}') - 1] \tag{6.73}$$

In particular, for a system of neutral hard spheres lacking an attractive interaction tail, for which $\phi_{\alpha\beta}(r) = 0$ when $r > \sigma_{\alpha\beta}$, the MSM is equivalent to the Percus–Yevick approximation and can be solved exactly by analytic means (Lebowitz, 1964). This model, although crude, still has the advantage of providing analytic expressions for the direct correlation functions and the thermodynamic properties in terms of parameters such as the hard-sphere diameters and the partial densities of the components.

We shall indicate the nature of the procedure for the solution of the model by discussing first the case of a one-component fluid of hard spheres, and shall then report the solutions for the two-component fluid. (Machine calculations on hard spheres are available; see Alder and Hecht, 1969.)

6.5.1 Mean spherical model (MSM) for one-component fluid of neutral hard spheres

Virial expansion results for the fluid of hard spheres (Nijboer and van Hove, 1952; Ashcroft and March, 1967) show that in the Percus–Yevick approximation, to second order in the packing fraction $\eta = (\pi/6)\rho\sigma^3$, the direct correlation function $c(r)$ inside the hard core diameter $\sigma(r < \sigma)$ is a simple third-order polynomial in r/σ. This form is preserved in the exact solution of the MSM (Wertheim, 1963; Thiele, 1963), which reads, as we already noted in chapter 2,

$$c(r) = A_0 + A_1(r/\sigma) + A_3(r/\sigma)^3 \quad \text{for} \quad r < \sigma \tag{6.74}$$

with

$$\left.\begin{aligned} A_0 &= -(1+2\eta)^2/(1-\eta)^4 \\ A_1 &= 6\eta(1+\tfrac{1}{2}\eta)^2/(1-\eta)^4 \\ A_3 &= -\tfrac{1}{2}\eta(1+2\eta)^2/(1-\eta)^4 \end{aligned}\right\} \tag{6.75}$$

We note that the function $c(r)$ has two singularities, at $r = 0$ and at $r = \sigma$. As a consequence, its Fourier transform $\tilde{c}(k)$ at large k will contain two types of terms:

(1) Terms behaving as k^{-4} and k^{-6}, with coefficients proportional to A_1 and A_3, respectively.

(2) Terms behaving as $\cos(k\sigma)/(k\sigma)^{2n}$ and $\sin(k\sigma)/(k\sigma)^{2n+1}$. Indeed, the Fourier transform $c(k)$ is easily evaluated to be

$$\tilde{c}(k)=24\eta\left\{-\frac{2A_1}{(k\sigma)^4}+\frac{24A_3}{(k\sigma)^6}-\frac{\cos(k\sigma)}{(k\sigma)^2}(A_0+A_1+A_3)+\frac{\sin(k\sigma)}{(k\sigma)^3}(A_0+2A_1+4A_3)\right.$$
$$\left.+\frac{\cos k\sigma}{(k\sigma)^4}(2A_1+12A_3)-24A_3\frac{\sin(k\sigma)}{(k\sigma)^5}-24A_3\frac{\cos(k\sigma)}{(k\sigma)^6}\right\} \qquad (6.76)$$

The coefficients of the polynomial for $c(r)$ inside σ are determined so as to satisfy the condition (6.71) that $g(r)=0$ for $r<\sigma$, which implies that $g(r)$ vanishes at the origin with all its derivatives. To this end we can calculate the structure factor $S(k)$ from $\tilde{c}(k)$ by the usual relation

$$S(k)=[1-\tilde{c}(k)]^{-1}$$
$$=1+\tilde{c}(k)+[\tilde{c}(k)]^2+[\tilde{c}(k)]^3+\cdots \qquad (6.77)$$

[since $\tilde{c}(k)<1$], and we must then require that its large-k expansion should not contain terms behaving as k^{-2n}, as these would herald terms in $[g(r)-1]$ proportional to $(r/\sigma)^{2n-3}$ for small r. By this technique it is possible to prove that $c(r)$ inside σ is indeed a third-order polynomial, and to find the following relations between the polynomial coefficients:

$$A_1=6\eta(A_0+A_1+A_3)^2 \qquad (6.78)$$

$$A_3=-\tfrac{1}{2}\eta(A_0{}^2-8A_3{}^2-16A_0A_3-12A_1A_3) \qquad (6.79)$$

It is easily verified that these relations are satisfied by the Wertheim–Thiele solution reported in eqn (6.75).

To obtain a complete specification of $c(r)$, however, we need a further condition to determine A_0. Here thermodynamic inconsistency arises as is well known in the P–Y solution. We could, for example, use the exact relation between $g(\sigma^+)$ and the thermodynamic pressure p (see, for instance, Longuet-Higgins and Pople, 1956), but this would *not* yield the Wertheim–Thiele solution. Alternatively we could require the internal energy to be only kinetic, to complete the solution, and this leads to the relation

$$1+c(r=0)=\tilde{c}(k=0) \qquad (6.80)$$

or

$$1+A_0=24\eta(\tfrac{1}{3}A_0+\tfrac{1}{4}A_1+\tfrac{1}{6}A_3) \qquad (6.81)$$

which is indeed satisfied by the Wertheim–Thiele solution. In fact, using the relation $S(0)=\rho k_B T K_T$ between the structure factor at long wavelength and the isothermal compressibility, we can rewrite (6.80) as

$$c(r=0)=-\frac{1}{S(0)}=-\beta\left(\frac{\partial p}{\partial\rho}\right)_T \qquad (6.82)$$

so that this procedure relates the coefficient A_0 to the isothermal compressibility. By integrating the expression of A_0 we then find the equation of state in the form

$$p=\rho k_B T\frac{1+\eta+\eta^2}{(1-\eta)^3} \tag{6.83}$$

a result previously obtained by Reiss, Frisch and Lebowitz (1959).

The conclusion from this argument is that, by a study of the singularities in $c(r)$, we can calculate the coefficients in the polynomial for $c(r)$ inside σ without ever having to explicitly impose the troublesome condition $g(r)=0$ inside σ by direct calculation in $\boldsymbol{r}$ space.

6.5.2 Mean spherical model for binary mixture of hard spheres

The exact solution of the MSM for a binary mixture of hard spheres of diameters σ_1 and $\sigma_2(\sigma_2>\sigma_1)$ has been obtained by Lebowitz (1964). The direct correlation functions inside the cores have again a simple form:

$$C_{\alpha\alpha}(r)=A_{0\alpha}+A_{1\alpha}(r/\sigma_\alpha)+A_{3\alpha}(r/\sigma_\alpha)^3\ (r<\sigma_\alpha) \tag{6.84}$$

$$C_{12}(r)=\begin{cases} A_{01} & (r\leqslant\lambda) \\ A_{01}+\dfrac{1}{r}[A_1(r-\lambda)^2+4\lambda A_3(r-\lambda)^3+A_3(r-\lambda)^4] & [\lambda\leqslant r<\tfrac{1}{2}(\sigma_1+\sigma_2)] \end{cases} \tag{6.85}$$

where $\lambda=\frac{1}{2}(\sigma_2-\sigma_1)$. Again the coefficients A_{01} and A_{02} are related to the long-wavelength limit of the partial structure factors, namely

$$-A_{0\alpha}=\frac{1}{k_B T}\left(\frac{\partial p(\rho_1,\rho_2)}{\partial\rho_\alpha}\right)_{T,\rho_{\bar\alpha}}=\frac{1}{k_B T}\sum_\beta\rho_\beta\left(\frac{\partial\mu_\beta(\rho_1,\rho_2)}{\partial\rho_\alpha}\right)_{T,\rho_{\bar\alpha}}$$

$$=v_\alpha/(k_B T K_T) \tag{6.86}$$

where v_α is the partial volume per molecule of the αth species and $\bar\alpha$ is the species different from α. The equation of state is found to have the form

$$p=\frac{k_B T}{(1-\eta)^3}\left\{(\rho_1+\rho_2)(1+\eta+\eta^2)-\frac{\pi}{2}\rho_1\rho_2(\sigma_2-\sigma_1)^2\times\left[\sigma_1+\sigma_2+\sigma_1\sigma_2\frac{\pi}{6}(\rho_1\sigma_1{}^2+\rho_2\sigma_2{}^2)\right]\right\} \tag{6.87}$$

with $\eta=\frac{1}{6}\pi(\rho_1\sigma_1{}^3+\rho_2\sigma_2{}^3)$, while the other coefficients in eqns (6.84) and (6.85) are given by

$$A_{1\alpha}=\pi\left[\rho_\alpha\sigma_\alpha{}^3g_{\alpha\alpha}^2+\rho_{\bar\alpha}\sigma_\alpha\left(\frac{\sigma_1+\sigma_2}{2}\right)^2g_{12}^2\right] \tag{6.88}$$

$$A_1=\pi\left(\frac{\sigma_1+\sigma_2}{2}\right)g_{12}[\rho_1\sigma_1g_{11}+\rho_2\sigma_2g_{22}] \tag{6.89}$$

and

$$A_{3\alpha}\sigma_\alpha^{-3}=A_3=-\frac{\pi}{12}(\rho_1 A_{01}+\rho_2 A_{02}) \tag{6.90}$$

with

$$g_{\alpha\alpha}=-c_{\alpha\alpha}(\sigma_\alpha)=\left[(1+\tfrac{1}{2}\eta)+\frac{\pi}{4}\rho_{\bar\alpha}\sigma_{\bar\alpha}^{2}(\sigma_1-\sigma_2)\right]\Big/(1-\eta)^2 \tag{6.91}$$

$$g_{12}=-c_{12}\left(\frac{\sigma_1+\sigma_2}{2}\right)=(\sigma_2 g_{11}+\sigma_1 g_{22})/(\sigma_1+\sigma_2) \tag{6.92}$$

Numerical evaluations of the partial structure factors in a mixture based on this model have been carried out by Ashcroft and Langreth (1967*a*) and Enderby and North (1968), and some of their results are reported in *figure 6.11.*

6.6 Hydrodynamic correlation functions in binary mixtures

We turn now to the dynamic correlation functions in a binary mixture. It should be remarked at the outset, and should be no surprise from the earlier

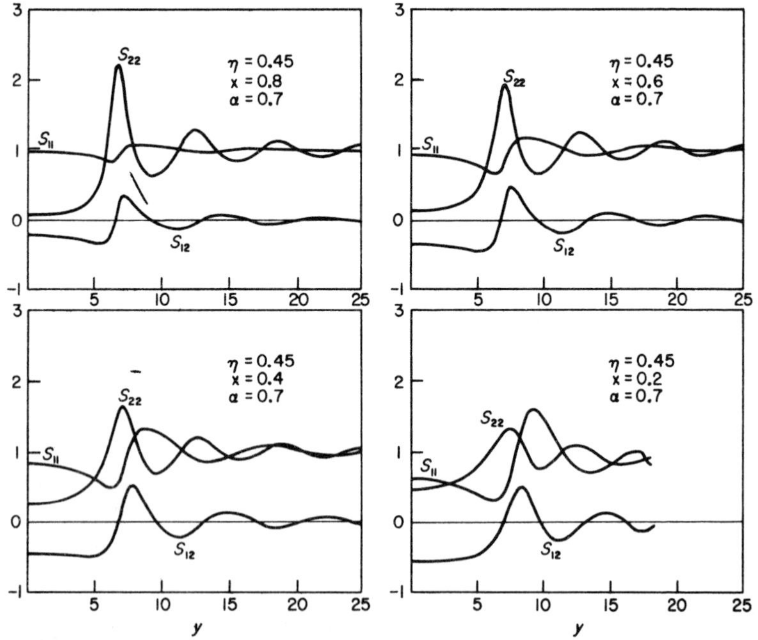

Figure 6.11 Partial structure factors for Percus–Yevick mixture of hard spheres with packing fraction $\eta=0.45$ and diameter ratio $\alpha=\sigma_1/\sigma_2=0.7$, for four values of the concentration x of the larger species (from Ashcroft and Langreth, 1967*a*)

discussion of the scattering function $S(k, \omega)$ of a one-component liquid, that only in relatively simple models, or in the hydrodynamic regime, can closed results be written down.

6.6.1 Partial van Hove functions

Essentially, the linearised equations of hydrodynamics now allow us to calculate the long wavelength, low frequency forms of the correlation functions in a binary mixture. The detailed arguments (Cohen, Sutherland and Deutch, 1971; Bhatia, Thornton and March, 1974) are set out in appendix 6.5 and we shall be content to summarise here the results, and comment on their physical interest.

The structure in k and ω of the concentration–concentration correlation function $S_{cc}(k, \omega)$ is, in fact,

$$S_{cc}(k, \omega)=\frac{Nk_BT}{Z}\left\{\frac{2A_7Xk^2}{\omega^2+X^2k^4}+\frac{2A_8Yk^2}{\omega^2+Y^2k^4}\right\} \tag{6.93}$$

the k and ω dependence being explicit. The other quantities can be obtained in terms of the physical parameters which enter the linearised equations of hydrodynamics.

Similarly the structure of the number–concentration correlation function in the same limit of small k and ω is given by

$$S_{Nc}(k, \omega)=Nk_BT\left\{\frac{2A_4Xk^2}{\omega^2+X^2k^4}+\frac{2A_5Yk^2}{\omega^2+Y^2k^4} + A_6\frac{k}{c_0}\left[\frac{\omega+c_0k}{(\omega+c_0k)^2+\Gamma^2k^4}-\frac{\omega_0-c_0k}{(\omega-c_0k)^2+\Gamma^2k^4}\right]\right\} \tag{6.94}$$

Thirdly, the number–number correlation function $S_{NN}(k, \omega)$, closely related to the density–density correlation function $S(k, \omega)$ in a one-component fluid, is given by

$$S_{NN}(k, \omega)=\frac{Nk_BTK_T}{V\gamma}\left\{\frac{2A_1Xk^2}{\omega^2+X^2k^4}+\frac{2A_2Yk^2}{\omega^2+Y^2k^4} + \left[\frac{\Gamma k^2}{(\omega+c_0k)^2+\Gamma^2k^4}+\frac{\Gamma k^2}{(\omega-c_0k)^2+\Gamma^2k^4}\right] + A_3\frac{k}{c_0}\left[\frac{\omega+c_0k}{(\omega+c_0k)^2+\Gamma^2k^4}-\frac{\omega-c_0k}{(\omega-c_0k)^2+\Gamma^2k^4}\right]\right\} \tag{6.95}$$

6.6.2 Kubo relations

It is now relatively straightforward to obtain Kubo formulae for the transport coefficients in a mixture, analogous to those obtained earlier for one-component fluids. The simplest result is for interdiffusion, namely

$$\lim_{\omega\to 0} \omega^2 \left[\lim_{k\to 0} \frac{1}{k^2} S_{cc}(k,\omega) \right] = 2Nk_B T\, D/Z \qquad (6.96)$$

For sound wave attenuation, however, it turns out to be more convenient to express $\frac{4}{3}\eta+\zeta$ in terms of the mass–mass correlation function $S_{MM}(k,\omega)$. It is again straightforward to construct the hydrodynamic form of the M–X correlation functions. We record here the Kubo relations that follow:

$$\lim_{\omega\to 0} \omega^4 \left[\lim_{k\to 0} \frac{1}{k^4} S_{MM}(k,\omega) \right] = \frac{2k_B T}{\rho} \left(\tfrac{4}{3}\eta+\zeta\right) \qquad (6.97)$$

$$\lim_{\omega\to 0} \omega^4 \left[\lim_{k\to 0} \frac{1}{k^4} S_{MX}(k,\omega) \right] = 2k_B T \frac{\gamma V D \Sigma}{Y K_T} \qquad (6.98)$$

$$\lim_{\omega\to 0} \omega^2 \left[\lim_{k\to 0} \frac{1}{k^2} S_{XX}(k,\omega \right] = 2mk_B T V\, D/Z_X \qquad (6.99)$$

Again all the quantities appearing are defined in appendix 6.5.

We shall give an example or two below for isotopic mixtures and for conformal solutions, in which these formulae can be used for actual evaluation of transport coefficients.

6.7 Dynamical properties of isotopic mixtures

We turn next to the simplest classical mixture—that of isotopes. Here the forces between the components are all the same, as are the static structure factors. Being an 'ideal solution' the static concentration fluctuations are independent of k and given by

$$S_{cc}(k) = c_1 c_2 \qquad (6.100)$$

The dynamical properties of such mixtures have been discussed by Parrinello, Tosi and March (1974*a*).

We shall first tackle the problem by setting up some physical models with the correct structure as dictated by the general theoretical framework of response functions, to understand the nature of the excitations which can occur in such an isotopic mixture. Secondly, we shall study the dynamical concentration fluctuations, in order to examine mass transport in the mixture. Computer simulation data for mass transport in Lennard–Jones isotopic mixtures are available (Ebbsjö et al., 1974).

6.7.1 Models of partial dynamical structure factors

For the ideal solution which an isotopic mixture constitutes, the dynamical structure factors $S_{NN}(k,\omega)$ and $S_{Nc}(k,\omega)$ have zero-order moments given by

$$\int \frac{d\omega}{2\pi} S_{Nc}(k, \omega) = S_{Nc}(k) = 0 \tag{6.101}$$

and

$$\int \frac{d\omega}{2\pi} S_{NN}(k, \omega) = a(k) \tag{6.102}$$

Before going into more details, let us set up the very simplest model for the partial dynamical structure factors. The idea (following Feynman—see chapter 8) is to assume well-defined excitations, with dispersion relations $\omega(k)$, and to determine $\omega(k)$ in terms of the static structure using the second moments. These are, in fact, given by

$$\int_{-\infty}^{\infty} \frac{d\omega}{2\pi} \omega^2 S_{\alpha\beta}(k, \omega) = \delta_{\alpha\beta} k_B T\, k^2/m_\alpha \tag{6.103}$$

which immediately yields

$$\int_{-\infty}^{\infty} \frac{d\omega}{2\pi} \omega^2 S_{Nc}(k, \omega) = c_1 c_2 k_B T\, k^2 \left(\frac{1}{m_1} - \frac{1}{m_2}\right) \tag{6.104}$$

$$\int_{-\infty}^{\infty} \frac{d\omega}{2\pi} \omega^2 S_{NN}(k, \omega) = k_B T\, k^2 \left(\frac{c_1}{m_1} + \frac{c_2}{m_2}\right) \tag{6.105}$$

and

$$\int_{-\infty}^{\infty} \frac{d\omega}{2\pi} \omega^2 S_{cc}(k, \omega) = c_1 c_2 m k_B T k^2 / m_1 m_2 \tag{6.106}$$

with $m = c_1 m_1 + c_2 m_2$.

Now for small mass differences, $\int d\omega \omega^2 S_{Nc}(k, \omega) \to 0$ with $m_1 - m_2$, whereas the other second moments remain finite. Thus, the number density $\rho(\mathbf{k}, t)$ and the concentration $c(\mathbf{k}, t)$ oscillate independently in this case, with frequencies $\omega_{10}(k)$ and $\omega_{20}(k)$, respectively. To satisfy the zeroth and second moments in this limit of small $(m_1 - m_2)$ we therefore adopt the forms

$$S_{NN}(k, \omega) \doteqdot \tfrac{1}{2} a(k)[\delta(\omega - \omega_{10}(k)) + \delta(\omega + \omega_{10}(k))] \tag{6.107}$$

and

$$S_{cc}(k, \omega) \doteqdot \tfrac{1}{2} c_1 c_2 [\delta(\omega - \omega_{20}(k)) + \delta(\omega + \omega_{20}(k))] \tag{6.108}$$

with $S_{Nc}(k, \omega) \doteqdot 0$. Then we find (compare Ascarelli, Paskin and Harrison, 1967 for a one-component liquid)

$$\omega_{10}^2(k) = \frac{k_B T\, k^2}{a(k)} \left(\frac{c_1}{m_1} + \frac{c_2}{m_2}\right) \tag{6.109}$$

and

$$\omega_{20}^2(k) = m k_B T k^2 / m_1 m_2 \tag{6.110}$$

The first mode, for equal masses, is the classical analogue of Feynman's result, and is the frequency with which the density oscillates. The second frequency ω_{20} describes the self-function, which in the same approximation and for equal masses is

$$S_s(k, \omega) \doteqdot \tfrac{1}{2}[\delta(\omega - k/(\beta m)^{1/2}) + \delta(\omega + k/(\beta m)^{1/2})] \tag{6.111}$$

Two limitations of such a treatment must be noted:

(1) The velocity of sound v_s in the isotopic mixture will only be correct if the ratio of the specific heats γ is near to unity. This is of little consequence for many liquid metals, for which the values of γ at the melting point are 1.2–1.3, but for argon at the triple point, with $\gamma = 2.2$, the error is a major one. Actually, in the ideal solution which the isotopic mixture represents we have the exact result

$$v_s{}^2 = \gamma k_B T/ma(0) \tag{6.112}$$

and the concentration dependence is only in $m = c_1 m_1 + c_2 m_2$. This exact formula agrees with the above model when $m_1 \simeq m_2$ and $\gamma \simeq 1$.

(2) The treatment is only valid for small mass differences.

In the case of arbitrary mass difference, $S_{Nc}(k, \omega)$ is non-zero and the pure modes ω_{10} and ω_{20} interact and mix. The most convenient formulation is then in terms of response functions. We shall simply summarise the main results here.

The new squared frequencies are

$$\omega^2(k) = \tfrac{1}{2}\left\{\omega_{10}^2 + \omega_{20}^2 \pm \left[(\omega_{10}^2 + \omega_{20}^2)^2 - \frac{4k^4}{m_1 m_2 \beta^2 a(k)}\right]^{1/2}\right\} \tag{6.113}$$

When the mass difference is small, we have

$$\frac{4k^2}{m_1 m_2 \beta^2 a(k)} \simeq 4\omega_{10}^2 \omega_{20}^2 \tag{6.114}$$

giving back the pure mode frequencies.

Evidently, given $a(k)$, this formula determines the frequencies explicitly for arbitrary concentration and masses. Therefore, only a brief discussion will be given of the nature of the solution for the limiting case when m_2/m_1 becomes large. Then we have

$$\omega_1{}^2 \doteqdot \frac{c_1}{m_1}\frac{k_B T k^2}{a(k)} \tag{6.115}$$

and

$$\omega_2{}^2 \doteqdot \frac{c_2}{m_1} k_B T k^2 \tag{6.116}$$

When c_1 becomes small, that is a light impurity in the host liquid, the two modes are well separated, $\omega_1{}^2$ becoming very small, although still reflecting the oscillations in $a(k)$ at large k.

Some discussion of the damping of such modes has been given by Parrinello, Tosi and March (1974*b*) and we refer to this paper for further details.

6.7.2 Atomic transport for mass difference $(m_1 - m_2)$ small

Having dealt with models for arbitrary mass differences, we shall discuss the dynamics carefully by perturbation theory, the perturbation being the deviations of m_1 and m_2 from some suitably chosen mass M. The theory can be posed precisely as follows. We take a reference liquid, with the same force law as the mixture, composed of $N = N_1 + N_2$ identical atoms of mass M. Then the difference H' between the Hamiltonian of the mixture and that of the reference liquid is

$$H' = -\left[\delta_1 \sum_{i=1}^{N_1} p_i^{\,2} + \delta_2 \sum_{i=1}^{N_2} p_i^{\,2}\right] \bigg/ 2M \tag{6.117}$$

with $\delta_i = (m_i - M)/m_i$. With the assumption that H' can be treated as a perturbation, it is possible to derive the following expression for the interdiffusion coefficient D:

$$D = c_2 D_1 + c_1 D_2 \tag{6.118}$$

where D_1 and D_2 are self-diffusion coefficients of the two isotopes in the mixture. This relation is correct to first order in the mass difference. Furthermore by choosing

$$M = \left(\frac{c_1}{m_1} + \frac{c_2}{m_2}\right)^{-1} \tag{6.119}$$

there follow the additional relations

$$\begin{aligned} c_1 D_1 + c_2 D_2 = D_0(M) &= D_1^{\,0}\left(c_1 + c_2 \frac{m_1}{m_2}\right)^{1/2} \\ &= D_2^{\,0}\left(c_2 + c_1 \frac{m_2}{m_1}\right)^{1/2} \end{aligned} \tag{6.120}$$

where $D_\alpha^{\,0}$ are the self-diffusion coefficients in the pure isotopes. It is also worth noting that, in the limit of vanishingly small concentration c_2, the method of radioactive tracers, which is essentially a measurement of the mutual diffusion coefficient D, actually measures $D_1^{\,0}$ (that is, the diffusion constant of the pure matrix) to order δ^2.

A fuller development of the perturbation theory for small mass differences allows one to express the dynamic correlation functions in the mixture in terms of correlation functions in the one-component reference liquid, thus exhibiting explicitly the dependence of $S_{\alpha\beta}(k, \omega)$ and $S^{s}_{\alpha\alpha}(k, \omega)$ on concentration and masses. Molecular dynamics calculations of the velocity autocorrelation functions (related to $S^{s}_{\alpha\alpha}(k, \omega)$) in an isotopic mixture with Lennard–Jones interactions have been reported by Ebbsjö et al. (1974). Simple scaling with concentration and masses, as given by the perturbative theory in the mass difference, is found to hold even for rather large mass differences. *Figure 6.12* reports the results obtained for mixtures of atoms with masses equal to 40 and 80 atomic units. Significant violations of simple scaling are observed only for very large mass differences ($m_1=40$ and $m_2=200$).

6.8 van Hove functions and transport in conformal solutions

We discussed in section 6.4 the static partial structure factors in a conformal solution. To develop a perturbative treatment of the dynamic correlation functions in such a mixture, we must, of course, supplement the assumptions of conformal solution theory on the inter-particle potentials by the assumption that the masses of the atomic constituents be sufficiently close. A perturbative treatment is then possible (Parrinello, Tosi and March, 1974*b*), which parallels the treatment given in the preceding section for isotopic mixtures.

Assuming, for simplicity, equal atomic masses of the two constituent species, the change in the intermediate scattering function for a pair of atoms, relative to the reference liquid, is given by

$$\Delta F_{ij}(k,t)=\tfrac{1}{2}(k_BT)^{-1}\sum_{\alpha,\beta=1}^{2}\sum_{l=1}^{N_\alpha}\sum_{l'=1}^{N_\beta}\Bigg\{\langle\Delta\phi_{\alpha\beta}(r_{ll'}(t))\exp\left(-i\boldsymbol{k}\cdot[\boldsymbol{r}_i(0)-\boldsymbol{r}_j(t)]\right)\rangle_c$$

$$+\int_{t_0}^{t}dt_1\left\langle\Delta\phi_{\alpha\beta}(r_{ll'}(t_1))\exp\left[(i\boldsymbol{k}\cdot\boldsymbol{r}_j(t)\right]\frac{d}{dt_0}\exp\left[-i\boldsymbol{k}\cdot\boldsymbol{r}_i(t_0)\right]\right\rangle_c$$

$$+k_BT\int_{t_0}^{t}dt_1\langle\Delta\phi_{\alpha\beta}(r_{ll'}(t_1))[\exp(i\boldsymbol{k}\cdot\boldsymbol{r}_j(t)),\exp(-i\boldsymbol{k}\cdot\boldsymbol{r}_i(t_0))]\rangle_c\Bigg\}_{t_0=0}$$

$$\tag{6.121}$$

where $\langle\,\rangle_c$ denotes the cluster part of the correlation function and

$$\Delta\phi_{\alpha\beta}(r)=A_{\alpha\beta}\phi(\lambda_{\alpha\beta}r)-\phi(r)$$

$$\doteqdot(A_{\alpha\beta}-1)\phi(r)+(\lambda_{\alpha\beta}-1)P(r) \tag{6.122}$$

Though we shall not derive eqn (6.121), it is clear that at $t=0$ this must be equivalent to the theory of the static partial structure factors in a conformal solution, given earlier in section 6.3. In fact, letting $t\to 0$ in eqn (6.121), only the first term on the right-hand side remains. Going into $\boldsymbol{r}$ space by Fourier transform, contact can be established with eqn (6.50). In that equation, the two-body terms arise from terms in the summations in eqn (6.121) for which $l=i$, $l'=j$, or vice versa. Similarly, three-body terms arise when only the index l (or l') coincides with i or j.

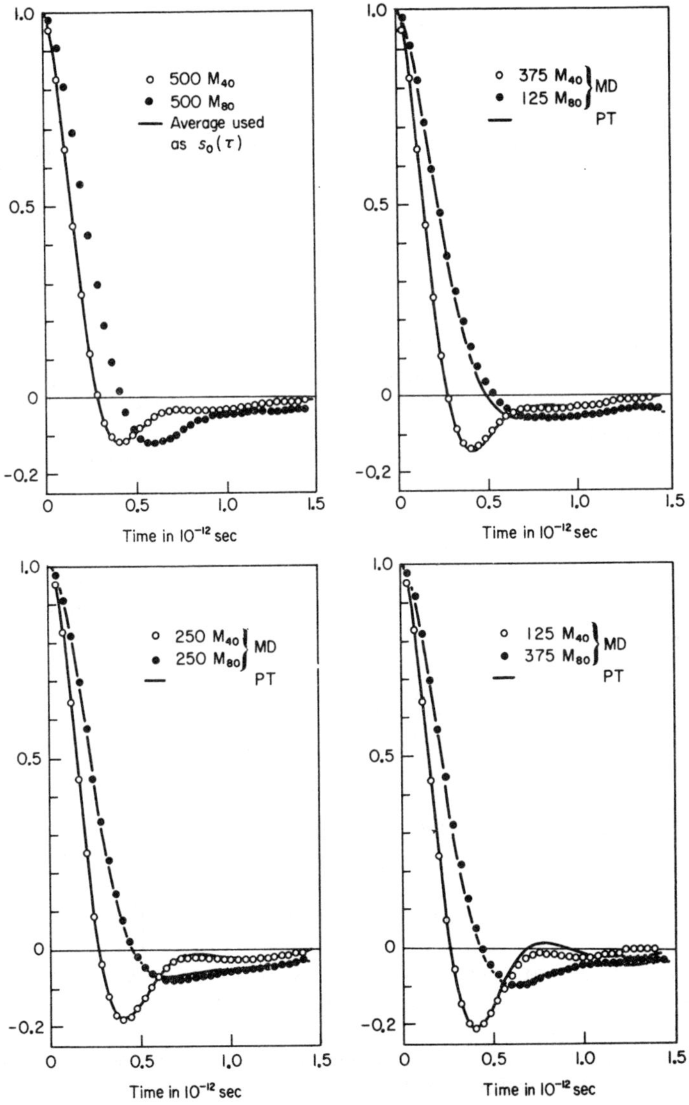

Figure 6.12 Velocity autocorrelation functions in isotopic mixtures of particles with atomic masses of 40 and 80 atomic units, in various concentrations (from Ebbsjo et al., 1974). The circles are the results of molecular dynamics experiments, while the curves are calculated by scaling through perturbation theory for small mass difference

Using eqn (6.122), the dynamic correlations in eqn (6.121) can be expressed solely in terms of dynamic correlations in the reference liquid.

6.8.1 Partial dynamic structure factors

The desired dynamic structure factors $S_{\alpha\beta}(k, \omega)$ in conformal solution theory are now simply obtained from the Fourier transforms of $F_{ij}(k, t)$ with respect to time. Specifically, the self-structure factor can be calculated.

One interesting consequence is that the dynamical concentration fluctuations take the form

$$S_{cc}(k\omega) = c_1 c_2 [c_2 S^s_{11}(k\omega) + c_1 S^s_{22}(k\omega) - c_1 c_2 \beta d M^{12}_{12}(k\omega)] \tag{6.123}$$

where $S^s_{\alpha\alpha}(k\omega)$ are the self-functions in the mixture while M^{12}_{12} is solely a property of the reference liquid. The S^s_{11} and S^s_{22} terms express the fact that self-motion of atoms 1 and 2 leads to concentration fluctuations in the solution. The presence of the correction M^{12}_{12} is associated with a kind of backflow of other particles accompanying this self-diffusion.

An expression for the mutual diffusion coefficient D has been given earlier in terms of $S_{cc}(k\omega)$ namely

$$D = \tfrac{1}{2}\beta \left(\frac{\partial^2 G}{\partial c_2{}^2}\right) \lim_{\omega \to 0} \omega^2 \left[\lim_{k \to 0} \frac{1}{k^2} S_{cc}(k, \omega)\right] \tag{6.124}$$

(see eqn (6.96)). In conformal solution theory the thermodynamics discussed previously yields

$$\frac{\partial^2 G}{\partial c_2{}^2} = \frac{k_B T}{c_1 c_2} - 2dE_0 \tag{6.125}$$

Using the expression given above for $S_{cc}(k, \omega)$ we can then write

$$D = c_2 D_1 + c_1 D_2 + c_1 c_2 \gamma \tag{6.126}$$

where D_1 and D_2 are the self-diffusion coefficients of the two components in the mixture and

$$\gamma = -\beta d \left[2E_0 D_0 + \tfrac{1}{2} \lim_{\omega \to 0} \omega^2 \lim_{k \to 0} \frac{1}{k^2} M^{12}_{12}(k, \omega)\right] \tag{6.127}$$

Although in the derivation of the above expression for D we have assumed equal masses, we have seen already in section 6.7 that the above result for D with $\gamma = 0$ is valid to first order in the mass difference.

The interest in the result for D resides in the concentration dependence $c(1-c)$ of the difference between D and $c_2 D_1 + c_1 D_2$. Obviously this is a symmetrical function around $c = \frac{1}{2}$. There is no easy route to estimating γ numerically from the above theory. However, computer simulation in Ar–Kr mixtures (Jacucci and McDonald, 1976) indicates that the correction $c_1 c_2 \gamma$ in eqn (6.126) for D is very small (i.e. less than 10% of $c_2 D_1 + c_1 D_2$).

Similarly, from the Kubo relation (6.97) using $S_{NN}(k, \omega)$ calculated from the perturbation theory for conformal solutions with equal masses, we find

$$\tfrac{4}{3}\eta+\zeta=c_1(\tfrac{4}{3}\eta+\zeta)_1+c_2(\tfrac{4}{3}\eta+\zeta)_2+c_1c_2\beta d\gamma' \tag{6.128}$$

where the dependence on concentration is explicit.

Neutron scattering experiments from an alloy system such as Na–K might well be used to test the partial van Hove functions derived here. Clearly more direct tests, however, are available by measurements of the transport coefficients in the mixture. Until now, no precise test of the above results is possible on a suitable system to which conformal solution theory should apply.

CHAPTER 7
CHARGED FLUIDS

Hitherto we have concerned ourselves exclusively with neutral fluids. This is a perfectly correct description for liquid argon, for example. But already, even for liquid sodium, it is not correct in all respects—the liquid is evidently formed from singly-charged positive ions and conduction electrons. More obviously, no one would describe liquid lithium iodide without starting out from an ionic picture.

Besides dealing in this chapter with ionic fluids, a short discussion is included of dipolar liquids and, in particular, of water. Ions in dipolar liquids, such as water, are briefly considered. In this connection, there is, of course, by now a vast literature on molecular fluids, to which we cannot do justice in a book of this scope. A recent review of the experimental situation in this area has been given by Allen and Higgins (1973). An interesting reference on the way in which scattering experiments, for example, must be interpreted in a manner which includes not only the internal structure of the molecules themselves but also the effect of the correlation of relative orientation of the molecules is that of Egelstaff, Page and Powles (1971). Specific examples considered there are liquid carbon tetrachloride and liquid germanium tetrabromide.

7.1 Classical one-component plasma

The classical one-component plasma is an assembly of identical, point-like charged particles, with charge e and mass m, embedded in a uniform background of charge and obeying the laws of classical statistical mechanics. The charge density of the background is chosen so that the whole system is electrically neutral. Even though the model can simulate a system of rigid ions (possibly floating in a uniform dielectric which scales e^2 by its dielectric constant) in the limit when the effects of the ionic cores become negligible, we shall call the plasma particles 'classical electrons'. The properties of a fluid of electrons in the fully degenerate limit will be discussed in the next section.

The simplified model of the classical one-component plasma already exhibits the characteristics which distinguish real Coulomb systems, such as plasmas and electrolytic solutions, from neutral fluids. These are the phenomena of screening and plasma oscillations, which arise from the long-range nature of the Coulomb interaction.

7.1.1 Screening and plasma oscillations

Each electron in the plasma interacts not just with a limited number of neighbours, as is the case for a particle in a neutral fluid, but with all the other electrons, so that its motion cannot easily be decoupled from that of the other electrons. As a consequence, the plasma exhibits a strong collective behaviour in long wavelength phenomena, where the Fourier transform of the Coulomb potential, $4\pi e^2/k^2$, diverges. A manifestation of this behaviour is the collective rearrangement of the plasma around a fixed charge inserted in it, which effects a complete screening of the added charge: the plasma density is locally depleted (or enhanced) so as to neutralise the impurity charge, leading to an impurity potential which decays much faster than r^{-1} at large distance (*static screening*). Another manifestation of the collective behaviour of the plasma is the existence of a collective mode of oscillation, in which the restoring force is provided by the mean self-consistent field of all the electrons moving in concert. The collective plasma motions effect a *dynamic screening* of the Coulomb potential of each electron in the plasma: as the electron moves through the fluid the other electrons move in a correlated fashion to restore local electrical neutrality, and one may say that each electron carries with itself a region of low electron density (a 'screening hole').

To illustrate these phenomena we consider the continuity equation relating the electron density $\rho(\mathbf{r}, t)$ to the current density $\mathbf{j}(\mathbf{r}, t)$:

$$\frac{\partial \rho(\mathbf{r}, t)}{\partial t} = -\nabla \cdot \mathbf{j}(\mathbf{r}, t) \tag{7.1}$$

We assume that the plasma contains some scattering centres, such as neutral impurities, so as to ensure a finite value for the conductivity σ†. The expression of the current density depends then on the magnitude of the frequency ω as compared to the inverse relaxation time τ^{-1} for scattering against the neutral impurities. If $\omega\tau \ll 1$, and for slow variations in space ($ka \ll 1$ where a is the mean distance between electrons) the plasma is in a collision-dominated regime, and considering small deviations from equilibrium we can use the linearised hydrodynamic expression,

$$e\mathbf{j}(\mathbf{r}, t) = \sigma\left[\mathbf{E}(\mathbf{r}, t) - \frac{1}{e}\nabla\mu(\mathbf{r}, t)\right] \tag{7.2}$$

Here, $\mathbf{E}(\mathbf{r}, t)$ is the electric field and $\mu(\mathbf{r}, t)$ is the local chemical potential, and we have taken the temperature as constant to avoid consideration of thermoelectric effects at this stage. The hydrodynamic behaviour of a two-component plasma is briefly discussed in section 7.3.3.

From the Poisson equation,

$$\nabla \cdot \mathbf{E}(\mathbf{r}, t) = 4\pi e[\rho_e(\mathbf{r}, t) + \delta\rho(\mathbf{r}, t)] \tag{7.3}$$

†In the absence of scattering centres the conductivity would be infinite, since collisions between the electrons conserve the total momentum of the electron fluid.

where $\rho_e(\boldsymbol{r}, t)$ describes a density of external charge inserted in the plasma, and writing

$$\delta\mu(\boldsymbol{r}, t) = \left(\frac{\partial\mu}{\partial\rho}\right)_T \delta\rho(\boldsymbol{r}, t) \tag{7.4}$$

we have

$$\frac{\partial\rho(\boldsymbol{r}, t)}{\partial t} = -\sigma\left\{4\pi[\rho_e(\boldsymbol{r}, t) + \delta\rho(\boldsymbol{r}, t)] - \frac{1}{e^2}\left(\frac{\partial\mu}{\partial\rho}\right)_T \nabla^2\rho(\boldsymbol{r}, t)\right\} \; (ka \ll 1, \; \omega\tau \ll 1) \tag{7.5}$$

Taking Fourier transforms we find that in static conditions ($\omega = 0$) an external charge density $e\rho_e(\boldsymbol{k})$ creates a density distortion in the plasma which is given by

$$\rho(\boldsymbol{k}) = -\frac{\rho_e(\boldsymbol{k})}{1 + k^2/k_s^2} \tag{7.6}$$

where

$$k_s = \left[4\pi e^2 \Big/ \left(\frac{\partial\mu}{\partial\rho}\right)_T\right]^{1/2} = (4\pi\rho^2 e^2 K_T)^{1/2} \tag{7.7}$$

is the 'screening wave vector' given in terms of the plasma density ρ and of the isothermal compressibility K_T. Equation (7.6) yields complete screening of the external charge, since

$$\lim_{k\to 0} \rho(\boldsymbol{k}) = -\lim_{k\to 0} \rho_e(\boldsymbol{k}) \tag{7.8}$$

or

$$\int \mathrm{d}\boldsymbol{r}\,\delta\rho(\boldsymbol{r}) = -\int \mathrm{d}\boldsymbol{r}\,\rho_e(\boldsymbol{r}) \tag{7.8'}$$

We also note from eqn (7.6) that in the absence of the external perturbation $[\rho_e(\boldsymbol{k}) = 0]$ static charge density fluctuations $\rho(\boldsymbol{k})$ can arise in the plasma provided that $k = ik_s$, namely, such fluctuations decay in space over a distance k_s^{-1}. Finally, the effective potential generated by a screened external charge is

$$V_{\mathrm{eff}}(\boldsymbol{k}) = \frac{4\pi e^2}{k^2}[\rho_e(\boldsymbol{k}) + \rho(\boldsymbol{k})] = \frac{4\pi e^2}{k^2 + k_s^2}\rho_e(\boldsymbol{k}) \tag{7.9}$$

which for a point-like external charge, $\rho_e(\boldsymbol{k}) = Z$, yields

$$V_{\mathrm{eff}}(r) = \frac{Ze^2}{r}\exp(-k_s r) \tag{7.10}$$

This latter result is, of course, not correct since eqn (7.9) holds only for small wave numbers, but will in the present case represent correctly the behaviour of the effective potential at large distance from the charged impurity.

Before proceeding to a discussion of the opposite limit, $\omega\tau \gg 1$, it is instructive to examine how the above behaviour differs from that of a neutral fluid. We are here considering changes of the plasma density relative to the inert background, and thus the proper analogue of $\delta\rho(\boldsymbol{r}, t)$ is the concentration fluctuation $\delta c(\boldsymbol{r}, t)$ in a two-component neutral fluid (this point will be further clarified in section 7.3). In such a fluid at constant temperature the linearised interdiffusion current is given by (compare eqn A6.5.5)

$$j(\boldsymbol{r}, t) = -\rho D[\nabla c(\boldsymbol{r}, t) + k_p \nabla p(\boldsymbol{r}, t)] \tag{7.11}$$

where D is the interdiffusion coefficient, $p(\boldsymbol{r}, t)$ is the pressure and $k_p = \left(\frac{\partial \mu}{\partial p}\right)_{T,c} \Big/ \left(\frac{\partial \mu}{\partial c}\right)_{T,p}$. The continuity equation consequently gives

$$\frac{\partial c(\boldsymbol{r}, t)}{\partial t} = D\nabla^2[c(\boldsymbol{r}, t) + k_p p(\boldsymbol{r}, t)] \tag{7.12}$$

which is to be compared with eqn (7.5) for the plasma. Since both terms on the right-hand side of (7.12) are of order k^2, static screening is absent in the neutral fluid, and its presence in the plasma is seen to derive from the form of the Poisson equation, namely from the nature of the Coulomb interaction.

Let us now consider the behaviour of the plasma in the collisionless regime, $\omega\tau >> 1$. We still restrict ourselves to slow variations in space so that we may take the force on an electron as given by $e\boldsymbol{E}(\boldsymbol{r}, t)$, and thus neglect local field corrections. We can now write, from Newton's second law,

$$\frac{m}{\rho}\frac{\partial j(\boldsymbol{r}, t)}{\partial t} = e\boldsymbol{E}(\boldsymbol{r}, t) \tag{7.13}$$

which it is interesting to contrast with eqn (7.2). Combining eqn (7.13) with the continuity equation and Poisson's equation we have

$$\frac{\partial^2 \rho(\boldsymbol{r}, t)}{\partial t^2} = -\omega_p{}^2[\rho_e(\boldsymbol{r}, t) + \delta\rho(\boldsymbol{r}, t)] \; (ka \ll 1, \omega\tau \gg 1) \tag{7.14}$$

where

$$\omega_p = (4\pi\rho e^2/m)^{1/2} \tag{7.15}$$

is called the plasma frequency. From (7.14) we see that, in the absence of external charges, the plasma can execute free oscillations of frequency ω_p at long wavelength. We also see that, in the presence of an external charge distribution given by $e\rho_e(\boldsymbol{k}, \omega)$, the plasma is polarised according to

$$\rho(\boldsymbol{k}, \omega) = \frac{\omega_p{}^2}{\omega^2 - \omega_p{}^2}\rho_e(\boldsymbol{k}, \omega) \tag{7.16}$$

This expression describes the dynamic screening of a time-dependent, long-wavelength disturbance of the plasma. Dynamic screening clearly has a resonance at the frequency ω_p, expressing the fact that the plasma oscillation can be excited by a disturbance of exactly its frequency.

In the very dense plasma formed by the conduction electrons in a metal ($\rho \sim 10^{23}\mathrm{cm}^{-3}$) the quantum of the plasma oscillation (the 'plasmon') has an energy $\hbar\omega_p \sim 10\,\mathrm{eV}$. But in the low density plasma, of course, the plasmon zero-point energy, proportional to $\rho^{1/2}$ is greatly reduced, and the plasma oscillation will be thermally excited at high temperatures.

7.1.2 Dielectric function

We can conveniently summarise the above results and prepare the ground for their extension to arbitrary k and ω by introducing the longitudinal† dielectric function of the plasma, $\varepsilon(k, \omega)$. We define this function by an obvious extension of the electrostatic definition of the dielectric constant,

$$\varepsilon(k, \omega) = \hat{k} \cdot \boldsymbol{D}(\boldsymbol{k}, \omega)/\hat{k} \cdot \boldsymbol{E}(\boldsymbol{k}, \omega) \tag{7.17}$$

where

$$\nabla \cdot \boldsymbol{D}(\boldsymbol{r}, t) = 4\pi e \rho_e(\boldsymbol{r}, t) \tag{7.18}$$

Combining these with the Poisson equation we have

$$\frac{1}{\varepsilon(k, \omega)} = 1 + \frac{\rho(\boldsymbol{k}, \omega)}{\rho_e(\boldsymbol{k}, \omega)} \tag{7.19}$$

Equation (7.5) therefore gives

$$\varepsilon(0, \omega) = 1 + \frac{4\pi i \sigma}{\omega} \qquad (\omega\tau \ll 1) \tag{7.20}$$

and

$$\varepsilon(k, 0) = 1 + \frac{k_s^{\,2}}{k^2} \qquad (ka \ll 1) \tag{7.21}$$

while equation (7.14) gives

$$\varepsilon(0, \omega) = 1 - \frac{\omega_p^{\,2}}{\omega^2} \qquad (\omega\tau \gg 1) \tag{7.22}$$

These results are exact even for a quantal one-component plasma. The divergence of $\varepsilon(k, 0)$ for $k \to 0$ is a consequence of the plasma being a perfect conductor and leads to complete screening of a static disturbance. We also note that the plasmon resonance at long wavelength appears as a zero of the dielectric function. In the quantal plasma (e.g. the alkali metals), the implication on the optical properties is that ω_p (in the ultraviolet) marks the transition from absorption to transparency (see for example Kittel, 1963).

†The longitudinal function is involved because we are dealing with the displaced charge density.

It is useful at this stage to relate the dielectric function to the van Hove function $S(k, \omega)$ of the plasma. We define the density response function of the plasma, $\chi(k, \omega)$, as in section 3.2.5 by

$$\rho(\boldsymbol{k}, \omega)=\chi(k, \omega)V_{\mathrm{e}}(\boldsymbol{k}, \omega) \tag{7.23}$$

where $V_{\mathrm{e}}(\boldsymbol{k}, \omega)$ is the externally applied potential, given by

$$V_{\mathrm{e}}(\boldsymbol{k}, \omega)=\frac{4\pi e^2}{k^2}\rho_{\mathrm{e}}(\boldsymbol{k}, \omega) \tag{7.24}$$

(see eqn (7.18)). From eqn (7.19) we have

$$\frac{1}{\varepsilon(k, \omega)}=1+\frac{4\pi e^2}{k^2}\chi(k, \omega) \tag{7.25}$$

this relation being obviously true in general. For a classical plasma the fluctuation–dissipation theorem, eqn (3.46), then yields

$$S(k, \omega)=-\frac{k^2 k_{\mathrm{B}}T}{2\pi\rho e^2\omega}\operatorname{Im}\frac{1}{\varepsilon(k, \omega)} \tag{7.26}$$

We also have for the classical plasma

$$\begin{aligned} S(k) &= -\frac{k_{\mathrm{B}}T}{\rho}\chi(k, 0) \\ &= \frac{k^2 k_{\mathrm{B}}T}{4\pi\rho e^2}\left[1-\frac{1}{\varepsilon(k, 0)}\right] \end{aligned} \tag{7.27}$$

The long-wavelength expressions for the static and dynamic structure factors in the classical plasma follow at once from eqns (7.20)–(7.22). Equation (7.21) yields

$$S(k)=\frac{k_{\mathrm{B}}T}{4\pi\rho e^2}k^2 \qquad (ka \ll 1) \tag{7.28}$$

which should be contrasted with the limiting values of the structure factors in classical neutral fluids, which are non-vanishing constants. Equation (7.20) yields

$$S(k, \omega)=\frac{k_{\mathrm{B}}T}{\rho e^2}\frac{2\sigma k^2}{\omega^2+(4\pi\sigma)^2} \tag{7.29}$$

Comparing this result with the concentration structure factor for a two-component neutral fluid in the hydrodynamic regime given in eqn (6.93) which upon neglect of heat fluctuations can be written

$$S_{\mathrm{cc}}(k, \omega)=\frac{k_{\mathrm{B}}T}{(\partial\mu/\partial c)_T}\frac{2Dk^2}{\omega^2+(Dk^2)^2} \tag{7.30}$$

we see that the finite quantity $4\pi\sigma$ replaces in the denominator of (7.29) the vanishing quantity Dk^2 in the denominator of (7.30). This implies that, whereas D can be obtained from $(\omega^2/k^2)S_{cc}(k, \omega)$ in the Kubo limit, no Kubo relation exists between σ and $(\omega^2/k^2)S(k, \omega)$ in the plasma (Martin, 1967).

From eqn (7.22) we have

$$\frac{1}{\varepsilon(0, \omega)} = 1 + \frac{\omega_p^{\,2}}{\omega^2 - \omega_p^{\,2} + i\eta} \qquad (\omega\tau \gg 1) \tag{7.31}$$

where we have introduced an infinitesimal imaginary part to account for causality of the response. Consequently,

$$\begin{aligned} S(k, \omega) &= \frac{k^2 k_B T \omega_p^{\,2}}{2\rho e^2 \omega}\, \delta(\omega^2 - \omega_p^{\,2}) \\ &= \frac{k^2 k_B T}{4\rho e^2}\, [\delta(\omega - \omega_p) + \delta(\omega + \omega_p)] (ka \ll 1,\ \omega\tau \gg 1) \end{aligned} \tag{7.32}$$

In the limit $\tau \to \infty$ this expression applies to all frequencies and then the plasmon excitation exhausts all the long-wavelength sum rules. In particular we find

$$S(k) = \int_{-\infty}^{\infty} \frac{d\omega}{2\pi}\, S(k, \omega) = \frac{k_B T}{4\pi\rho e^2}\, k^2 \qquad (ka \ll 1) \tag{7.33}$$

in agreement with (7.28), and

$$\int_{-\infty}^{\infty} \frac{d\omega}{2\pi}\, \omega^2 S(k, \omega) = \frac{k_B T}{m}\, k^2 \qquad (ka \ll 1) \tag{7.34}$$

in agreement with the exact second-moment sum rule (3.56).

Finally, we complete the general discussion of the dielectric properties of the plasma by introducing the 'screened' response function, which is akin to the polarisability constant in electrostatics in that it relates the polarisation of the plasma not to the external potential $V_e(\mathbf{k}, \omega)$ but to the macroscopic internal potential. The latter is the sum of $V_e(\mathbf{k}, \omega)$ and of the polarisation potential

$$V_p(\mathbf{k}, \omega) = \frac{4\pi e^2}{k^2}\, \rho(\mathbf{k}, \omega) \tag{7.35}$$

and we write

$$\rho(\mathbf{k}, \omega) = \tilde{\chi}(k, \omega)[V_e(\mathbf{k}, \omega) + V_p(\mathbf{k}, \omega)] \tag{7.36}$$

which defines the screened response function $\tilde{\chi}(k, \omega)$. By comparison with (7.23) we have

$$\chi(k, \omega) = \frac{\tilde{\chi}(k, \omega)}{1 - \dfrac{4\pi e^2}{k^2}\, \tilde{\chi}(k, \omega)} \tag{7.37}$$

and

$$\varepsilon(k, \omega) = 1 - \frac{4\pi e^2}{k^2}\, \tilde{\chi}(k, \omega) \tag{7.38}$$

The definition (7.36) has allowed us to account at one stroke for the long-range effects of the Coulomb interactions, and the screened response function is, in fact, analogous in character to the density response function $\chi(k, \omega)$ of a neutral fluid. In particular, from eqns (7.20)–(7.22) we have

$$\tilde{\chi}(k, 0) = -\rho^2 K_{\mathrm{T}} \qquad (ka \ll 1) \tag{7.39}$$

which is the equivalent of the Ornstein–Zernike formula relating the long-wavelength structure factor of a neutral fluid to its compressibility, and

$$\tilde{\chi}(k, \omega) = -i\frac{\sigma k^2}{e^2 \omega} \qquad (ka \ll 1,\ \omega\tau \ll 1) \tag{7.40}$$

$$\tilde{\chi}(k, \omega) = \frac{\rho k^2}{m\omega^2} \qquad (ka \ll 1,\ \omega\tau \gg 1) \tag{7.41}$$

Equation (7.40) implies that the conductivity σ can be obtained from the screened response by a Kubo-like formula,

$$\sigma = -e^2 \lim_{\omega \to 0} \omega \left[\lim_{k \to 0} \frac{1}{k^2} \operatorname{Im} \tilde{\chi}(k, \omega) \right] \tag{7.42}$$

7.1.3 Structure of the plasma

The earliest theory of the structure of the classical plasma was presented in the work of Debye and Hückel (1923), which is rightly celebrated as an early example of self-consistent field theory. The theory determines self-consistently the effective potential $V_{\mathrm{eff}}(r)$ generated by an electron known to reside at $\boldsymbol{r} = 0$, by requiring that it satisfy the Poisson equation

$$\nabla^2 V_{\mathrm{eff}}(r) = -4\pi e^2 [\delta(\boldsymbol{r}) - \rho + \rho g(r)] \tag{7.43}$$

where the three contributions to the charge density, respectively, arise from the electron at the origin, from the inert background and from the distribution of electrons around the given electron. By relating this distribution to $V_{\mathrm{eff}}(r)$ through a Maxwell–Boltzmann exponential,

$$g(r) = \exp\left[-V_{\mathrm{eff}}(r)/k_{\mathrm{B}}T\right] \tag{7.44}$$

and expanding this exponential to linear terms the Poisson equation can be integrated to find

$$V_{\mathrm{eff}}(r) = \frac{e^2}{r} \exp(-k_0 r), \qquad k_0 = \left(\frac{4\pi\rho e^2}{k_{\mathrm{B}}T} \right)^{1/2} \tag{7.45}$$

Comparison with (7.10) shows that the Debye–Hückel theory assumes the result (7.9) as valid for all wave numbers and approximates k_s by its free-particle value ($K_T=(\rho k_B T)^{-1}$ for free particles). We therefore expect the theory to become valid for the long-wavelength structural properties of a low-density, high-temperature plasma.

Alternatively, the Debye–Huckel theory may be characterised by its approximation on the static structure factor of the plasma,

$$S_{DH}(k)=\frac{k^2}{k^2+k_0{}^2} \tag{7.46}$$

which implies the form

$$\tilde{c}_{DH}(k)=-\frac{4\pi\rho e^2}{k_B T k^2} \tag{7.47}$$

for the Fourier transform of the direct correlation function. Thus the theory neglects all short-range correlations between the electrons, and becomes in fact completely invalid at short distance, as can be seen from the result for the radial distribution function,

$$g_{DH}(r)=1-\frac{e^2}{k_B T r}\exp(-k_0 r) \tag{7.48}$$

which diverges for $r\to 0$.

A careful analysis of short-range correlations and their effect on the behaviour of $g(r)$ at large and small r has been given by O'Neil and Rostoker (1965) by means of perturbation theory. The appropriate perturbative parameter is the so-called plasma parameter Γ,

$$\Gamma=\frac{e^2}{k_B T a} \tag{7.49}$$

where $a=(4\pi\rho/3)^{-1/3}$ is the radius of a sphere whose volume equals the volume per electron. Thus Γ measures the ratio between the average potential energy and the average kinetic energy per particle. O'Neil and Rostoker consider the perturbation expansion of the three-body function $g^{(3)}(1,2,3)$ entering the force equation (2.11) for $g(r)$, and show that the Debye–Hückel result follows by neglecting the triplet correlation function $t(1,2,3)$ defined in eqn (2.8). Their result for the short-range correlations in perturbation theory may then be stated in terms of decoupling approximations for $t(1,2,3)$ through products of the hole functions $h(r)=g(r)-1$, which read (Abramo and Tosi, 1972)

$$t(1,2,3)=h(1,3)h(3,2)+h(3,1)h(1,2)+h(1,2)h(2,3)+\rho\int d\mathbf{r}_4 h(1,4)h(2,4)h(3,4) + 0(\Gamma^3) \tag{7.50}$$

for large r_{12} and

$$t(1,2,3)=h(1,3)h(3,2)+h(3,1)h(1,2)+h(1,2)h(2,3)+h(1,2)h(2,3)h(3,1)+0(\Gamma^3\ln\Gamma) \tag{7.51}$$

for small r_{12}. The structure of the perturbation expansion is clearly quite different in the two different ranges of r_{12}, changing from a Kirkwood superposition form at small r_{12} where close encounters of three electrons are dominant, to a convolution-type form at large r_{12} where collective effects are dominant. The integral term in (7.50) effectively serves to screen the term $h(1,2)h(2,3)$, which would otherwise lead to a divergence at small k.

Various approximate theories of structure have been applied to the evaluation of $g(r)$ in the classical plasma for $\Gamma \leqslant 2$ (Carley, 1963), a common feature of the results being fair agreement with the Debye–Hückel theory at large distance ($r/a \gtrsim 1$) and the removal of the divergence at short distance. Extensive computer simulation experiments have been carried out by the Monte Carlo method by Brush, Sahlin and Teller (1966) and have lately been refined and extended up to $\Gamma = 160$ by Hansen (1973). A remarkable feature of the Monte Carlo results is the appearance of oscillations in $g(r)$ for $\Gamma \gtrsim 2.5$, which are indicative of the incipient formation of short-range order in the Coulomb fluid. These oscillations become increasingly more marked as Γ increases (see *figure 7.1*) until the fluid undergoes a transition to a spacially ordered state for $\Gamma \simeq 155$. Recent calculations of $g(r)$ by liquid structure theories for $\Gamma \leqslant 10$ (Hirt, 1967; Cooper, 1973) indicate that over this limited range of Γ the hypernetted chain approximation, based on the assumption (cf. eqn 2.38)

$$c(r) = -\frac{e^2}{rk_BT} + h(r) - \ln g(r) \tag{7.52}$$

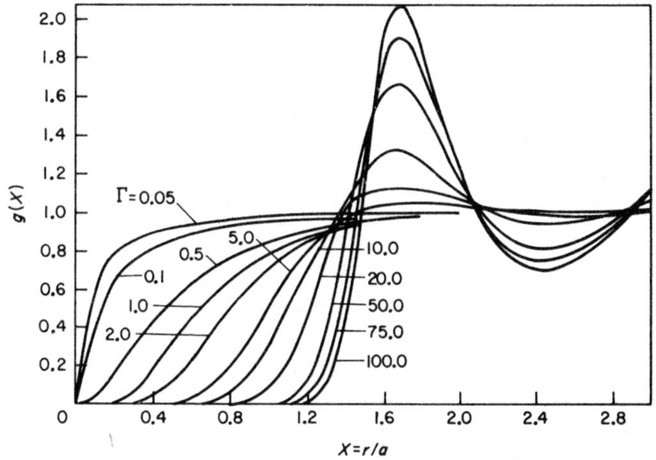

Figure 7.1 Radial distribution function of the classical one-component plasma as a function of distance (in units of $a = (4\pi\rho/3)^{-1/3}$) for a series of values of the plasma parameter $\Gamma = e^2/(k_BTa)$ (from Brush, Sahlin and Teller, 1966)

is in best agreement with the Montė Carlo data. It should also be noted that the mean spherical model discussed in section 6.5 has also been modified to apply to the classical plasma by Gillan (1974), good agreement being obtained with computer experiments.

7.1.4 Dynamics of the plasma

The dynamic extension of the Debye–Hückel theory for the classical plasma is known as the random phase approximation (RPA). The RPA form of the density response function can be derived in several alternative ways, for instance by introducing collective coordinates pertaining to the plasmon excitation (Bohm and Pines, 1953) or by appropriate decouplings in the equation of motion for the density fluctuations (Ehrenreich and Cohen, 1959)†, but in view of the detailed discussion given in section 7.1.2 it may here be simply stated as assuming that $\tilde{\chi}(k, \omega)$ coincides with the density response function $\chi_0(k, \omega)$ of a perfect gas. Namely,

$$\chi_{\mathrm{RPA}}(k, \omega) = \frac{\chi_0(k, \omega)}{1 - \dfrac{4\pi e^2}{k^2}\chi_0(k, \omega)} \tag{7.53}$$

The RPA therefore includes the interactions between the electrons only in accounting for the polarisation potential $V_{\mathrm{Pf}}(k, \omega) = (4\pi e^2/k^2)\rho(\mathbf{k}, \omega)$ created by a density fluctuation, but neglects all short-range correlations.

We have already discussed the evaluation of $\chi_0(k, \omega)$ for a classical perfect gas in section 3.4.1 and in Appendix 5.2 (see eqns (A5.2.11) and (A5.2.12)). We immediately find

$$\begin{aligned} S_{\mathrm{RPA}}(k, \omega) &= -\frac{2k_{\mathrm{B}}T}{\rho\omega} \frac{\operatorname{Im}\chi_0(k, \omega)}{\left[1 - \dfrac{4\pi e^2}{k^2}\operatorname{Re}\chi_0(k, \omega)\right]^2 + \left[\dfrac{4\pi e^2}{k^2}\operatorname{Im}\chi_0(k, \omega)\right]^2} \\ &= \frac{2\pi^{1/2}}{k(2k_{\mathrm{B}}T/m)^{1/2}} \\ &\times \frac{\exp(-x^2)}{\left[1 + \dfrac{k_0^2}{k^2} - 2\dfrac{k_0^2}{k^2}x\exp(-x^2)\displaystyle\int_0^x \exp(t^2)\,\mathrm{d}t\right]^2 + \left[\pi^{1/2}\dfrac{k_0^2}{k^2}x\exp(-x^2)\right]^2} \end{aligned} \tag{7.54}$$

where $k_0^2 = 4\pi\rho e^2/k_{\mathrm{B}}T$ and $x = \omega/k(m/2k_{\mathrm{B}}T)^{1/2}$. Comparison with eqn (5.53) shows that this expression can be obtained by the Vlasov equation approach if one assumes the Debye–Hückel form for the direct correlation function, $\tilde{c}_{\mathrm{DH}}(k) = -k_0^2/k^2$. It is easily seen that the classical RPA reduces to the Debye–Hückel theory in the static limit.

†These decouplings can be thought of as equivalent to time-dependent self-consistent field theory.

We can now use the above expression for $S(k, \omega)$ to derive the long-wavelength properties of the plasmon. The damping of the excitation vanishes for $k \to 0$, and its frequency follows by requiring that the first term in the denominator of (7.54) vanish. Using the expansion

$$2x \exp(-x^2) \int_0^x \exp(t^2)\,\mathrm{d}t = 1 + (2x^2)^{-1} + 3(4x^4)^{-1} + \cdots \tag{7.55}$$

we find

$$\begin{aligned} \omega_p^{\,2}(k) &= \omega_p^{\,2}\left[1 + \frac{3k^2 k_B T}{m\omega_p^{\,2}(k)} + \cdots\right] \\ &= \omega_p^{\,2} + \frac{3k_B T}{m} k^2 + \cdots \end{aligned} \tag{7.56}$$

The dispersion curve of the plasmon at long wavelength is therefore rising and is determined by free-gas properties, specifically by $3k_B T$ which is twice the kinetic energy per particle. This RPA result is clearly valid only for a low-density, high-temperature plasma. We may in fact look upon the coefficient of k^2 in (7.56) as defining a 'high-frequency' elastic constant in the sense of Schofield (1966), which in RPA equals $3\rho k_B T$ and therefore differs from the ordinary, low-frequency bulk modulus, $K_T^{\,-1} = \rho k_B T$. We shall return to this point when discussing the zero-sound excitation in liquid He^3.

The RPA calculation of $S(k, \omega)$ can be improved to account for short-range correlations between the electrons by the same techniques that we have discussed for neutral liquids in sections 5.4 and 5.5. For instance, use of the approximation developed by Pathak and Singwi (1970) for liquid argon, reviewed in section 5.5, leads to a dispersion curve of the form (Abramo and Tosi, 1974)

$$\omega_p^{\,2}(k) = \omega_p^{\,2} + \frac{k^2}{m}(3k_B T + \tfrac{4}{15}u) + \cdots \tag{7.57}$$

where u is the mean potential energy per particle. This result is most easily obtained by requiring that the plasmon exhaust the fourth moment sum rule at long wavelengths. Since u is negative and increases in magnitude with increasing plasma parameter Γ, the dispersion of the plasmon decreases with increasing Γ, until it becomes negative for an estimated value of $\Gamma \simeq 10$. On the other hand, as k increases at fixed Γ, damping mechanisms associated with the decay of the plasmon excitation into disordered particle motions become increasingly important (see *figure 7.2*).

These results are in good agreement with computer simulation experiments performed by Hansen, Pollock and McDonald (1974). Their results for $\Gamma = 110.4$ are reported in *figure 7.3*, and show a very sharp peak in $S(k, \omega)$ which shifts to lower frequency and progressively broadens as k increases, until at sufficiently large k the plasmon peak disappears and the spectrum takes a Gaussian shape peaked at $\omega = 0$, which is characteristic of a perfect gas. Upon decreasing Γ at

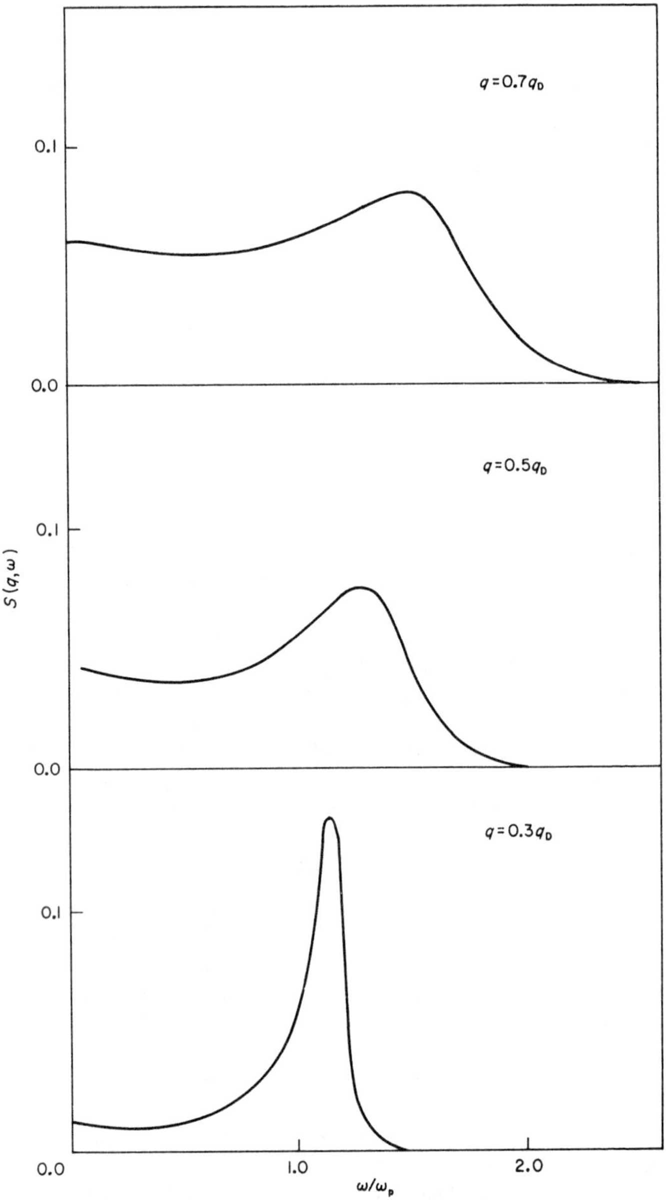

Figure 7.2 Dynamic structure factor (in units of $k_B T/\rho\omega_p$) of the classical one-component plasma with $\Gamma = 1$ for increasing values of the wave number (in units of $q_D = (4\pi\rho e^2/k_B T)^{1/2}$), as obtained by mean-field theory (from Abramo and Tosi, 1974)

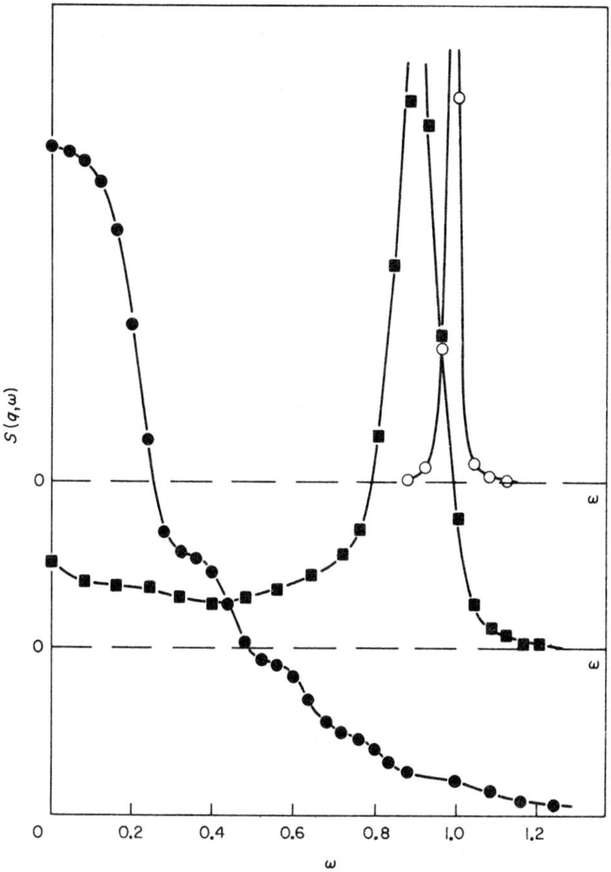

Figure 7.3 Dynamic structure factor of the classical one-component plasma with $\Gamma = 110.4$ for three values of the wave number, as obtained by computer simulation (from Hansen, Pollock and McDonald, 1974)

constant k, the collective peak is found to broaden and its dispersion flattens, becoming positive for $\Gamma \lesssim 10$.

7.2 Degenerate electron liquid and the jellium model

We now turn to a discussion of the properties of a quantum plasma of charged fermions in the fully degenerate limit ($k_B T \ll \varepsilon_F$). This model mainly serves to simulate the sea of conduction electrons in a simple metal†. Our discussion of

†For a discussion of this subject, and departures from this simple model, see section 7.4 below.

the hydrodynamic and dielectric properties of the classical plasma in sections 7.1.1 and 7.1.2 has, in fact, been sufficiently general that many of the results hold for the quantum plasma as well. The results that require modifications for quantum effects are, as we shall indicate in the following, the relations between the dielectric response and the structure factors, eqns (7.26) and (7.27), and the long-range form of the screened static potential, eqn (7.10). The book of Pines and Nozières (1966) is recommended for a more detailed understanding of Fermi liquids.

7.2.1 Dielectric response in the random-phase approximation (RPA)

The evaluation of the free-particle response function $\chi_0(k, \omega)$ entering the RPA expression for $\chi(k, \omega)$, eqn (7.53), must now take into account the Fermi character of the statistics. The single-particle states, which are plane waves of energy $\varepsilon_k = \hbar^2 k^2/2m$, are not occupied according to Maxwell's distribution, but those of lowest energy are occupied, each by two electrons, up to a maximum wave vector—the Fermi wave vector k_F. This is fixed so as to accommodate all the available electrons:

$$2V \int_0^{k_F} \frac{d\mathbf{k}}{(2\pi)^3} = N \quad \rightarrow \quad k_F = (3\pi^2 \rho)^{1/3} \tag{7.58}$$

The maximum electron energy is the Fermi energy $\varepsilon_F = \hbar^2 k_F{}^2/2m$, and the mean kinetic energy per electron is easily found to be $\frac{3}{5}\varepsilon_F$. Since the mean potential energy is of order $e^2\rho^{-1/3}$, the ratio of the kinetic to the potential energy per particle varies as $\rho^{1/3}$, and thus the Fermi system approaches ideality at *high* density.

The Fermi–Thomas approximation (see for example Schiff, 1959; March, 1974) is the quantum extension of the Debye–Hückel theory. Again the theory considers the variation in density of the plasma around a particle known to reside at the origin, and takes account of the fact that the maximum kinetic energy will be locally varying in space according to

$$\varepsilon_F(r) = \frac{\hbar^2}{2m} [3\pi^2 \rho(r)]^{2/3} \simeq \varepsilon_F \left[1 + \frac{2}{3} \frac{\delta\rho(r)}{\rho}\right] \tag{7.59}$$

The self-consistency requirement relating $\delta\rho(r)$ to the effective potential $V_{\mathrm{eff}}(r)$ generated by the particle at the origin is that the local chemical potential, given by $\varepsilon_F(r) + V_{\mathrm{eff}}(r)$, should equal the chemical potential at large distance, which is a constant given by ε_F. This yields

$$\delta\rho(r) = -\frac{3\rho}{2\varepsilon_F} V_{\mathrm{eff}}(r) \tag{7.60}$$

and integration of the Poisson equation leads to

$$V_{\mathrm{eff}}(r) = \frac{e^2}{r} \exp(-k_{FT} r) \tag{7.61}$$

with

$$k_{\mathrm{FT}}=\left(\frac{6\pi\rho e^2}{\varepsilon_{\mathrm{F}}}\right)^{1/2}=(4\pi\rho^2 e^2 K_0)^{1/2} \tag{7.62}$$

$K_0=3(2\rho\varepsilon_{\mathrm{F}})^{-1}$ being the compressibility of the ideal Fermi gas.

The evaluation of $\chi_0(k,\omega)$ can be carried out from the general expression for the density response function given in eqn (3.43), and is detailed in appendix 7.1 (Linhard, 1954). The main results of the calculation are summarised in the following.

The static dielectric function takes the form

$$\varepsilon_{\mathrm{RPA}}(k,0)=1+\frac{k_{\mathrm{FT}}^2}{2k^2}\left\{1+\frac{k_{\mathrm{F}}}{k}\left[1-\left(\frac{k}{2k_{\mathrm{F}}}\right)^2\right]\ln\left|\frac{1+k/2k_{\mathrm{F}}}{1-k/2k_{\mathrm{F}}}\right|\right\} \tag{7.63}$$

reducing for $k \ll k_{\mathrm{F}}$ to the Fermi–Thomas form,

$$\varepsilon_{\mathrm{FT}}(k,0)=1+\frac{k_{\mathrm{FT}}^2}{k^2} \tag{7.64}$$

This equivalence of the RPA with the simplest self-consistent theory is, however, no longer true at finite wave vector, contrary to the classical case. In particular, the RPA dielectric function (7.63) contains not only the perfect-screening singularity at $k=0$, but also a singularity in its derivative at $k=2k_{\mathrm{F}}$. This latter singularity arises from the sharp change of the occupancy of the single-particle states at $k=k_{\mathrm{F}}$; clearly, $2k_{\mathrm{F}}$ gives then the largest possible momentum change for an electron in an elastic scattering process, from one point to the diametrically opposite point on the Fermi surface. The singularity in the dielectric function at $k=2k_{\mathrm{F}}$ dominates the behaviour of the polarisation charge $\delta\rho(r)$ and the effective potential $V_{\mathrm{eff}}(r)$ at large distance: in the quantum plasma these do not decay exponentially with distance, but have instead an oscillatory behaviour of the type

$$\delta\rho(r)\sim\frac{\cos(2k_{\mathrm{F}}r)}{(k_{\mathrm{F}}r)^3} \tag{7.65}$$

(Friedel, 1958; Langer and Vosko, 1959). Furthermore, since the Fermi surface remains sharp in the presence of electron–electron interactions (which are neglected in the evaluation of $\chi_0(k,\omega)$), the above effect is a characteristic of real Fermi plasmas. Such oscillations in the charge density around a static charged impurity have, in fact, very small amplitude (see *figure 7.4*) but have been revealed in nuclear magnetic resonance experiments, as they give rise to electric field gradients over several shells of neighbours of an impurity in a metal (Rowland, 1960; Kohn and Vosko, 1960). The Kohn singularities in the dispersion curves of phonons in metals (Kohn, 1959) also derive from the same effect, as the effective ion–ion interaction involves the electronic dielectric function (see section 7.4.1).

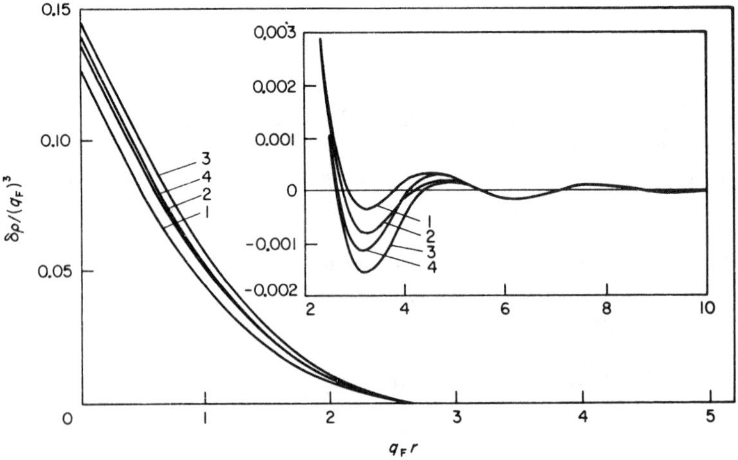

Figure 7.4 Displaced charge density around a static positive impurity in the degenerate electron liquid, as obtained by various theories of correlations treating the impurity-electron interaction by linear response (from Singwi et al., 1970). In particular, curve 1 is based on the RPA discussed in detail in the text. The long-range Friedel oscillations in the displaced charge density are displayed in the inset on an enlarged scale

The static structure factor of the quantum plasma cannot be obtained directly from the static response: eqn (7.27) is no longer valid and one must explicitly perform the integration of the van Hove function,

$$S(k, \omega) = -\frac{2\hbar}{\rho}\,\mathrm{Im}\,\chi(k, \omega)$$
$$= -\frac{\hbar k^2}{2\pi\rho e^2}\,\mathrm{Im}\,\frac{1}{\varepsilon(k, \omega)} \tag{7.66}$$

over the frequency. However, it remains generally true that the plasmon excitation exhausts all the long-wavelength sum rules in the pure plasma, so that we can write

$$\lim_{k\to 0} S(k, \omega) = \frac{\hbar\omega_p k^2}{4\rho e^2}\,\delta(\omega - \omega_p) \tag{7.67}$$

having used (7.31). Consequently,

$$\lim_{k\to 0} S(k) = \frac{\frac{1}{2}\hbar\omega_p}{4\pi\rho e^2}\,k^2 \tag{7.68}$$

which has the form of eqn (7.28), with the thermal energy $k_B T$ being replaced by the zero-point energy of the plasmon. The RPA satisfies this general result, but the numerical evaluations of the pair correlation function (Glick and Ferrell, 1960; Hedin, 1965) show that, even if the Debye–Hückel divergence for $r \to 0$ is not present, still $g_{\mathrm{RPA}}(r)$ becomes negative at small values of r as the density decreases from the ideal gas limit. This deficiency is a consequence of the neglect of short-range correlations between the electrons, except for those correlations between electrons with parallel spins which arise from the ideal Fermi statistics.

Finally, and in contrast with the classical case, $\operatorname{Im} \chi_0(k, \omega)$ is now different from zero only in a limited part of the (ω, k) plane, determined by the inequalities

$$\frac{\hbar}{m}(-k_{\mathrm{F}}k + \tfrac{1}{2}k^2) \leqslant \omega \leqslant \frac{\hbar}{m}(k_{\mathrm{F}}k + \tfrac{1}{2}k^2) \tag{7.69}$$

(see *figure 7.5*). This fact is again a consequence of the Fermi distribution of the electrons in the ground state, which allows excitations to occur with conservation of momentum and energy only if the available energy $\hbar\omega$ and the available momentum $\hbar \boldsymbol{k}$ are so related. Therefore, the plasmon excitation, which starts at a finite frequency ω_{p} for $k \to 0$, is rigorously undamped in the RPA over a considerable range of wave vector, until its dispersion curve meets the continuum of single-particle excitations reproduced in *figure 7.5*, when the decay of a plasmon into an electron-hole pair becomes possible with conservation of momentum and energy. This result is, however, an artefact of the RPA: in the real quantum plasma, simultaneous excitations of two or more electrons in the same process are possible and are not restricted to any special part of the (ω, k) plane. In particular, the damping of the plasmon at long wavelength is

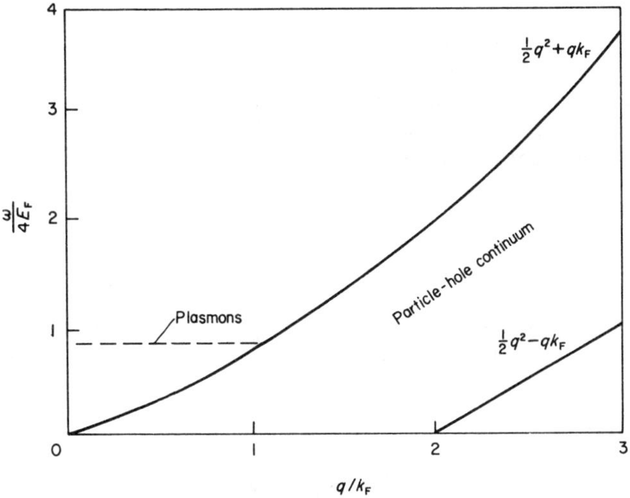

Figure 7.5 Continuum of single electron-hole excitations in a degenerate electron plasma (schematical) (from March, Young and Sampanthar, 1967)

dominated by decay into two electron-hole pairs (Dubois, 1959), which leads to a plasmon lifetime proportional to k^{-2}. We should stress that the situation is quite different in solid or liquid metals: decay of a plasmon into a single electron-hole pair may be allowed even for $k \to 0$, with the ions providing the excess momentum needed for this process.

To obtain the dispersion of the plasmon in the RPA we write from eqn (A7.1.1)

$$\mathrm{Re}\,\chi_0(k,\omega)=\frac{2}{\hbar}\sum_{\substack{k' \\ (k'<k_F,\,|k'+k|>k_F)}}\left\{\frac{1}{\omega-\dfrac{\hbar}{m}\boldsymbol{k}\cdot\boldsymbol{k}'-\dfrac{\hbar k^2}{2m}}-\frac{1}{\omega+\dfrac{\hbar}{m}\boldsymbol{k}\cdot\boldsymbol{k}'+\dfrac{\hbar k^2}{2m}}\right\}$$

$$\underset{(k\to 0)}{\longrightarrow}\ \frac{2k^2}{m\omega^2}\sum_{\substack{k' \\ (k'<k_F)}}\left[1+3\left(\frac{\hbar\boldsymbol{k}\cdot\boldsymbol{k}'}{m\omega}\right)^2\right]=\frac{\rho k^2}{m\omega^2}\left[1+\frac{6}{5}\frac{\varepsilon_F k^2}{m\omega^2}\right] \tag{7.70}$$

and determine the frequency $\omega_p(k)$ of the plasmon resonance by requiring that the RPA dielectric function vanish at this frequency. This gives

$$\omega_p{}^2(k)=\omega_p{}^2+\frac{6}{5}\varepsilon_F\frac{k^2}{m}+\cdots \tag{7.71}$$

where, again, the high-frequency elastic constant determining the dispersion is twice the kinetic energy per particle. This expression is in fair agreement with the measured dispersion of the plasmon excitation in metals (Raether, 1965; see *figure 7.6*).

7.2.2 Extensions of the RPA

Early attempts at improving the RPA description of a degenerate plasma by inclusion of short-range correlations between the electrons were made by Hubbard (1957) and by Nozières and Pines (1958). Hubbard considered the leading exchange corrections to the RPA calculation of the screened response, with the results

$$\tilde{\chi}_H(k,\omega)=\frac{\chi_0(k,\omega)}{1+\dfrac{4\pi e^2}{k^2}G_H(k)\chi_0(k,\omega)} \tag{7.72}$$

and

$$\chi_H(k,\omega)=\frac{\chi_0(k,\omega)}{1-\dfrac{4\pi e^2}{k^2}[1-G_H(k)]\chi_0(k,\omega)} \tag{7.73}$$

where

$$G_H(k)\simeq\frac{1}{2}\frac{k^2}{k^2+k_F{}^2} \tag{7.74}$$

gives a short-range exchange correction, structurally entering the response

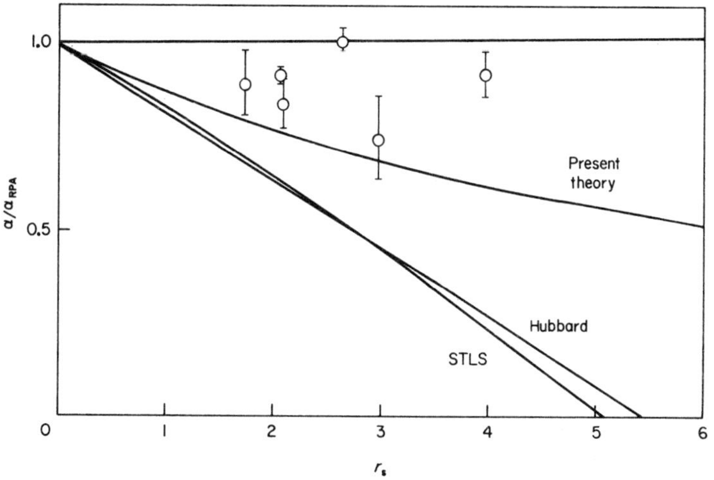

Figure 7.6 Coefficient of the leading term of the plasmon dispersion in the degenerate electron plasma, in units of its RPA value given in eqn (7.71), as a function of r_s extending over the range of metallic densities. The results of various theories of correlation are shown (from Singwi et al., 1970). The experimental values for a few simple metals (from Raether, 1965) are obtained from the angular dependence of the intensity of inelastically scattered high-energy electrons

function (7.73) as a modification of the collective potential in a Vlasov-type approach. Nozières and Pines evaluated instead the leading corrections to the ground state energy coming from processes at large momentum transfers, their results for this property being very similar to those of Hubbard. *Table 7.1* collects various numerical results for the correlation energy of the degenerate plasma over the range of electronic densities encountered in metals: this energy is defined by writing the ground state energy per electron as

$$E_{gs} = \left(\frac{2.21}{r_s^2} - \frac{0.916}{r_s} + \varepsilon_{corr}\right) \text{ryd} \tag{7.75}$$

where the first two terms are the kinetic energy and the potential energy obtained by an exact Hartree–Fock treatment, and $r_s = [(4\pi/3)\rho a_0^3]^{-1/3}$ is the radius of the electronic sphere in units of the Bohr radius a_0. It is remarkable that the calculations of the correlation energy subsequent to the RPA are in fair agreement with the very early estimate of Wigner (1938),

$$\varepsilon_{corr} = -\frac{0.88}{r_s + 7.8} \tag{7.76}$$

This was based on an interpolation between an estimate on the fluid for $r_s = 1$

Table 7.1. Correlation energy in the electron fluid at metallic densities (from Vashishta and Singwi, 1972)

r_s	1	2	3	4	5	6
Wigner	−0.10	−0.090	−0.082	−0.075	−0.069	−0.064
RPA	−0.157	−0.124	−0.105	−0.094	−0.085	−0.078
Hubbard	−0.131	−0.102	−0.086	−0.076	−0.069	−0.064
Nozières–Pines	−0.115	−0.094	−0.081	−0.072	−0.065	−0.060
Vashishta–Singwi	−0.112	−0.089	−0.075	−0.065	−0.058	−0.052

and a calculation for very large r_s, where Wigner proposed that the increasing predominance of the interactions over the kinetic effects would lead to a crystallisation of the electron fluid (for a review, see Care and March, 1975). For many numerical purposes, the simple analytic expression of Nozières and Pines (1958),

$$\varepsilon_{\text{corr}} = -0.115 + 0.031 \ln r_s \tag{7.77}$$

is still recognised as sufficiently accurate over the range of metallic densities ($2 \lesssim r_s \lesssim 6$).

Improvements to the Hubbard dielectric function have been developed in recent years along two main directions. On the one hand, the calculation of the screened response by perturbative techniques has been extended to include short-range Coulomb correlations. In particular, a careful evaluation of the static dielectric function, with special attention to the value of the compressibility that determines screening at long wavelength (see eqn (7.21)), has been presented by Geldart and Taylor (1970), while the properties of the response at high frequencies, with special attention to plasmon dispersion and damping and the effects of multi-pair excitations, have been examined by Ôsaka (1962), Dubois and Kivelson (1969), and Glick and Long (1971). On the other hand, the Vlasov-type approach has also been improved to include short-range correlations in the collective potential. By expressing these correlations through the radial distribution function $g(r)$, a self-consistent scheme which determines simultaneously the dielectric function and the static structure factor through frequency integration of eqn (7.66) can be developed (Singwi et al., 1968). The simplest approximation is to write the response function in the form (7.73) with the short-range correlation factor given by

$$G(k) = -\frac{1}{\rho} \int \frac{d\boldsymbol{k}'}{(2\pi)^3} \frac{\boldsymbol{k} \cdot \boldsymbol{k}'}{k'^2} [S(\boldsymbol{k} - \boldsymbol{k}') - 1] \tag{7.78}$$

This would in fact correspond, in the classical case, to the inclusion of the first term of eqn (7.50), while in the degenerate case it reduces to the Hubbard expression, eqn (7.74), when the static structure factor is replaced by its Hartree–Fock value. A subsequent, more refined, approximation,

$$G(k) = -\frac{1}{\rho}\left(1 + a\rho\frac{\partial}{\partial\rho}\right)\int\frac{d\boldsymbol{k}'}{(2\pi)^3}\frac{\boldsymbol{k}\cdot\boldsymbol{k}'}{k'^2}\left[S(\boldsymbol{k}-\boldsymbol{k}') - 1\right] \tag{7.79}$$

allows one to satisfy nearly exactly the compressibility relation given in eqn (7.39) over the range of metallic densities (Vashishta and Singwi, 1972) through a choice of the parameter $a = \frac{2}{3}$. Available results for the radial distribution function of the degenerate plasma at a density corresponding to metallic sodium, $r_s = 4$, are reported in *figure 7.7*.

The reader who wishes to have a full quantitative account of the properties of the quantal electron fluid is referred to the excellent review article by Hedin and Lundqvist (1969).

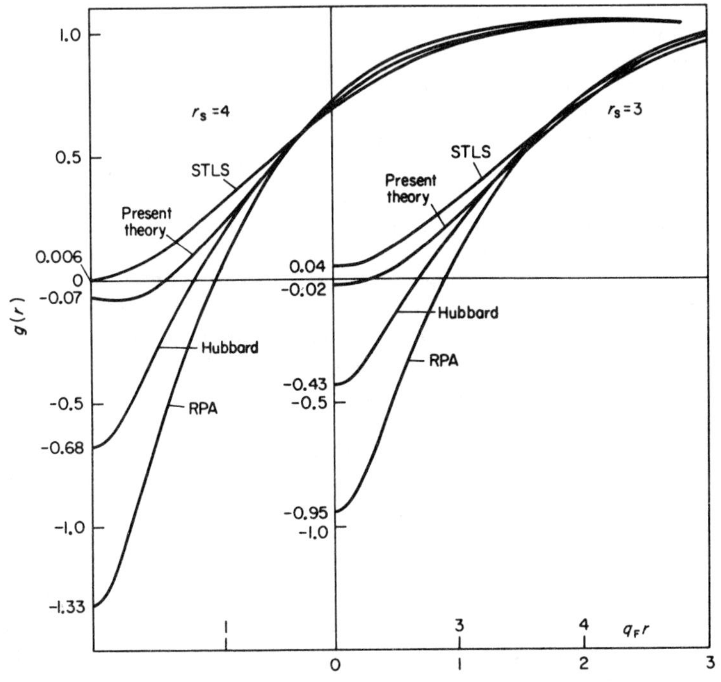

Figure 7.7 Pair correlation function in the degenerate electron plasma for two values of r_s in the range of metallic densities, according to various theories of correlations (from Singwi et al., 1970)

7.3 Ionic melts

7.3.1 Potential models

We begin our discussion of two-component charged fluids by reviewing the growing field of molten salts. Theoretical studies of these systems have so far been based almost exclusively on a rigid-ion model for the interactions, which neglects polarisation of the closed electron shells of the ions. The potential between two ions is written as

$$\varphi_{ij}(r)=\frac{Z_i Z_j e^2}{r}+\phi_{ij}(r) \tag{7.80}$$

and the short-range potential $\phi_{ij}(r)$ is usually taken as the sum of a van der Waals attractive term and an ion-overlap repulsive term having an exponential dependence on distance,

$$\phi_{ij}(r)=-\frac{c_{ij}}{r^6}+b_{ij}\exp\left[(r_i+r_j-r)/\rho_{ij}\right] \tag{7.81}$$

r_i and r_j being the radii of the two ions. The exponential form of the repulsive potential is qualitatively based on the fact that this term arises from the overlap of closed-shell ionic configurations, whose wave functions decay approximately exponentially. Of as great practical importance as the precise functional form of $\phi_{ij}(r)$ is, however, the accurate determination of the intervening parameters.

A long-established procedure for the determination of the parameters, developed mainly by Born and Mayer, is based on the use of optical crystal data for the van der Waals coefficients and of thermodynamic crystal data for the repulsive parameters. A detailed review has been given by Tosi (1964). The thermodynamic data (equilibrium volume and compressibility) must be corrected for contributions from lattice vibrations, by means of approximate forms of the crystal equation of state, in order to reduce them to values appropriate to a static crystal where lattice symmetry suppresses polarisation effects. In practice, the values of the ionic radii are now rather well established (Fumi and Tosi, 1964), also by comparison with X-ray maps of the electronic distribution in a few salts, whereas the accuracy of the other parameters and of the van der Waals coefficients is more questionable and is certainly not uniformly good for the various salts.

Results of computer simulation of the thermodynamic properties of both solid and molten alkali halides in a rigid-ion model have been reported by Adams and McDonald (1974). The model yields good values for the equilibrium volume and the internal energy of the room-temperature solid, but the values of the compressibility and the thermal expansivity are somewhat poorer; furthermore, the results worsen as the temperature is raised and the system melts. Thermal effects in the thermodynamic properties of the solid arise, of course, from lattice vibrations, that the rigid-ion model is not adequate to describe with quantitative accuracy. Preliminary calculations on the liquid by Romano and Margheritis (1974) indicate that the inclusion of electronic polarisation reduces the discrepancy with experiment.

Considerable information on polarisable-ion models for small deviations from the perfect lattice is available from detailed studies of the phonon dispersion curves in ionic crystals, and the importance of these effects has been underlined by the vast work on point defects in these crystals. It is well established that a primary point for a reliable theory is a good account of the static dielectric constant of the crystal, which contains not only contributions arising from the polarisation of the electronic shells and from the displacements of the ions from their lattice sites, but also contributions associated with a coupling between these two effects which is not purely electrostatic in origin. The shell model (see for example Dick and Overhauser, 1958; Cochran, 1963) has been introduced to account for this coupling and is now accepted as the basic theoretical model for both lattice vibrations and lattice defects in ionic crystals. Its adoption and adaptation for liquid state properties will probably be necessary to obtain accurate quantitative results.

To illustrate the simplest version of the shell model in a framework appropriate for extension to the liquid state, let us consider a simple alkali halide molecule (Tosi and Doyama, 1967). We write the energy of the molecule relative to the free-ion state, as a function of the internuclear distance r and of the electronic dipoles carried by the two ions, as

$$E(r, \mu_+, \mu_-) = -\frac{e^2}{r} + \phi(r) - (\mu_+ + \mu_-)\frac{e}{r^2} - \frac{2\mu_+\mu_-}{r^3} + \frac{\mu_+{}^2}{2\alpha_+} + \frac{\mu_-{}^2}{2\alpha_-} + q\mu_-\phi_{\text{rep}}(r) \quad (7.82)$$

where the first two terms are the rigid-ion contribution, the subsequent four terms represent the polarisation contributions as given by electrostatic theory in the linear dipole approximation, and the last term represents a short-range coupling between the dipoles and the overlap, limited for simplicity to the (usually more polarisable) negative ion. The coupling parameter q can be determined from the static dielectric constant of the crystal, while the electronic polarisabilities, α_+ and α_-, enter the high-frequency dielectric constant. The last term in (7.82), which is characteristic of the shell model, physically accounts for the fact that the mutual repulsion of closed electronic shells induces an electronic dipole moment opposite in sign to that induced by the electric field, as is apparent from the expression

$$\mu_-(r) = \alpha_-\left[\frac{e}{r^2} + 2\frac{\mu_+(r)}{r^3} - q\phi_{\text{rep}}(r)\right] \quad (7.83)$$

obtained by minimising E with respect to μ_-.

The above simple model, when evaluated with parameters entirely determined from solid-state properties, already accounts rather well for the basic properties of the alkali halide molecules (binding energy, equilibrium internuclear distance, dipole moment, vibrational frequency). It should be stressed, however, that the assumption of constant polarisabilities breaks down for large interpenetrations of the electronic shells of the two ions, as signalled by the polarisation catastrophe presented by the model. Similar results in a quantum mechanical framework have been reported by Brumer and Karplus (1973) and Brumer

(1974) who have analysed the molecular properties to derive additional information on the force law. Experimental data are available also on alkali halide dimers, which could be useful for the same purpose.

7.3.2 Static structure factors

Partial structure factors for two-component charged liquids, and linear combinations thereof, are introduced by the same definitions that we used for mixtures (see section 6.1). It will be noted, however, that in the present case the average concentrations are fixed by the overall neutrality condition,

$$c_1 Z_1 = c_2 Z_2 \tag{7.84}$$

Z_1 and Z_2 being now the absolute valences of the two types of ions. It is then easily seen that the concentration–concentration structure factor introduced in eqns (6.4) and (6.7) becomes, aside from a trivial factor, the charge–charge structure factor $S_{qq}(k)$ in a rigid-ion model, defined as

$$S_{qq}(k) = Z_1 S_{11}(k) + Z_2 S_{22}(k) - 2\sqrt{Z_1 Z_2}\, S_{12}(k) \tag{7.85}$$

the partial structure factors being defined as in (6.2). It is this latter structure factor that we have been discussing in the two preceding sections on one-component Coulomb fluids in an inert background.

The relations of the partial structure factors to the thermodynamic properties derived by Kirkwood and Buff (1951) for mixtures in the long wavelength limit and discussed in section 6.2 still hold, but are considerably simplified by the requirement of screening in the Coulomb fluid. Around a cation, say, positive charges are distributed at distance r with charge density $Z_1 e\rho g_{11}(r)$ and negative charges are distributed at distance r with charge density $-Z_2 e\rho g_{22}(r)$. We impose perfect screening by requiring that the total charge density seen from the cation integrate to $-Z_1 e$, namely

$$-Z_1 e = e\rho \int \mathrm{d}\mathbf{r}\, [Z_1 g_{11}(r) - Z_2 g_{22}(r)] \tag{7.86}$$

or

$$\lim_{k\to 0}\, [Z_1 S_{11}(k) = \sqrt{Z_1 Z_2}\, S_{12}(k)] \tag{7.87}$$

By a similar argument applied to the anion, and using the Kirkwood–Buff compressibility relation, eqn (6.18), we find

$$\lim_{k\to 0} [Z_1 S_{11}(k) = \sqrt{Z_1 Z_2}\, S_{12}(k) = Z_2 S_{22}(k)] = \rho k_B T\, K_T \tag{7.88}$$

The charge–charge structure factor, $S_{qq}(k)$, correspondingly behaves as k^2 in the long wavelength limit, in accord with eqn (7.27). With reference to eqn (A6.1.19), we see that the 'osmotic compressibility' vanishes in a Coulomb fluid, as a consequence of the fact that charge separation requires plasmon excitation. The inclusion of polarisation effects in the long wavelength structure factors has been discussed by Abramo, Parrinello and Tosi (1973*a*) in the case of low conductivity.

The formal solution of the mean spherical model for an ionic liquid with hard-core short-range interactions, based on the assumptions

$$g_{ij}(r)=0 \qquad (r<\sigma_{ij})$$
$$c_{ij}(r)=-(1)^{i+j}\frac{Z_iZ_je^2}{rk_BT} \qquad (r>\sigma_{ij}) \tag{7.89}$$

has been presented by Waisman and Lebowitz (1970; 1972). As a consequence of the powerful analytic restrictions imposed by the condition that $g_{ij}(r)$ vanish inside the cores, the direct correlation functions $c_{ij}(r)$ inside the cores have again simple forms which, in fact, coincide with those reported in section 6.5.2 for a mixture of neutral hard spheres. An explicit solution for the coefficients in terms of density, valences and core radii has so far been obtained, however, only in the case where the ions have equal radii (for more recent progress on the general case, see Gillan et al., 1976). In this case the hard-core terms and the Coulomb terms are completely independent, and one finds in the case $Z_1=Z_2=1$

$$c_{ij}(r)=c_{ij}^0(r)+(-1)^{i+j}(\alpha_0+\alpha_1 r/\sigma) \qquad (r<\sigma) \tag{7.90}$$

where σ is the hard-sphere diameter, $c_{ij}^0(r)$ is the neutral hard-sphere solution for a one-component fluid given in eqns (6.74) and (6.75), and the new constants α_0 and α_1 are given by

$$2\pi\rho\sigma^3\alpha_0=Bx^2 \tag{7.91}$$

and

$$2\pi\rho\sigma^3\alpha_1=\tfrac{1}{2}B^2x^2 \tag{7.92}$$

with $x^2=4\pi\rho e^2\sigma^2/k_BT$ and

$$B=-\frac{1}{x}[1+x-(1+2x)^{1/2}] \tag{7.93}$$

Furthermore, the internal energy per particle associated with the Coulomb interactions, which is

$$E=\tfrac{1}{2}\rho\int d\mathbf{r}\frac{e^2}{r}[g_{11}(r)+g_{22}(r)-2g_{12}(r)] \tag{7.94}$$

is soon shown, using (7.89), to be determined only by the constant α_0 and to be specifically

$$E=Be^2/\sigma \tag{7.95}$$

Integration of E over inverse temperature yields the Coulomb free energy, and differentiation of the free energy yields the equation of state in analytic form,

$$P=P_0+\frac{k_BT}{4\pi\sigma^3}[x+x(1+2x)^{1/2}-\tfrac{2}{3}(1+2x)^{3/2}+\tfrac{2}{3}] \tag{7.96}$$

By comparison with Monte Carlo data, the mean spherical model has been found to be a useful first approximation to the pair correlation functions of a liquid of charged, equi-sized hard spheres, at least when the charges are small (Larsen, 1974). More reliable results are provided, however, by the hypernetted chain approximation. The extension of the approach of Andersen, Weeks and Chandler (1971; see also section 2.6.6) to ionic liquids has been effected by Schofield (1974).

Monte Carlo determinations of the pair correlation functions, based on rigid-ion models of the type (7.80), have been reported for several molten salts. As shown in *figure 7.8* for molten KCl (Woodcock and Singer, 1971), the model predicts a rather sharp coordination between first-neighbour unlike ions and good charge ordering, in that the peaks of $g_{11}(r)$ and $g_{22}(r)$ systematically lie in correspondence to the valleys of $g_{12}(r)$. A typical set of partial structure factors for the same model is reported in *figure 7.9* for KBr (from data of Lewis, Woodcock and Singer, 1975): again, the fairly sharp peaks are indicative of good short-range ordering of the various types of ion pairs.

The experimental situation, as revealed by neutron diffraction measurements based on isotopic substitution, appears to be somewhat more complex in detail. The general features of the Monte Carlo $g_{KCl}(r)$ in the rigid-ion model for molten KCl are in accordance with the diffraction data of Derrien and Dupuy (1975), but some differences emerge for the correlation functions of like ions. The neutron data (see *figure 7.10*) indicate a first-neighbour like-ions peak at 4.8 Å, against the Monte Carlo value of 4.3 Å; indeed, by comparison with the first-neighbour distances in the solid near the melting point, the data suggest a tendency in the melt for contraction of the unlike-ions distance and for dilation of the like-ions distance. Similar data on molten CsCl (Derrien and Dupuy, 1975) indicate that the strongest correlations between Cs ions, say, are indeed due to their participating in the coordination sphere of a Cl ion.

Formation of molecular-ion complexes has been invoked for a qualitative explanation of neutron diffraction data on molten CuCl by Page and Mika (1971). The measured partial structure factors in this system, reported in *figure 7.11*, indicate good short-range ordering for Cu–Cl and Cl–Cl pairs, but practically no ordering for the Cu–Cu pairs. Since the Cu ion is known to have two valence states rather close in energy, Page and Mika suggest formation of $(CuCl_2)^-$ complexes, leaving very loose correlations between the Cu ions. It is, of course, doubtful whether simple polarisable-ion models are capable of simulating such a situation.

7.3.3 Dynamic structure factors

The linearised hydrodynamic equations for a two-component perfect conductor have been solved by Giaquinta (1974) to obtain the various correlation functions between mass, charge and temperature fluctuations in the hydrodynamic limit. We simply comment here on the main results of physical interest.

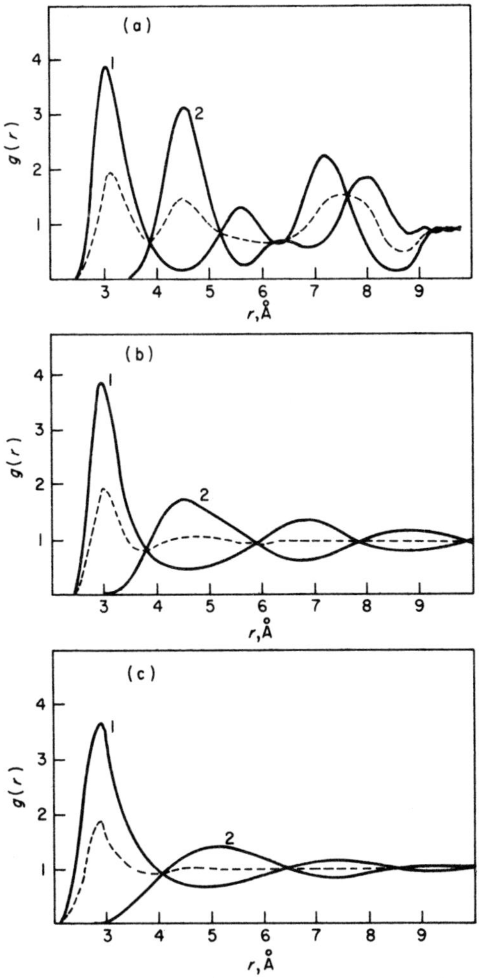

Figure 7.8 Pair correlation functions for KCl in (*a*) the solid state at $T = 1045$ *K* and $V = 41.48$ cm^3/mole; (*b*) the liquid state at T = 1045 K and $V = 48.80$ cm^3/mole; and (*c*) the liquid state at $T = 2874$ K and $V = 97.60$ cm^3/mole (from Monte Carlo determinations by Woodcock and Singer, 1971). In each figure, curves 1 and 2 give the pair functions for unlike and for like ions, respectively, while the broken curve gives $g_{NN}(r)$

First, the charge–charge correlation function $S_{qq}(k, \omega)$ is a superposition of two Lorentzian peaks centred at $\omega = 0$ and associated with electrical and thermal conduction, reducing, however, to the simple expression (7.29) when only terms

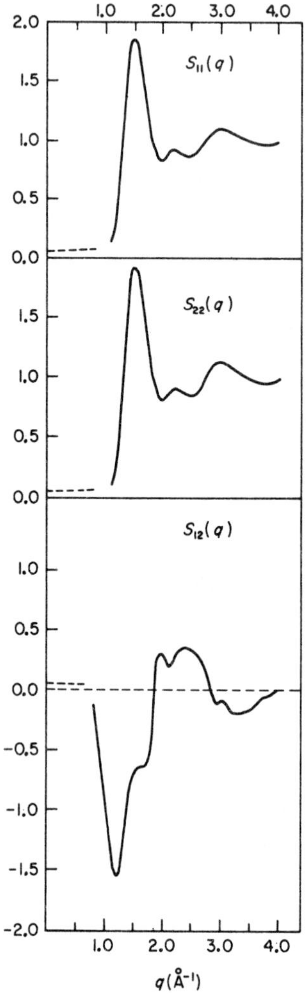

Figure 7.9 Partial structure factors in molten KBr near the freezing point, based on Monte Carlo determinations of the pair correlation functions by Lewis, Woodcock and Singer, 1975 (see Abramo, Parrinello and Tosi, 1973*a*)

of order k^2 are retained. Correspondingly, the mass–mass correlation function $S_{MM}(k, \omega)$ presents a Rayleigh peak containing a contribution from electric conduction, while the Brillouin doublet associated with sound waves is broadened by electric conduction in addition to viscosity and thermal conduction. A

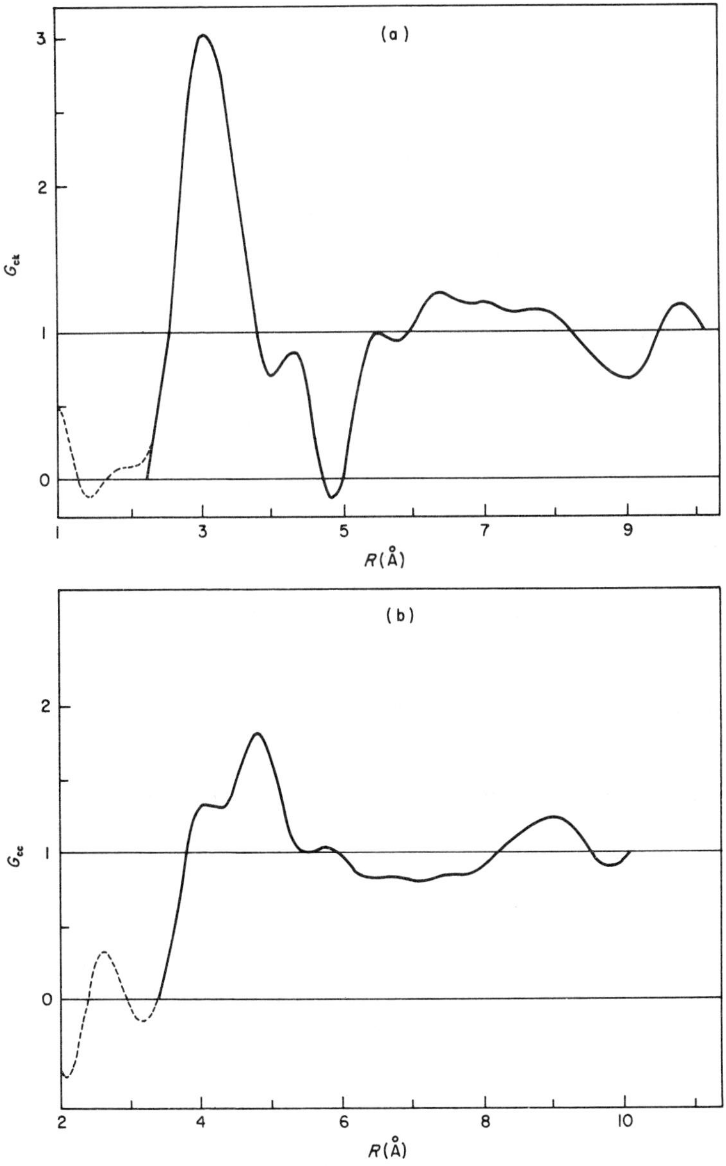

Figure 7.10 (See over for caption)

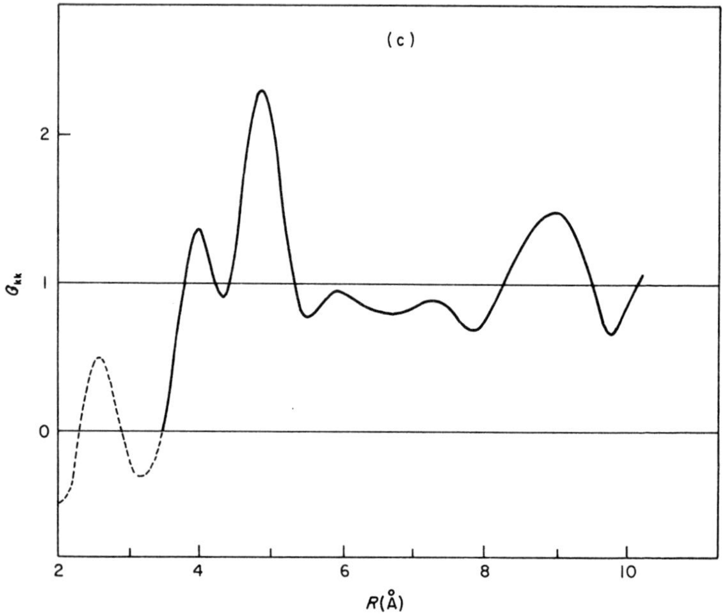

Figure 7.10 Pair correlation functions in molten KCl at 800°C, as obtained by neutron diffraction (from Derrien and Dupuy, 1975)

similar structure is presented by the mass–charge correlation function, $S_{Mq}(k, \omega)$, and by the charge–temperature correlation function, $S_{qT}(k, \omega)$. These two functions obviously describe electrostrictive and thermoelectric effects, respectively.

It is straightforward to derive Kubo relations for the transport coefficients in the two-component charged fluid from the hydrodynamic correlation functions. As usual, we find

$$\lim_{\omega\to 0} \omega^4 \left[\lim_{k\to 0} \frac{1}{k^4} S_{MM}(k, \omega) \right] = \frac{2k_B T}{\rho} \left(\tfrac{4}{3}\eta + \zeta\right) \tag{7.97}$$

while, as we noted in section 7.1, the correlation functions involving charge fluctuations must be 'screened' before proceeding to the Kubo limit. Precisely, defining 'screened' response functions $\tilde{\chi}_{AB}(k, \omega)$ by

$$\tilde{\chi}_{AB}(k, \omega) = \varepsilon(k, \omega)\chi_{AB}(k, \omega) \tag{7.98}$$

we find the relations

$$\lim_{\omega\to 0} \omega \left[\lim_{k\to 0} \frac{1}{k^2} \operatorname{Im} \tilde{\chi}_{qq}(k, \omega) \right] = \sigma/e^2 \tag{7.99}$$

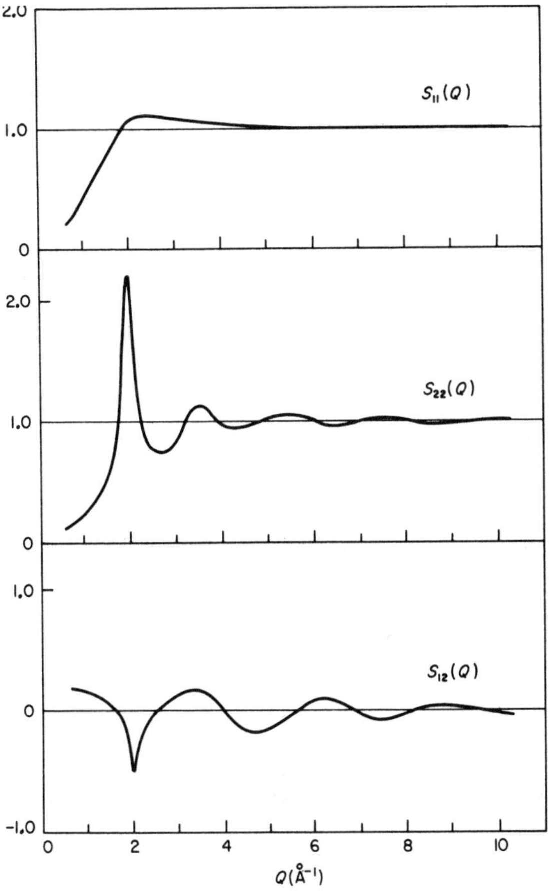

Figure 7.11 Partial structure factors in molten CuCl, as obtained by neutron diffraction (from Page and Mika, 1971). Cu and Cl are indicated by 1 and 2 respectively

$$\lim_{\omega \to 0} \omega^3 \left[\lim_{k \to 0} \frac{1}{k^4} \operatorname{Im} \tilde{\chi}_{\mathrm{Mq}}(k, \omega) \right] = \frac{\gamma\sigma}{K_{\mathrm{T}}} \left(\frac{\partial \mu}{\partial \rho_{\mathrm{q}}} \right)_{\rho_{\mathrm{m}}, \mathrm{T}} P \tag{7.100}$$

and

$$\lim_{\omega \to 0} \omega \left[\lim_{k \to 0} \frac{1}{k^2} Im\, \tilde{\chi}_{\mathrm{qT}}(k, \omega) \right] = \frac{1}{C_{\mathrm{p}}} \left[\alpha + \frac{\sigma T}{e} \left(\frac{\partial \mu}{\partial T} \right)_{\rho_{\mathrm{m}}, \rho_{\mathrm{q}}} \right] \tag{7.101}$$

Equation (7.99) is clearly equivalent to eqn (7.40), while eqns (7.100) and (7.101) give Kubo-like expressions for the electrostrictive coefficient P and for the thermoelectric coefficient α.

Outside the hydrodynamic regime we will make the usual approximation of neglecting the coupling of density fluctuations to temperature fluctuations. The variables mass density and charge density obviously remain the most convenient at long wavelength, where (*a*) the charge fluctuations are completely independent of mass fluctuations to order k^2, and (*b*) the mass fluctuations are coupled to charge fluctuations at order k^4 through an electrostrictive coefficient, which we may, however, suppose to be small at least when the two types of ions have similar sizes. In this situation we thus expect that the dynamics of the fluid may be described to a good approximation by two independent modes, namely a sound wave damped by viscosity and a plasmon-like 'optical phonon' damped by electric conduction (or, more precisely, by high-frequency electric conduction). The situation is, however, bound to be quite different at finite wavelengths. To realise this it is sufficient to notice that, in the opposite limit of short wavelength, the system should behave as a collection of independent particles, so that the most convenient dynamical variables become the density fluctuations of positive and of negative ions. We thus expect that at intermediate wavelengths the mass and charge modes will interact and mix.

This analysis is confirmed by the explicit calculation of the fourth moments of the van Hove functions (Abramo, Parrinello and Tosi, 1973*b*), reported in *figure 7.12* for molten NaCl. At long wavelengths, the fourth moment of $S_{MM}(k,\omega)$ behaves as k^4, as for a neutral fluid, while the fourth moment of $S_{qq}(k, \omega)$ behaves as k^2, as for a plasma. At $k \sim 1$ Å^{-1}, however, these two moments cross and the fourth moment of $S_{Mq}(k, \omega)$ correspondingly reaches a large value.

In extending the memory function approach described in section 5.5 to a two-component fluid it is, of course, convenient to choose dynamical variables which are as much as possible independent, in view of our present insufficient knowledge of the physical processes underlying the relaxation times $\tau(k)$ introduced in eqn (5.65). A choice of such variables for a molten salt has been proposed by Abramo, Parrinello and Tosi (1974), in the form

$$A_1(\mathbf{k}, t) = m_+{}^{1/2} \sin \vartheta_k \rho_+(\mathbf{k}, t) + m_-{}^{1/2} \cos \vartheta_k \rho_-(\mathbf{k}, t)$$

$$A_2(\mathbf{k}, t) = m_+{}^{1/2} \cos \vartheta_k \rho_+(\mathbf{k}, t) - m_-{}^{1/2} \sin \vartheta_k \rho_-(\mathbf{k}, t) \tag{7.102}$$

where the mixing angle ϑ_k is chosen so as to diagonalise the matrix

$$\frac{\mu_{ij}^{(4)}(k)}{k_B T k^2} + k_B T k^2 S_{ij}(k)/[S_{11}(k)S_{22}(k) - S_{12}^2(k)] \tag{7.103}$$

for each value of k. This is equivalent to the diagonalisation of the dynamical matrix that is needed in lattice dynamics of ionic crystals to arrive at independent acoustic and optic phonons, and leads to the expected limiting behaviours

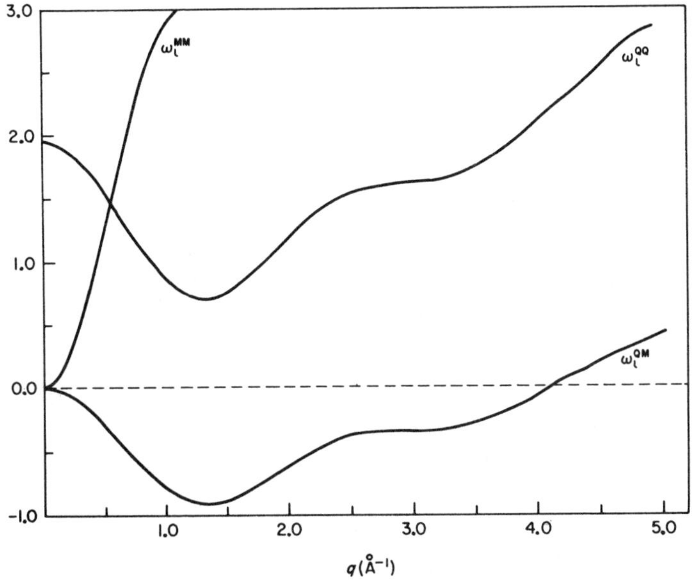

Figure 7.12 Fourth moments of the mass–charge dynamic structure factors in molten NaCl near its freezing point, in units of $4\pi\rho e^2 k_B T/[(m_+ + m_-)k^2]$ (from Abramo, Parrinello and Tosi, 1973*b*)

$$\lim_{k\to 0} A_1(\boldsymbol{k}, t) = m^{-1/2} M(\boldsymbol{k}, t)$$

$$\lim_{k\to 0} A_2(\boldsymbol{k}, t) = \mu^{1/2} \rho_q(\boldsymbol{k}, t) \tag{7.104}$$

where $m = m_+ + m_-$ and $\mu = m_+ m_-/m$, and

$$\lim_{k\to\infty} A_1(\boldsymbol{k}, t) = m_-{}^{1/2} \rho_-(\boldsymbol{k}, t)$$

$$\lim_{k\to\infty} A_2(\boldsymbol{k}, t) = m_+{}^{1/2} \rho_+(\boldsymbol{k}, t) \tag{7.105}$$

The relaxation times appropriate to these dynamical variables are then fixed according to the criterion adopted by Lovesey (1971) for liquid argon, neglecting the cross relaxation time. As noted in section 5.5 this procedure should be improved, as it has been for liquid argon, to make contact with the values of the relaxation times in the hydrodynamic regime. The numerical results for the longitudinal current correlation functions in molten NaCl are reported in *figure 7.13* for wave vectors corresponding to the first peak of the static structure factors and higher. The spectra of these extensions of acoustic and optic phonons to the liquid state are clearly very broad at these wavelengths. At longer wave-

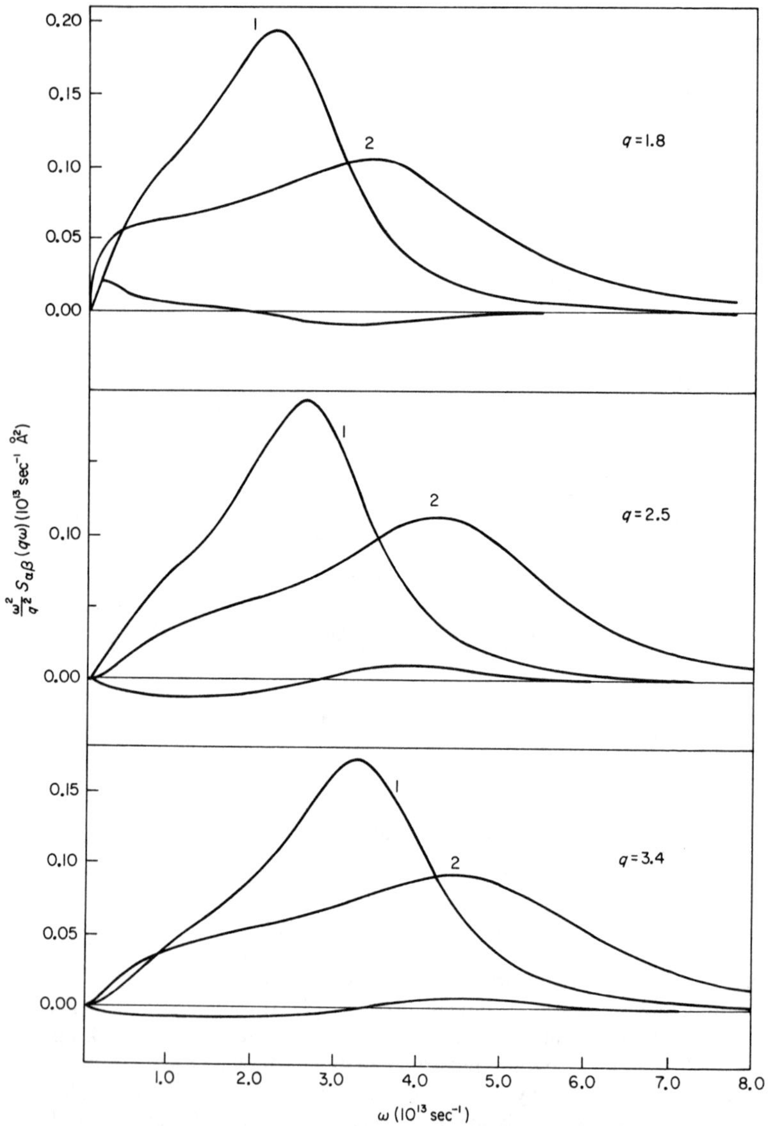

Figure 7.13 Longitudinal current spectra for the variables $A_i(\boldsymbol{k}, t)$ defined in the text, in molten NaCl near its freezing point (from Abramo, Parrinello and Tosi, 1974)

lengths ($k < 1$ Å^{-1}), however, the 'acoustic' spectrum (spectrum 1 in *figure 7.13*) will shift to lower frequencies, and the low-frequency part of the 'optic' spectrum (spectrum 2) will be depressed and its peak enhanced, approaching a behaviour similar to that depicted in *figure 7.2* for the plasmon in the classical plasma (with a negative dispersion, however). These theoretical results appear to be in accord with simulation work carried out on molten NaCl by Singer and coworkers (to be published) and on molten RbBr by Rahman and Copley (1974), although a better knowledge of the static structure factors and spectral moments entering the theory would be needed for a quantitative comparison.

Finally, it may be noted that simulation work on the velocity correlation functions in ionic melts has also been carried out, particularly by Rahman (1974*c*) who reports a variety of behaviours as compared with the behaviour of this function in liquid argon that we discussed in section 4.1. Thus, in molten LiI the velocity autocorrelation function for the heavy iodine ion does not show a negative region, so that these ions appear to approach the ideal diffusor behaviour over a very short time interval. On the other hand, the beryllium ions in molten BeF_2, which sit in a tight tetrahedral cage of fluorine ions, appear to be able to execute a considerable number of slightly damped oscillations before ultimately diffusing.

7.4 Liquid metals

We shall now briefly discuss some aspects of the theory of liquid metals that are intrinsically dependent on their nature as two-component electron–ion systems. Although the discussion will be mostly based on the assumption of weak electron–ion interactions, we shall briefly review the experimental evidence which indicates the need of transcending this assumption even for simple metals and the initial theoretical steps taken in this direction.

7.4.1 Electron theory of interionic potentials in metals

One of the most important uses of the static dielectric function of the degenerate electron plasma is the evaluation of the effective ion–ion interaction in metals. It is recognised that the conduction electrons will rearrange themselves around each ion so as to screen its potential at long distance, and such an effect can be calculated through $\varepsilon(k, 0)$ if the electron–ion interaction, $V_{\mathrm{ie}}(\boldsymbol{k})$, is weak. Each ion is then viewed as a weak, static external disturbance on the electron fluid, creating a polarisation

$$\rho(\boldsymbol{k}) = \frac{k^2}{4\pi e^2} V_{\mathrm{ie}}(\boldsymbol{k}) \left[\frac{1}{\varepsilon(k, 0)} - 1 \right] \tag{7.106}$$

(see eqns (7.23) and (7.25)). The interaction energy of the polarisation with another ion is given in Fourier transform by $\rho(\boldsymbol{k}) V_{\mathrm{ie}}(-\boldsymbol{k})$, and the effective ion–

ion potential then is

$$V_{\text{eff}}(\boldsymbol{k}) = V_{\text{ii}}(\boldsymbol{k}) + \rho(\boldsymbol{k}) V_{\text{ie}}(-\boldsymbol{k})$$

$$= V_{\text{ii}}(\boldsymbol{k}) + \frac{k^2}{4\pi e^2} |V_{\text{ie}}(\boldsymbol{k})|^2 \left[\frac{1}{\varepsilon(k, 0)} - 1 \right] \qquad (7.107)$$

Here, $V_{\text{ii}}(\boldsymbol{k})$ is the direct potential of interaction between the ions, which in simple metals with small ionic cores can usually be represented by $4\pi(Ze)^2/k^2$, the Born–Mayer repulsion between the cores being, in practice, negligible.

This simple theory requires, in addition to the dielectric function of the homogeneous plasma, the bare electron–ion potential $V_{\text{ie}}(\boldsymbol{k})$, and considerable effort has been devoted to its determination—see, for example, Heine (1970). We note, however, that this interaction is simply Coulombic at large distance, so that we can write $V_{\text{ie}}(k) = -4\pi Ze^2/k^2$ for $k \to 0$, and consequently

$$\lim_{k \to 0} V_{\text{eff}}(\boldsymbol{k}) = \lim_{k \to 0} \frac{4\pi(Ze)^2}{k^2 \varepsilon(k, 0)} = \frac{4\pi(Ze)^2}{k_s{}^2} \qquad (7.108)$$

as a result of complete screening. The vibrational frequency of the ions, which in this limit would be the ion plasma frequency $\omega_{\text{pi}} = [4\pi\rho_{\text{i}}(Ze)^2/M]^{1/2}$ in the absence of screening, is thereby reduced to an acoustic-type phonon frequency,

$$\omega_{\text{ph}}^2(k) = \lim_{k \to 0} \frac{\omega_{\text{pi}}^2}{\varepsilon(k, 0)} = \omega_{\text{pi}}^2 \frac{k^2}{k_s{}^2} = \frac{m}{M} Zs^2 k^2 \qquad (7.109)$$

where $s = (\rho_e m K)^{-1/2}$ is the sound velocity in the plasma in terms of its compressibility K. Using the free particle value $K_0 = 3(2\rho_e \varepsilon_F)^{-1}$ given in eqn (7.62), we find the result of Bohm and Staver (1952),

$$\omega_{\text{ph}}(k) = \left(\frac{1}{3} Z \frac{m}{M} \right)^{1/2} v_F k \qquad (7.110)$$

It should be stressed that, while the dielectric function entering the simple theory refers to the *homogeneous* electron plasma, the inclusion of the effects of the electron–ion interactions on the dielectric function becomes necessary in some situations even for simple metals (Pethick, 1970). For instance, an important effect in the sound-wave limit is the modification of the electrons' chemical potential due to the electron–ion interaction. Such effects become paramount for *d*-type or tight-binding electrons. The whole dielectric approach is, of course, invalid if the electron–ion interaction is strong.

7.4.2 Partial structure factors in electron–ion liquids

The arguments used in section 7.3.2 to evaluate the long wavelength structure factors in a two-component rigid-ion conductor are, of course, immediately applicable to a liquid metal, where they lead to the results (Watabe and Hasegawa, 1973; Chihara, 1973; Gray, 1973)

$$\lim_{k\to 0}\,[ZS_{\rm ii}(k)=Z^{1/2}S_{\rm ie}(k)=S_{\rm ee}(k)]=\rho k_{\rm B}T\;K_{\rm T} \tag{7.111}$$

and (Tosi and March, 1973)

$$\lim_{k\to 0}\,[ZS_{\rm ii}(k)+S_{\rm ee}(k)-2Z^{1/2}S_{\rm ie}(k)]=\frac{\frac{1}{2}\hbar\omega_{\rm p}}{4\pi\rho_{\rm e}e^2}\,k^2 \tag{7.112}$$

with $\omega_{\rm p}{}^2=4\pi\rho_{\rm e}e^2(Zm+M)/mM$. These arguments have been extended in two directions:

(1) By considering a binary alloy as a three-component electron–ion system, the long wavelength electron–ion and electron–electron structure factors as functions of concentration can be related to the ion–ion structure factors (March, Tosi and Bhatia, 1973), the latter being determined from directly measurable thermodynamic properties of the alloy according to the discussion given in section 6.2. The results for the Na–K alloy, reported in *figure 7.14*, show a strong variation with concentration of the total electronic charge seen by either type of ion in the alloy.

(2) Analogous screening conditions hold for pairs of particles in a pure liquid metal (Parrinello and Tosi, 1973), and one can then relate the ion–ion–electron triplet function, integrated over the electron coordinates, to the three-ion function as determined from measurements of the ion–ion structure factor under pressure according to the discussion of section 2.4. *Figure 7.15* reports these results for liquid rubidium.

A simple relation between the electron–ion and the ion–ion structure factors in a pure liquid metal persists at finite wavelengths in the case of weak electron–ion interaction (Tosi and March, 1973). Consider the response of the liquid metal to a weak external potential $V_{\rm e}(\boldsymbol{k},\omega)$ acting on the electrons only: we can write the polarisation of the electron and of the ion fluid in terms of the appropriate response functions as

$$\rho_{\rm e}(\boldsymbol{k},\omega)=\chi_{\rm ee}(k,\omega)V_{\rm e}(\boldsymbol{k},\omega) \tag{7.113}$$

and

$$\rho_{\rm i}(\boldsymbol{k},\omega)=\chi_{\rm ie}(k,\omega)V_{\rm e}(\boldsymbol{k},\omega) \tag{7.114}$$

On the other hand, for weak electron–ion interaction we can think of $\rho_{\rm i}(\boldsymbol{k},\omega)$ as the polarisation of the ion fluid in response to the electron polarisation potential acting directly on the ions as an external potential, namely we can write

$$\rho_{\rm i}(\boldsymbol{k},\omega)=\chi_{\rm ii}(k,\omega)V_{\rm ie}(\boldsymbol{k})\rho_{\rm e}(\boldsymbol{k},\omega) \tag{7.115}$$

whence

$$\chi_{\rm ie}(k,\omega)=\chi_{\rm ee}(k,\omega)V_{\rm ie}(\boldsymbol{k})\chi_{\rm ii}(k,\omega)$$

$$\simeq\chi_{\rm ee}(k,0)V_{\rm ie}(\boldsymbol{k})\chi_{\rm ii}(k,\omega) \tag{7.116}$$

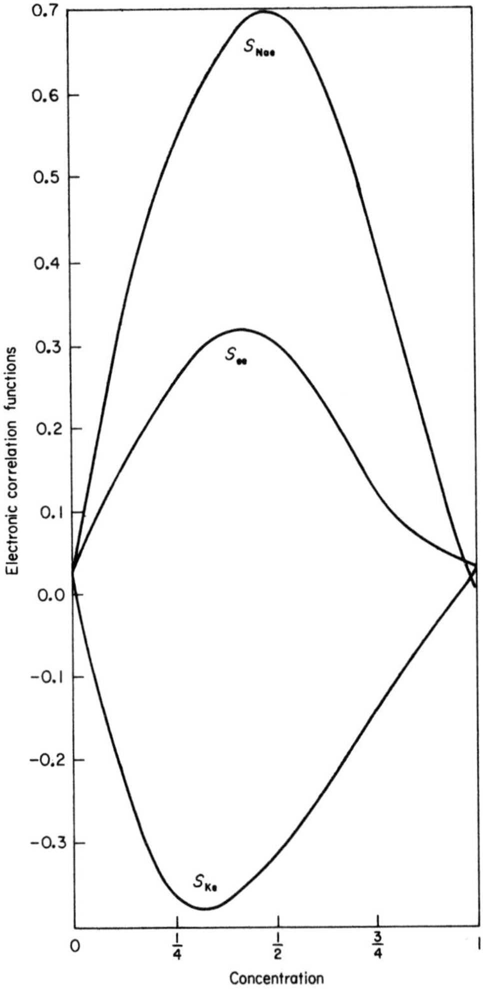

Figure 7.14 Electronic structure factors at long wavelength in the Na–K liquid alloy, as functions of concentration (from March, Tosi and Bhatia, 1973)

where, in the last step, we have made use of the adiabatic approximation. Use of the fluctuation–dissipation theorem finally gives

$$Z^{1/2}S_{ie}(k)=\chi_{ee}(k,0)V_{ie}(k)S_{ii}(k)=\tilde{\chi}_{ee}(k,0)\frac{V_{ie}(k)}{\varepsilon(k,0)}S_{ii}(k) \tag{7.117}$$

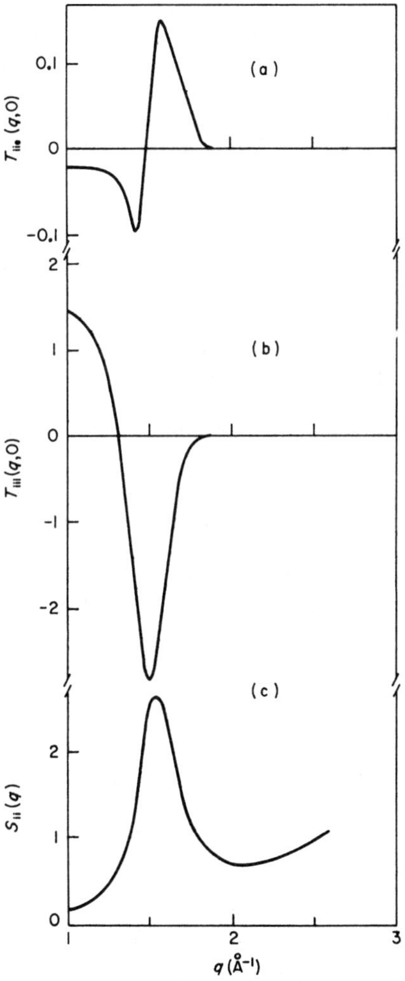

Figure 7.15 Ion–ion–electron (*a*) and ion–ion–ion (*b*) triplet correlations in liquid Rb at 333 K, as derived from the ion–ion structure factor under pressure (*c*) of Egelstaff, Page and Heard, 1971 (from Parrinello and Tosi, 1973)

Using eqns (7.39) and (7.38), this result is soon shown to reduce to (7.111) for $k \to 0$. We now note that the screened electron–ion potential in metals is negative for $k \to 0$ but usually decreases in absolute magnitude as k increases and changes sign for $k \sim k_F$. Thus, according to eqn (7.117) we expect that if $(a)k_F$ is smaller than the position of the first peak in $S_{ii}(k)$ and (*b*) the screened electron–ion potential is positive and still sizable for $k>k_F$, then a sizable negative dip in $S_{ie}(k)$ may be present at wave vectors comparable with the position of the first peak of $S_{ii}(k)$. We saw this to be the case for $S_{12}(k)$ in molten salts (see *figure 7.9*). Such a behaviour of $S_{ie}(k)$ would imply good charge ordering in the liquid metal, as is apparent from the definition of $S_{qq}(k)$ given in eqn (7.85). We also expect that $S_{ie}(k)$ will rapidly vanish, in general, for $k>2k_F$, because of the quantum behaviour of $\tilde{\chi}_{ee}(k, 0)$ discussed in section 7.2.1.

Approximate self-consistent calculations of the partial structure factors for a model liquid metal have been reported by Chihara (1972), but it appears that the chosen system does not have the characteristics indicated above. Indeed, the calculated $S_{ie}(k)$ vanishes for $k \sim k_F$ without having a negative region, and $S_{ee}(k)$ is qualitatively indistinguishable from the electron structure factor in the homogeneous plasma. On the other hand, Egelstaff, March and McGill (1974) have recently emphasised that the differences between experimentally determined X-ray and neutron diffraction intensities from liquid metals require, for their explanation, that the conduction electrons exhibit some short-range order, and that the range of this order be longer than the range of ionic order, as observed by the neutrons. The oscillations anticipated in $g_{ee}(r)$ are, in fact, relatively small, perhaps with $g_{ee}(r)$ going up to 1.1 as a maximum, but the range of such oscillations is of main interest. A parallel suggestion to the same effect has been put forward for conduction electrons in *solid* metals by Platzman and Eisenberger (1974), whose experiments on inelastic X-ray scattering show the plasmon excitation to be still visible as a distinct spectral peak, with negative dispersion, for wave vectors between k_F and $2k_F$.

7.4.3 Cohesive properties of metals

In the standard treatment of the thermodynamic properties of a metal, the effective potential energy of the ionic system is written in the form of a density-dependent term which is independent of structure, and a pair potential term which is dependent on density. Namely, the internal energy is written as

$$U=\tfrac{3}{2}Nk_BT+Nu_0(\rho)+\tfrac{1}{2}\sum_{i\neq i'}\phi(r_{ii'},\rho) \qquad (7.118)$$

the sum being extended to all ion pairs. Correspondingly, the expression (1.15) for the pressure has to be generalised to read (see Ascarelli, Paskin and Harrison, 1967)

$$P=\rho k_BT+\rho^2\frac{du_0}{d\rho}-\tfrac{1}{2}\rho^2\int d\mathbf{r}\,g(r)\left[\tfrac{1}{3}r\frac{\partial\phi(r,\rho)}{\partial r}-\rho\frac{\partial\phi(r,\rho)}{\partial r}\right] \qquad (7.119)$$

The above results, with a precise identification of the functions u_0 and ϕ, follow easily under the assumption of weak electron–ion interactions. In full generality, we should write the internal energy density as

$$u=\tfrac{3}{2}\rho_i k_B T+t_e+\tfrac{1}{2}\int d\mathbf{r}[\rho_e{}^2 g_{ee}(r)V_{ee}(r)+\rho_i{}^2 g_{ii}(r)V_{ii}(r)+2\rho_i\rho_e g_{ie}(r)V_{ie}(r)] \quad (7.120)$$

where t_e is the electronic kinetic energy density in the metal and all the pair functions refer to the metallic state. Under the assumptions already used in the preceding section to derive eqn (7.117), we can consider the switching on of the electron–ion interaction as a process of polarisation of the homogeneous electron fluid by the ions in their fixed configuration as described by $g_{ii}(r)$. By a standard result of linear response theory, one-half of the energy gained by the interaction of the polarisation with the polarising field cancels against the 'quasi-elastic' energy spent in creating the polarisation. The former energy in the present problem is

$$\tfrac{1}{2}\int d\mathbf{r}\,2\rho_i\rho_e[g_{ie}(r)-1]V_{ie}(r)$$

while the latter is

$$t_e+\tfrac{1}{2}\int d\mathbf{r}\,\rho_e{}^2[g_{ee}(r)-1]v_{ee}(r)-u_h$$

where $u_h(\rho)$ is the energy density of the homogeneous electron fluid at the electronic density obtaining in the metal. Using also the relation (7.117) to re-express $g_{ie}(r)$ in terms of $g_{ii}(r)$, we arrive at the result

$$u=\tfrac{3}{2}\rho_i k_B T+u_h+\tfrac{1}{2}\rho_i[V_{\text{eff}}(r)-V_{ii}(r)]_{r=0}$$
$$+\tfrac{1}{2}\int d\mathbf{r}[\rho_i{}^2V^s_{ii}(r)+2\rho_i\rho_e V^s_{ie}(r)-\rho_i{}^2V_{\text{eff}}(r)]+\tfrac{1}{2}\rho_i{}^2\int d\mathbf{r}\,g_{ii}(r)V_{\text{eff}}(r) \quad (7.121)$$

where $V^s_{ii}(r)$ and $V^s_{ie}(r)$ are the short-range (non-Coulombic) parts of the ion–ion and electron–ion interaction, and $V_{\text{eff}}(r)$ is the effective ion–ion potential defined in eqn (7.107). We thus see, by comparison with (7.118), that in this approximation $\phi(\text{r})$ is indeed the effective ion–ion potential as given by dielectric theory, and we identify u_0 with all the remaining terms of (7.121).

This formulation has been used by Price (1971; see also Hartman, 1971) to evaluate the thermodynamic properties (equilibrium volume, melting energy and compressibility) of liquid sodium, finding good agreement with experiment. The same formulation had been earlier applied to a number of simple metals in the solid state by Ashcroft and Langreth (1967*b*), again with satisfactory results.

In this connection, Ivanov et al. (1974) have reported accurate measurements of relative change of volume under pressure, and of the melting curve of sodium at pressures up to about 20 kilobar. This data has been employed by them to calculate the changes of various thermodynamic functions on melting. The parameter Γ in eqn (7.49), measuring the ratio of Coulomb to thermal energy, is demonstrated by these workers to tend along the melting curve to the value

155 ± 10 obtained for a transition to a crystalline state of the classical one-component plasma (see section 7.1.3).

7.4.4 Electronic transport and hydrodynamics

As we have seen in the preceding discussion, the assumption of weak electron–ion interactions and the adiabatic approximation turn out to be very useful for a semiquantitative understanding of the static properties of simple liquid metals. The same assumptions, usually supplemented by a random phase approximation on the electron fluid, form the basis for the calculation of electronic transport properties in metals, such as the electric resistivity and the thermal conductivity.

In this scheme, we note with Baym (1964) that the process of scattering of the electrons by the ions can be regarded as fully equivalent to a process of scattering of a beam of external particles in the Born approximation. We can thus appeal to the results of section 3.2.3 and write for the probability per unit time that an electron give up momentum $\hbar\boldsymbol{k}$ and energy $\hbar\omega$ to the ion system the expression

$$P(\boldsymbol{k}, \omega) = N|V_{ie}(\boldsymbol{k})|^2 S_{ii}(\boldsymbol{k}, \omega) \tag{7.122}$$

The same probability for the whole system of conduction electrons is obtained by multiplying $P(\boldsymbol{k}, \omega)$ by the dynamic structure factor $S'_{ee}(-\boldsymbol{k}, -\omega)$ for the electrons in uniform flow, and the total rate of momentum change of the flowing electron liquid is

$$-\frac{\mathrm{d}\boldsymbol{p}_e}{\mathrm{d}t} = N\sum_{\boldsymbol{k}} \hbar\boldsymbol{k}|V_{ie}(\boldsymbol{k})|^2 \int \frac{\mathrm{d}\omega}{2\pi} [S_{ii}(\boldsymbol{k}, \omega)S'_{ee}(-\boldsymbol{k}, -\omega) - S_{ii}(-\boldsymbol{k}, -\omega)S'_{ee}(\boldsymbol{k}, \omega)] \tag{7.123}$$

This quantity is obviously proportional to the inverse relaxation time for electric conduction, τ^{-1}.

Introducing the random phase approximation for the electronic structure factors,

$$S'_{ee}(\boldsymbol{k}, \omega) = S'_0(\boldsymbol{k}, \omega)/|\varepsilon(k, \omega)|^2 \tag{7.124}$$

where $S'_0(\boldsymbol{k}, \omega)$ is the van Hove function for a free electron gas in uniform flow, we can rewrite (7.123) as

$$-\frac{\mathrm{d}\boldsymbol{p}_e}{\mathrm{d}t} = N\sum_{\boldsymbol{k}} \hbar\boldsymbol{k} \int \frac{\mathrm{d}\omega}{2\pi} |\tilde{V}_{ie}(\boldsymbol{k}, \omega)|^2 [S_{ii}(\boldsymbol{k}, \omega)S'_0(-\boldsymbol{k}, -\omega) - S_{ii}(-\boldsymbol{k}, -\omega)S'_0(\boldsymbol{k}, \omega)] \tag{7.125}$$

where $\tilde{V}_{ie}(\boldsymbol{k}, \omega) = V_{ie}(\boldsymbol{k})/\varepsilon(k, \omega)$ is the screened electron–ion potential, that we can as usual replace by its static limit, $\tilde{V}_{ie}(\boldsymbol{k}) = V_{ie}(\boldsymbol{k})/\varepsilon(k, 0)$. Also, we have

$$S_0'(\boldsymbol{k}, \omega) = 2\pi \sum_{p} f(\boldsymbol{p})[1 - f(\boldsymbol{p}+\boldsymbol{k})]\delta(\omega - \varepsilon_{p+k} + \varepsilon_p) \tag{7.126}$$

where $f(\boldsymbol{p})$ is the usual 'displaced' Fermi distribution,

$$f(\boldsymbol{p}) = f_{\text{eq}}(\varepsilon_p - \boldsymbol{p} \cdot \boldsymbol{p}_{\text{e/m}}) \tag{7.127}$$

We can thus write eqn (7.125) to linear terms in $\boldsymbol{p}_e$ as

$$\frac{d\boldsymbol{p}_e}{dt} = -\boldsymbol{p}_e/\tau \tag{7.128}$$

where the relaxation time for electric conduction is given by (Baym, 1964)

$$\tau^{-1} = \frac{m}{12\pi^3\hbar^3} \int_0^{2k_F} k^3 \, dk |\tilde{V}_{\text{ie}}(\boldsymbol{k})|^2 \int_{-\infty}^{\infty} \frac{d\omega}{2\pi} S_{\text{ii}}(k, \omega) \frac{\hbar\beta\omega}{\exp(\hbar\beta\omega) - 1} \tag{7.129}$$

The Ziman (1961; see also Krishnan and Bhatia, 1945) expression for τ^{-1} follows by taking the high-temperature, classical limit ($\hbar\beta\omega \ll 1$), namely

$$\tau^{-1} = \frac{m}{12\pi^3\hbar^3} \int_0^{2k_F} k^3 \, dk |\tilde{V}_{\text{ie}}(\boldsymbol{k})|^2 S_{\text{ii}}(k) \tag{7.130}$$

Similar treatments can be developed for the electronic contributions to other transport properties of the liquid metal (Baym, 1964; see also Faber, 1973).

It is, in practice, extremely difficult to transcend the basic assumptions involved in the above theory of electronic transport (weak electron–ion and electron–electron interactions, and Born–Oppenheimer approximation). A formal discussion of transport coefficients and electronic effects in the dynamical structure of liquid metals has been given, however, by Tosi, Parrinello and March (1974). This elucidates the general structure of the dynamic correlation functions and gives general expressions for transport coefficients whose electronic terms reduce to the Baym theory when the appropriate simplifying assumptions are made.

7.5 Electron-hole liquids in semiconductors

Following an early suggestion by Keldysh (1968), considerable experimental evidence has accumulated that electrons and holes in Ge and Si undergo a gas-to-liquid phase transition at low temperature and at the high density attainable under high excitation power, condensing into droplets surrounded by a dielectric gas of excitons. The most direct evidence is perhaps provided by experiments on p–n junction noise revealing high current pulses as the drops are destroyed in reaching the junction (Asnin, Rogachev and Sablina, 1970; Benoît à la Guillaume et al., 1971), and by direct light scattering experiments from the droplets (Pokrovsky and Svistunova, 1971). Because of the indirect gap in these semiconductors, the recombination time is of the order of several microseconds, sufficiently longer than the thermalisation time that the drops can be observed under conditions of thermal equilibrium. Considerable experimental information is indeed available on the thermodynamic properties of the drops in Ge, including the ground state energy and the compressibility, the temperature dependence of the equilibrium density and the chemical potential, the critical temperature and density, and the 'enhancement factor' $g_{eh}(r=0)$. In particular, $E_b \sim 2$ meV and $\rho \sim 2 \times 10^{17}$ cm^{-3}.

Since the effective Bohr radius is sizably larger than the lattice constant, the electron-hole liquid can be regarded as a two-component, degenerate plasma with interactions given by Coulomb's potential screened by the static dielectric constant of the crystalline matrix considered as a uniform dielectric. An important role in determining the binding energy of the liquid relative to free excitons is already played by the detailed band structure of the matrix determining the free-particle kinetic energy and the exchange energy (Brinkman et al., 1972; Combescot and Nozières, 1972). Indeed, for simple parabolic bands one would write the ground state energy per electron-hole pair, as in section 7.2.2, in the form

$$E_{gs} = \left(\frac{2.21}{r_s^2} - \frac{1.832}{r_s} + \varepsilon_{corr} \right) \text{ryd} \tag{7.131}$$

which, for reasonable estimates of the correlation energy, is found to be smaller in absolute magnitude than the binding energy of -1 ryd of a free exciton (here, the unit of energy is the effective exciton rydberg, $\mu e^4/2\varepsilon^2\hbar^2$, and r_s is the mean distance in units of the effective Bohr radius, $\varepsilon\hbar^2/\mu e^2$). On the other hand, the electrons in Ge are in four ellipsoidal bands at the L point of the Brillouin zone with widely different longitudinal and transverse masses, and the holes are in two degenerate valence bands at the Γ point, and an approximate calculation yields

$$E_{gs} = \left(\frac{0.468}{r_s^2} - \frac{1.136}{r_s} + \varepsilon_{corr} \right) \text{ryd} \tag{7.132}$$

The decrease of the kinetic energy overcompensates for the increase of the exchange energy, and the liquid is bound by approximately 2 meV. The results are illustrated in *figure 7.16* (from Brinkman et al., 1972), eqns (7.131) and (7.132) being represented, respectively, by curves (*a*) and (*b*). Detailed calculations of the thermodynamic properties (Vashishta, Das and Singwi, 1974) are in good semiquantitative agreement with the experimental data.

Curve (*c*) in *figure 7.16* refers to Ge under a large $\langle 111 \rangle$ stress, which removes the degeneracies between the four electron ellipsoids and between the two valence bands at the zone centre, leaving just one electron and one hole ellipsoid. The theory indicates that the electron-hole liquid in this system may still be bound relative to free excitons, albeit only slightly and by an amount crucially dependent on the correlation energy. Also, because of the relative simplicity of

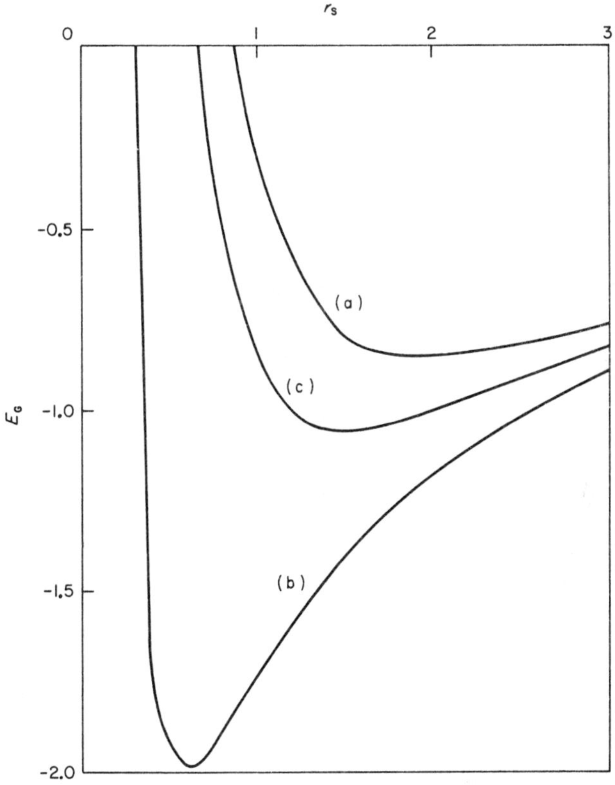

Figure 7.16 Ground state energy of the electron-hole liquid as a function of density for (*a*) a parabolic band model with equal electron and hole masses; (*b*) real germanium; and (*c*) germanium under a large ⟨111⟩ strain (from Brinkman et al., 1972)

its band structure, strained Ge has been the subject of very detailed theoretical studies as a testing ground of many-body theories of correlation. Some of the results for the pair correlation functions are reported in *figures 7.17* and *7.18* (from Vashishta, Bhattacharyya and Singwi, 1974). The fully self-consistent results (curves marked FSC) show that the two-component plasma is considerably more ordered than indicated by simple RPA-type theories, treating the interactions as essentially weak. In particular, the theory indicates a strong pile-up of electrons on holes, and marked oscillations in $g_{ee}(r)$ as the electrons around the electron at the origin try to distribute themselves in phase with the holes so as to maintain charge neutrality even on a local scale.

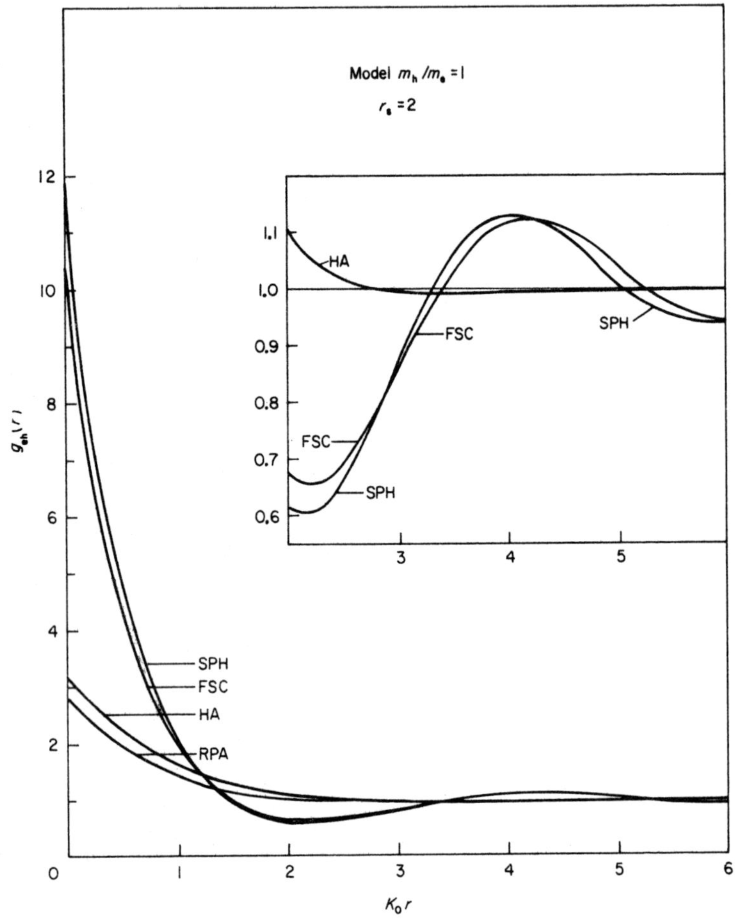

Figure 7.17 Electron-hole correlation function in an electron-hole liquid (from Vashishta, Bhattacharyya and Singwi, 1974). The fully self-consistent (FSC) results, based on an extension of the self-consistent theories described in section 7.2.2, transcend the linear treatment of the electron-hole interaction implicit in RPA-type treatments. These results should also be compared with the linear-response results of *figure 7.4*

7.6 Water and electrolyte solutions

The development of the theory of electrolyte solutions and electrochemical processes is one of the most important and long-studied topics in statistical mechanics. For a detailed understanding of these processes it appears necessary

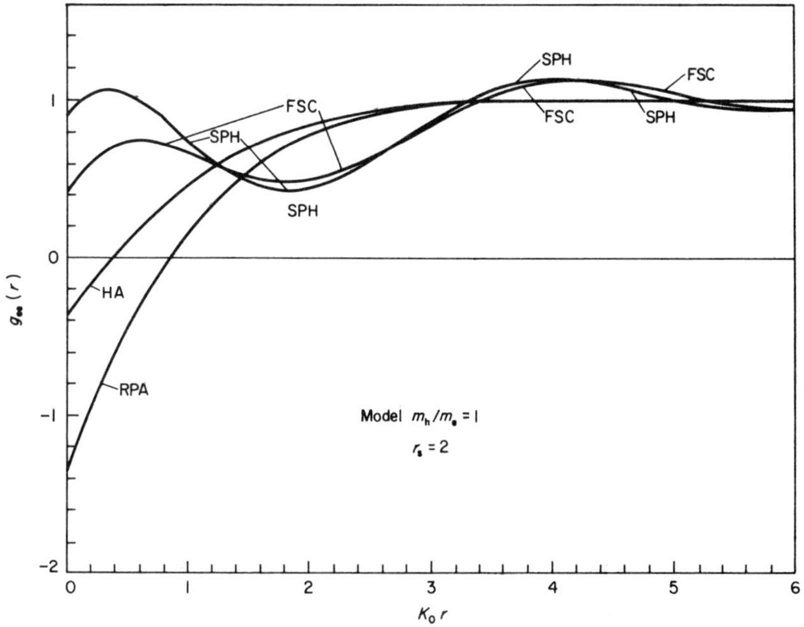

Figure 7.18 Electron–electron correlation function in an electron-hole liquid (from Vashishta, Bhattacharyya and Singwi, 1974)

to transcend the so-called primitive model of electrolytes, which treats the solution as a fluid of ions immersed in a uniform dielectric. Progress in this direction depends on a detailed understanding of pure water at the molecular level—a topic on which considerable progress has been made—and on a molecular treatment of the interactions between solutes and solvent.

7.6.1 Structure of water

One of the striking characteristics of water is its high dielectric constant. Each water molecule carries a permanent electric dipole, and the high dielectric constant is indicative of strong orientational correlations between the dipoles. The classical calculation of this property by Onsager (1936) serves to introduce several concepts that we shall need in the subsequent discussion.

The model considers each molecule as a small spherical cavity in a dielectric with dielectric constant ε_0, the cavity being filled with polarisable material (to simulate electronic polarisation effects, described through a 'high-frequency' dielectric constant ε_∞) and carrying a permanent dipole moment whose magnitude *in vacuo* we denote by μ_v. If $\boldsymbol{E}$ is the electric field at large distance from the cavity, then the field seen by the dipole in the cavity is the sum of (*a*) the 'cavity field' G which is determined by the applied field E as locally modified by the presence of the cavity, and (*b*) the reaction field R, which arises from the polarisation created in the dielectric by the dipole itself. It is a standard electrostatic calculation (see, for example, Fröhlich, 1958) to show that

$$G = \frac{3\varepsilon_0}{2\varepsilon_0 + \varepsilon_\infty} E \tag{7.133}$$

and

$$R = \frac{8\pi}{3} \frac{\varepsilon_0 - \varepsilon_\infty}{\varepsilon_\infty[2\varepsilon_0 + \varepsilon_\infty]} \mu \tag{7.134}$$

Here, μ is the effective dipole moment carried in the cavity, which on account of the polarisability of the material filling the cavity is enhanced over the dipole *in vacuo* by

$$\mu = \frac{\varepsilon_\infty + 2}{3} \mu_v \tag{7.135}$$

The energy of the dipole in the field is $-\mu \cdot (G+R)$, and the average dipole moment in the direction of E is calculated as

$$\langle \mu \rangle = \int_0^\pi (\mu \cos \vartheta) \exp[\mu \cdot (G+R)/k_B T] \sin \vartheta \, d\vartheta \bigg/ \int_0^\pi \exp[\mu \cdot (G+R)/k_B T] \sin \vartheta \, d\vartheta$$

$$\simeq \frac{\varepsilon_0}{2\varepsilon_0 + \varepsilon_\infty} \frac{\mu^2 E}{k_B T} \tag{7.136}$$

By imposing the self-consistency condition

$$\varepsilon_0 - \varepsilon_\infty = 4\pi\rho \langle \mu \rangle / E \tag{7.137}$$

one finds the Onsager formula,

$$\varepsilon_0 - \varepsilon_\infty = \frac{4\pi\rho\mu^2}{k_B T} \frac{\varepsilon_0}{2\varepsilon_0 + \varepsilon_\infty} \tag{7.138}$$

The result, with $\mu_v = 1.9 \times 10^{-18}$ for water, is $\varepsilon_0 \approx 30$ at room temperature.

We may look upon the discrepancy between the above calculation and the experimental value $\varepsilon_0 \approx 80$ as indicative of strong short-range correlations, neglected in the Onsager model. According to the early model of Bernal and Fowler (1933), there is a strong tendency for the four nearest neighbours of each water molecule to arrange themselves in a tetrahedral configuration, which allows the hydrogens in each molecule to point towards the oxygens of two neighbouring molecules. Inclusion of this effect in the calculation of the dielectric constant (see, for example, Fröhlich, 1958) augments Onsager's value by a factor 7/3, thus leading to satisfactory agreement with experiment.

We may expect, on very general grounds, that the dielectric constant of a dipolar fluid will be related to its pair correlation function, at least for non-polarisable molecules (or for some uniform-dielectric models of the electronic polarisation). Indeed, it is yet another aspect of the fluctuation–dissipation theorem that the static, wave-vector dependent dielectric function, which gives the response of the rigid-molecule fluid to an applied electric field, is directly related in the classical limit to the static structure factor expressing correlations between charge density fluctuations. We saw an example of this general property in eqn (7.27) for the classical one-component plasma. This question has received considerable attention in the recent literature, and in particular Nienhuis and Deutch (1971) and Ramshaw (1972) have discussed the properties of the pair correlation function which ensure the existence of a dielectric constant independent of the size and shape of the sample. Ramshaw (1972) and Wertheim (1973) give an explicit relation between the dielectric function and the direct correlation function $c(12)$ for non-polarisable molecules, in the form

$$\mathsf{Y}(12)=\frac{9k_{\mathrm{B}}T}{\mu^4\rho^2}\,\mu(\Omega_1)[\rho_1\delta(12)-\rho_1{}^2c(12)]\mu(\Omega_2) \tag{7.139}$$

where $\mathsf{Y}(12)$ is a tensor relating the electric field at $\boldsymbol{r}_1$ to the polarisation at $\boldsymbol{r}_2$, and the μ's are the dipole moments at the two points. The extension of the theory to polarisable molecules has been developed by Wertheim (1973).

As an example of the relation between dielectric constant and structure in an exactly soluble model, we consider the exact solution given by Wertheim (1971) for the mean spherical model of a fluid of hard spheres carrying permanent dipoles. The solution, for both the direct correlation function $c(12)$ inside the core and the total correlation function $h(12)$ outside the core, is the sum of a spherically symmetric term and of two terms with different dependences on the orientations of the dipoles. Specifically,

$$h(12)=h_{\mathrm{s}}(r)+h_{\mathrm{D}}(r)D(12)+h_{\Delta}(r)\Delta(12) \tag{7.140}$$

where

$$D(12)=\frac{3(\boldsymbol{s}_1\cdot\boldsymbol{r})(\boldsymbol{s}_2\cdot\boldsymbol{r})}{r^2}-\boldsymbol{s}_1\cdot\boldsymbol{s}_2 \tag{7.141}$$

and

$$\Delta(12)=\boldsymbol{s}_1\cdot\boldsymbol{s}_2 \tag{7.142}$$

These two functions, together with the identity $I(12)=1$, form a closed set, in that their various convolutions can be expressed through the same triplet of functions. Furthermore, Wertheim shows that $h_{\mathrm{s}}(r)$ coincides with the MSM solution for non-polar hard spheres (see section 6.5.1), and that $h_{\mathrm{D}}(r)$ and $h_{\Delta}(r)$ can be related to the same MSM solution for positive and negative values of the packing fraction η. The calculation of the dielectric constant, with a careful analysis of sample finiteness effects, leads to the result

$$\varepsilon_0=\frac{q(2\xi)}{q(-\xi)}=\frac{(1+4\xi)^2(1+\xi)^4}{(1-2\xi)^6} \tag{7.143}$$

where

$$q(\eta)=1-4\pi\rho\int_0^\sigma c_s(r;\rho)r^2\,dr=\frac{(1+2\eta)^2}{(1-\eta)^4} \tag{7.144}$$

and ξ is determined by the condition

$$q(2\xi)-q(-\xi)=\frac{4\pi\rho\mu^2}{3k_BT} \tag{7.145}$$

The above result is in agreement, for small values of $\rho\mu^2/k_BT$, with the earlier work of Jepsen and Friedman (1963) and Jepsen (1966).

Unfortunately, it appears that the mean spherical model for a point-dipole, hard-sphere fluid is not very successful in providing a semiquantitative estimate of the thermodynamic and structural properties of dipolar fluids, by comparison with Monte Carlo results (Patey and Valleau, 1973). In particular, the model leads to a serious underestimate of the dielectric constant of water ($\varepsilon_0\approx 15$ at room temperature) and it seems likely that an altogether more sophisticated model is necessary to account for hydrogen bonding and the consequent tendency to tetrahedral coordination. However, it may be noted that the Wertheim solution can be extended to treat hard-sphere particles carrying multipoles (Blum and Torruella, 1972; Blum, 1972; Blum, 1973*a*) and that, on the other hand, a perturbative scheme has been presented by Andersen and Chandler (1971) to improve the calculation of liquid properties starting from a hard-sphere solution (see also section 2.6.6). For related work in this area, using perturbation theory, reference should be made to Verlet and Weis (1974).

In parallel with these theoretical developments in the understanding of dipolar fluids, specific advances in the detailed description of the structure of water have recently been made both by neutron scattering methods (Page and Powles, 1971; Powles, Dore and Page, 1972) and by molecular dynamics methods (Rahman and Stillinger, 1971; 1972). The neutron experiments were carried out on heavy water, normal water and a mixture chosen so that the protons and deuterons give no contribution to the coherent scattering. The latter experiment gives a structure factor which corresponds very well to the oxygen-nucleus structure factor obtained from X-ray data. The analysis of the data is complicated by the high incoherent-scattering contribution from the protons in normal water, but the progress already achieved indicates that, by more precise measurements and more sophisticated separations of the coherent and incoherent scattering, the method of isotopic substitution should lead to a full experimental determination of the three partial structure factors.

The molecular dynamics experiments of Rahman and Stillinger, carried out on an assembly of 216 water molecules over a temperature range, start from the observation that there is a tendency when water molecules interact towards formation of linear hydrogen bonds between neighbours disposed in space in a tetrahedral coordination pattern. The model potential adopted in the experiments was proposed by Ben-Naim and Stillinger (1972) and consists of:

(1) A central Lennard–Jones component appropriate for the isoelectronic Ne–Ne interaction.

(2) An orientation dependent part based on a four-point charge complex within each water molecule.

Quantitatively this can be written

$$\phi_{\mathrm{eff}}(\boldsymbol{r}_i, \boldsymbol{r}_j) = \phi_{LJ}(r_{ij}) + S(r_{ij})v(\boldsymbol{r}_i, \boldsymbol{r}_j) \tag{7.146}$$

where r_{ij} is the radial distance between oxygen nuclei, while v is the sum of all sixteen Coulombic charge-pair interactions between the molecules. Each molecule contains two charges $+0.19e$, to be thought of as shielded protons, and two charges $-0.19e$ (unshared electrons), placed at the vertices of a regular tetrahedron with radius 1 Å, centred on the 0 nucleus. Finally S is an interpolation function, joining 0 at small r_{ij} to 1 at large r_{ij}. The specific form adopted for S is

$$\begin{aligned} S(r_{ij}) &= 0 && (0 \leqslant r_{ij} \leqslant R_L) \\ &= \frac{(r_{ij} - R_L)^2(3R_u - R_L - 2r_{ij})}{(R_u - R_L)^3} && (R_L \leqslant r_{ij} \leqslant R_u) \\ &= 1 && (R_u \leqslant r_{ij}) \end{aligned} \tag{7.147}$$

where $R_L = 2.04$ Å and $R_u = 3.19$ Å.

The above interaction has the merit that it incorporates the linear hydrogen-bonding tendency between neighbours in a tetrahedral pattern, although the force field thus constructed is not completely realistic as it is pairwise additive. Quantum mechanical calculations (Delbene and Pople, 1970; Hankins, Moscowitz and Stillinger, 1970) reveal that the intermolecular interactions have significant non-additive components.

The calculations of Rahman and Stillinger confined their 216 molecules to a cubical cell of side 18.62 Å, leading to the same density 1 gm/cm^3 at each temperature. Although the variation of density is of interest, all the results reported below are therefore for this fixed density.

Figures 7.19 and *7.20* show the O–O pair correlation function thus obtained, at two temperatures T_1 and T_2, the former corresponding to slightly supercooled water at 265 K while T_2 is 307.5 K. The most noteworthy feature (which is confirmed by even higher temperature data) is the decreasing amplitude of the oscillations with increasing temperature. *Figure 7.21* shows the O–H correlation functions at the same temperatures, while *figure 7.22* gives the H–H correlations.

Rahman and Stillinger also discuss molecular orientational correlations as a function of temperature in considerable detail in their work, and we refer the reader to their paper for this illuminating discussion.

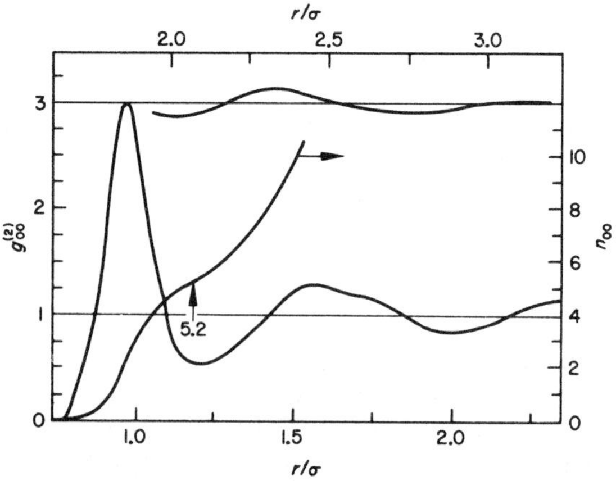

Figure 7.19 Radial pair correlation function $g_{OO}(r)$ for water at $T = 265$ K and density 1 gm/cm^3. The unit of distance is $\sigma = 2.82$ Å. The running coordination number is also shown, with scale on the right (from Rahman and Stillinger, 1972)

Their results indicate that the above potential is somewhat too directional and the next stage in this work is, pretty clearly, the refinement of the interaction.

Some quantities having dynamical information, such as the velocity autocorrelation function, were also calculated by Rahman and Stillinger. By a modest (6%) scaling of the interaction, both the magnitude of the self-diffusion coefficient, and its observed rapid temperature variation follow nicely from the model.

Watts (1974) by Monte Carlo computer simulation, has compared the results

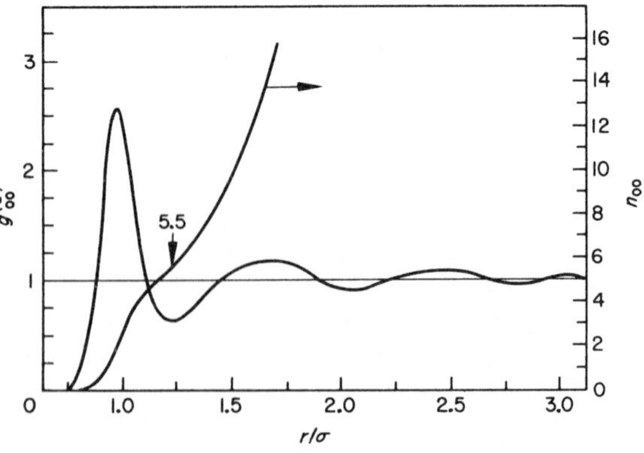

Figure 7.20 Same as *figure 7.19*, at $T = 307.5$K (from Rahman and Stillinger, 1972)

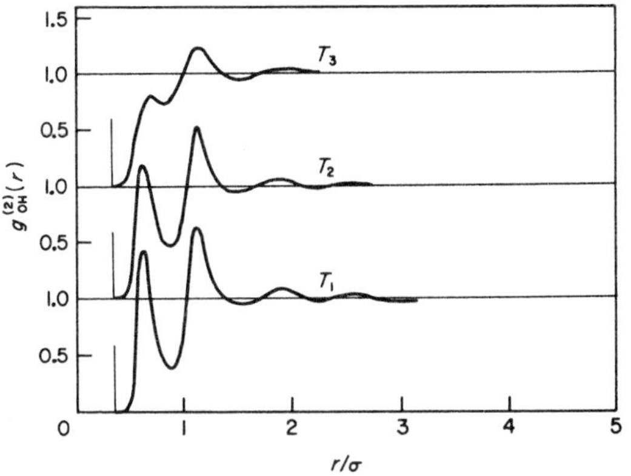

Figure 7.21 Radial cross-correlation function $g_{OH}(r)$ in water at three temperatures ($T_1 = 265$ K, $T_2 = 307.5$ K, $T_3 = 588$ K) and density 1 gm/cm^3. The vertical line indicates the intramolecular O–H distance (from Rahman and Stillinger, 1972)

of a Ben-Naim–Stillinger interaction with another due to Rowlinson (1951). While the latter interaction seems more accurate, neither is able to account for the high dielectric constant of water.

7.6.2 Dynamics of dipolar systems

Various dynamic extensions of the Onsager theory of the dielectric constant of water, to describe dielectric relaxation occurring through rotational Brownian motion of the dipoles, have been presented in the literature. The simplest extension (Cole, 1938) includes the cavity and the reaction field by replacing the static dielectric function in favour of its dynamic generalisation $\varepsilon(\omega)$, and represents the mean-square effective dipole moment by a simple relaxation formula through a relaxation time τ, thus leading to the following expression for the dielectric function in the long-wavelength limit:

$$\frac{[\varepsilon(\omega)-\varepsilon_\infty][2\varepsilon(\omega)+\varepsilon_\infty]}{3\varepsilon(\omega)} = \frac{(\varepsilon_\infty+2)^2}{9}\frac{4\pi\rho}{3k_BT}\frac{\mu_v{}^2}{1-i\omega\tau} \tag{7.148}$$

The electronic polarisation is still included through the constant ε_∞, the theory thus being applicable at low frequencies at which real electronic transitions can be neglected.

Fatuzzo and Mason (1967) have discussed in more detail the processes of 'dielectric friction', that is, the energy loss by a dipole as it librates around its mean direction (fixed by the field E) through its coupling with the surrounding dielectric. The calculation of the extra torque acting on the dipole from dielectric

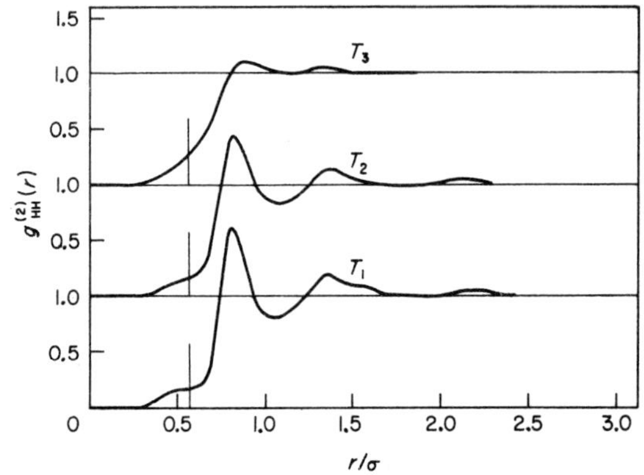

Figure 7.22 Radial correlation function $g_{HH}(r)$ in water, under the same thermodynamic conditions as in *figure 7.21*. The vertical line indicates the intermolecular HH distance (from Rahman and Stillinger, 1972)

friction leads to the following expression for the dielectric function,

$$\frac{\varepsilon_0[\varepsilon(\omega)-\varepsilon_\infty][2\varepsilon(\omega)+\varepsilon_\infty]}{(2\varepsilon_0+\varepsilon_\infty)(\varepsilon_0-\varepsilon_\infty)\varepsilon(\omega)}=\left\{1-i\omega\tau_0-\frac{(\varepsilon_0-\varepsilon_\infty)[\varepsilon(\omega)-\varepsilon_0]}{\varepsilon_0[2\varepsilon(\omega)+\varepsilon_\infty]}\right\}^{-1} \tag{7.149}$$

where τ_0 is a relaxation time accounting for ordinary viscous friction. Formally, this expression reduces to the Onsager–Cole formula (with $\tau_0 \to \tau$ and μ expressed in terms of ε_0 through (7.138)) when the last term in the brackets is omitted.

We shall here report in detail the discussion of dipolar motions given by Nee and Zwanzig (1970), which leads back to the Fatuzzo–Mason formula for spherically symmetrical Brownian rotation. Their model is a straightforward extension of the Onsager model, in that each molecule is thought of as a cavity bearing a permanent dipole μ and filled uniformly by a dielectric ε_∞ in a dielectric with dielectric function $\varepsilon(\omega)$. The magnitude of the dipole moment is taken as $\mu=\frac{1}{3}(\varepsilon_\infty+2)\mu_v$, so as to recover the Onsager result in the static limit. Then, if $\boldsymbol{E}(t)$ is the field far from the cavity, the field inside the cavity is

$$G(\omega)=\frac{3\varepsilon(\omega)}{2\varepsilon(\omega)+\varepsilon_\infty}E(\omega) \tag{7.150}$$

and the interaction of the dipole with this field can be treated as a perturbation on the natural motion of the dipole in the absence of external field, $\mu^{(0)}(t)$. Linear response theory then gives for the average dipole in the presence of the field

$$\begin{aligned}\langle\mu(t)\rangle &=(3k_BT)^{-1}\int_0^t ds\langle\dot{\mu}^{(0)}(0)\cdot\mu^{(0)}(s)\rangle G(t-s)\\ &=-(3k_BT)^{-1}\int_0^t ds\langle\mu^{(0)}(0)\cdot\dot{\mu}^{(0)}(s)\rangle G(t-s)\end{aligned} \tag{7.151}$$

having used translational invariance in time for the equilibrium correlation function entering the integral. By imposing the self-consistency condition that the above dipole moment give the polarisation in the field $E(t)$, namely

$$\langle \mu(\omega) \rangle = \frac{\varepsilon(\omega) - \varepsilon_\infty}{4\pi\rho} E(\omega) \tag{7.152}$$

in Fourier transform, one finds

$$\frac{[\varepsilon(\omega) - \varepsilon_\infty][2\varepsilon(\omega) + \varepsilon_\infty]}{3\varepsilon(\omega)} = -\frac{4\pi\rho\mu^2}{3k_B T} \int_0^\infty dt \exp(i\omega t) \frac{d}{dt} \phi(t) \tag{7.153}$$

with

$$\phi(t) = \langle \mu(0) \cdot \mu(t) \rangle / \langle \mu(0) \cdot \mu(0) \rangle \tag{7.154}$$

or (eliminating μ^2 in favour of ε_0)

$$\frac{\varepsilon_0[\varepsilon(\omega) - \varepsilon_\infty][2\varepsilon(\omega) + \varepsilon_\infty]}{(2\varepsilon_0 + \varepsilon_\infty)(\varepsilon_0 - \varepsilon_\infty)\varepsilon(\omega)} = -\int_0^\infty dt \exp(i\omega t) \frac{d}{dt} \phi(t) \tag{7.155}$$

The problem is thus to evaluate the time-dependent correlation function $\phi(t)$ in the absence of external field. The Onsager–Cole formula follows immediately if $\phi(t)$ is taken to decay exponentially.

To estimate $\phi(t)$ Nee and Zwanzig consider the rotational Brownian motion of the dipole. The frictional torque $N(t)$ arising from the coupling of the dipole with the surrounding dielectric is determined by the reaction field $R(t)$ as

$$N(t) = \mu(t) \times R(t) \tag{7.156}$$

where

$$R(\omega) = r(\omega)\mu(\omega), \qquad r(\omega) = \frac{8\pi\rho}{3} \frac{\varepsilon(\omega) - \varepsilon_\infty}{\varepsilon_\infty[2\varepsilon(\omega) + \varepsilon_\infty]} \tag{7.157}$$

The angular velocity of the dipole, $\Omega(\omega)$, is related to this torque through a generalised friction coefficient $\zeta(\omega)$ by

$$N(\omega) = -\zeta(\omega)\Omega(\omega) \tag{7.158}$$

with the quantity $D(\omega) \equiv k_B T/\zeta(\omega)$ having the meaning of a rotational diffusion coefficient of the dipole. Precisely, the rotational diffusion equation for the space density of the ensemble of particles, $\rho(t)$, has the form

$$\left.\frac{d\rho(t)}{dt}\right|_\omega = D(\omega)(u \times \nabla_u)^2 \rho(\omega) \tag{7.159}$$

where u is a unit vector specifying orientation, and the operator $u \times \nabla_u$ replaces the gradient operator entering the usual diffusion equation for translational Brownian motion. These equations are solved to give

$$\zeta(\omega)=\frac{\mu^2}{i\omega}[r(\omega)-r(0)]$$

$$=\frac{2k_{\mathrm{B}}T}{i\omega}\frac{\varepsilon_0-\varepsilon_\infty}{\varepsilon_0}\frac{\varepsilon(\omega)-\varepsilon_0}{2\varepsilon(\omega)+\varepsilon_\infty} \tag{7.160}$$

Adding to the rotational friction $\zeta(\omega)$ a term accounting for ordinary Stokes law friction and described by a relaxation time τ_0, one finally finds

$$-\int_0^\infty \mathrm{d}t\exp(i\omega t)\frac{\mathrm{d}\phi(t)}{\mathrm{d}t}=\left\{1-i\omega\tau_0-\frac{(\varepsilon_0-\varepsilon_\infty)[\varepsilon(\omega)-\varepsilon_0]}{\varepsilon_0[2\varepsilon(\omega)+\varepsilon_\infty]}\right\}^{-1} \tag{7.161}$$

an expression appropriate for spherically symmetrical rotational drift. Clearly, one regains the expression (7.149) for the dielectric function.

Lobo, Robinson and Rodriguez (1973) have used this approach to evaluate the dynamical properties of water and other dipolar fluids, at long wavelengths. Precisely, their expression for the dielectric function is

$$\varepsilon(\omega)-\varepsilon_\infty=\frac{\varepsilon(\omega)}{2\varepsilon(\omega)+\varepsilon_\infty}\frac{4\pi\rho\mu^2}{k_{\mathrm{B}}T}\left\{1-\frac{I^*\omega^2}{2k_{\mathrm{B}}T}-\frac{i\omega\tau_0}{1-i\omega\tau}-\frac{(\varepsilon_0-\varepsilon_\infty)[\varepsilon(\omega)-\varepsilon_0]}{\varepsilon_0[2\varepsilon(\omega)+\varepsilon_\infty]}\right\}^{-1} \tag{7.162}$$

where I^* is a suitably chosen moment of inertia. With the introduction of the inertial term, the above expression not only reduces to Onsager's formula in the static limit and reproduces the standard theory of dielectric relaxation at low frequencies, but also satisfies the exact sum rule (Lobo, Rodriguez and Robinson, 1967)

$$\frac{2}{\pi}\int_0^\infty \mathrm{d}\omega\,\omega\,\mathrm{Im}\,\varepsilon^{-1}(\boldsymbol{k},\omega)=-4\pi\rho\mu^2\left\{\langle(\hat{k}\times\hat{\mu})\cdot I^{-1}\cdot(\hat{k}\times\hat{\mu})\rangle+\frac{k^2}{m}\langle(\hat{k}\cdot\hat{\mu})^2\rangle\right\} \tag{7.163}$$

where I^{-1} is the inverse inertial tensor. This is equivalent to the second-moment sum rule given in (3.56) for a simple fluid and allows the theory to be applied at high frequencies. Similarly, a second relaxation time, τ is introduced in the Stokes term so as to obtain its correct behaviour both at low and at high frequency. In practice, while τ_0 can be related to the Debye relaxation time at low frequency, the ratio τ_0/τ is a parameter estimated to have the value $\tau_0/\tau\approx 13$ at room temperature.

By searching for the zeros and poles of the above dielectric function, Lobo et al. show that it exhibits a zero at a frequency

$$\omega_{\mathrm{p}}{}^2\approx\frac{4\pi\rho\mu^2}{I^*\varepsilon_\infty} \tag{7.164}$$

which corresponds to a longitudinal collective mode ('dipolar plasmon'), and a pole at a frequency

$$\omega_{\mathrm{t}}{}^2\approx\frac{k_{\mathrm{B}}T}{I^*}\left(1+\frac{\varepsilon_\infty}{\varepsilon_0}\right) \tag{7.165}$$

which corresponds to a transverse mode. The plasma frequency actually turns out in the model to be simultaneously a resonance for collective motions and for single-particle motions, so that the infrared resonances observed in dipolar fluids and commonly ascribed to hindered rotation or libration could also be interpreted according to the model in terms of dipolar plasmons.

Considering, in particular, the comparison with experiment for water, the available data (Eisenberg and Kauzmann, 1969; see also the extensive review work on water by Franks, 1973) show a broad peak in the infrared absorption spectrum at an angular frequency of 1.3×10^{14} sec^{-1} (0.95 in D_2O) and a peak in the neutron incoherent scattering at 1.1×10^{14} sec^{-1} (0.85 in D_2O), while the dipolar plasmon is found to lie at 1.18×10^{14} sec^{-1} for $\tau_0/\tau \approx 10$ (0.81 in D_2O). The comparisonwith the molecular dynamics results of Rahman and Stillinger is gratifying in that they show the existence of oscillations in the polarisation charge density, characteristic of a longitudinal collective mode, at a frequency of 1.6×10^{14} sec^{-1}.

Comparison of the theory with experiment for other dipolar fluids is presented in the paper of Lobo, Rodriguez and Robinson, 1973. For a review of experimental work on molecular fluids, we refer again to the article of Allen and Higgins (1973).

7.6.3 Ions in water

The statistical mechanical study of deviations from the primitive model of electrolytes due to the granularity of the solvent is still a relatively undeveloped subject, in spite of its relevance to phenomena such as solvation and ion hydration, which are of great interest to electrochemistry (see, for example, Hinton and Amis, 1971; Eliezer and Krindel, 1972). Early work applicable at low density was carried out by Jepsen and Friedman (1963) and a number of phenomenological approaches stemming from the cosphere concept of Gurney (1953) and Frank (1966) have been developed. The situation has been reviewed recently by Stell (1973), who has analysed the potentials of mean force between solute ions in a dipolar solvent. The primitive model assumes for this potential for a pair of ions the form

$$U(12) = \phi(12) + \frac{z_1 z_2 e^2}{\varepsilon r} \tag{7.166}$$

where $\phi(12)$ is the short-range potential between the two ions *in vacuo* and ε is the dielectric constant of the solvent.

Stell notes that the dielectric screening in (7.166) arises from the charge–dipole interaction between the solutes and the solvent. If this interaction were absent one would expect the ionic distribution in space to have pronounced wiggles at high density as in a rigid-ion model of molten salts, while these wiggles are moderated by dielectric screening, to which the primitive model gives primary attention. One may thus expect that the real ionic distribution will be qualitatively intermediate between that provided by the primitive model and that provided by a model omitting charge–dipole interactions.

Stell's analysis is then deepened by considering a model in which the solvent is allowed to become a continuous dielectric as far as the solvent–solvent interactions are concerned, but retains its granular nature in the solute–solvent interactions. This allows him to isolate, among the long-range part of the potential of mean force between two ions, not only the screened-Coulomb term of the primitive model, but also a 'cavity term' of the form A/r^4 already discussed by Jepsen and Friedman (1963). The remaining terms are essentially short range and included in the cosphere concept. A description of this region requires a truly microscopic picture: not only are the short-range potentials at play, but one may expect that, given the strong electric field in the immediate vicinity of an ion, there will be a substantial structural reorganisation and electronic distortion of the dipolar solvent. Molecular dynamics is clearly the appropriate tool, and one may expect considerable progress in the near future, in view of the recent progress in the understanding of pure water on the one hand and of molten salts on the other. It may here be noted that a series-expansion solution of the mean spherical model for a mixture of equi-sized hard spheres carrying charges and dipoles has been given recently by Blum (1974; see also Adelman et al. (1974)).

On the experimental side, the method of neutron scattering with isotopic substitution has been used by Howe, Howells and Enderby (1974) on concentrated solutions of $NiCl_2$ in D_2O to derive the partial structure factor pertaining to the Ni–Ni correlations. This is found to contain pronounced structure, and in particular a strong Ni–Ni first-neighbour peak, corresponding to a mean distance of 6 Å in the most concentrated solutions, the peak being comparable in height to the analogous peak observed in molten alkali halides at cation–cation distances of 4–5 Å by both neutron scattering and computer simulation techniques (see section 7.3.2). This finding suggests a remarkable degree of ordering for ions in concentrated solutions, at least in the case of transition metal ions. Such ions are known from diffraction experiments on hydrated salts in the crystalline state to have a strong tendency to formation of complexes with water molecules. A commonly found complex involves a divalent metal ion surrounded by four water molecules and two halogen ions in an octahedral configuration, the various complexes being locked in a crystalline structure via halogen–hydrogen–oxygen bonds and also via more complex hydrogen bonds involving water molecules in interstices of the lattice. It has been suggested (March & Tosi, 1974) that similar bonds may be operative in concentrated solutions of transition metal ions to produce ionic ordering.

Finally, we should record that an analysis relating the partial structure factors at long wavelength in an aqueous solution to thermodynamic properties has been given by Beeby (1973). On account of neutrality conditions analogous to the ones discussed for molten salts in section 7.3.2, the various structure factors are all related to the mean square fluctuations in the numbers of solute and solvent molecules, and their cross-correlation, these in turn being then related, via arguments similar to the Kirkwood–Buff argument discussed in appendix 6.1, to thermodynamic properties of the solution such as its compressibility and the partial molar volumes of the components.

CHAPTER 8
HELIUM LIQUIDS

8.1 Feynman theory of liquid helium-four

The theory of liquid He^4 given by Landau (see for example March, Young and Sampanthar, 1967) was based on the idea of well-defined elementary excitations which were phonon-like for small wave vector and of a more complex (roton) character at larger wave vectors, around the peak of the static structure factor $S(k)$, as needed to account for superfluid properties. This description was made fairly precise by the work of Feynman (1954), which we shall first briefly summarize.

Following the discussion in chapter 5, we can express the Hamiltonian of a non-viscous compressible fluid in the approximate form

$$H=\frac{1}{2}\sum_k \frac{m}{Nk^2}\left[\dot{\rho}_k\dot{\rho}_{-k}+\omega_k{}^2\rho_k\rho_{-k}\right] \tag{8.1}$$

where ρ_k, as usual, are the density fluctuations, while $\omega_k=v_s k$ for small k, v_s being the velocity of sound. In this approximation, it can be seen that H describes independent oscillators.

Now we have seen earlier that the structure factor $S(k)$ is essentially the expectation value of $N^{-1}\rho_k\rho_{-k}$ and since the average values of kinetic and potential energies are equal, from the virial theorem, for a harmonic oscillator, we can write

$$\frac{mv_s{}^2}{N}\langle\rho_k\rho_{-k}\rangle=\langle E_k\rangle \tag{8.2}$$

$\langle E_k\rangle$ being the average energy of the oscillator representing sound of wave number k. At absolute zero, all the oscillators are in their ground states and hence

$$\langle E_k\rangle=\tfrac{1}{2}\hbar\omega_k=\tfrac{1}{2}\hbar v_s k \tag{8.3}$$

The result for the structure factor $S(k)$ then follows immediately as

$$S(k)=\frac{\langle E_k\rangle}{mv_s{}^2}=\frac{\hbar k}{2mv_s} \tag{8.4}$$

This result has been proved by Gavoret and Nozieres (1964) to be exact for sufficiently small k.

Using Lighthill's (1958) work on Fourier transforms it is easy to show that the asymptotic form of the total correlation function $h(r)=g(r)-1$ follows almost immediately as

$$h(r)=-\hbar(2\pi^2\rho m v_s r^2)^{-1} \tag{8.5}$$

The above result for the structure factor $S(k)$ of the ground state is drastically modified in going to elevated temperatures. For then the probability of finding the oscillator representing phonons of wave number k in its nth excited state is proportional to $\exp(-E_n/k_BT)$ and it follows straightforwardly that

$$\begin{aligned} S(k) &= \frac{\hbar k}{2mv_s}\coth\left(\tfrac{1}{2}\hbar v_s k/k_BT\right) \\ &\cong \frac{k_BT}{mv_s{}^2}+\frac{\hbar^2k^2}{12mk_BT}+\cdots \end{aligned} \tag{8.6}$$

so that as $k\to 0$ we find

$$S(k)=\rho k_BTK_T \tag{8.7}$$

(provided we put $c_p/c_v\cong 1$) which is once again the result of classical fluctuation theory. It should be noted that the term linear in k has disappeared from the above expansion valid at elevated temperatures.

Since the above model describes independent phonons, we expect that $S(k, \omega)$ will simply show delta function behaviour at the phonon mode and we can write in the low temperature limit

$$S(k,\omega)=2\pi S(k)\delta(\omega-v_sk) \tag{8.8}$$

which, of course, immediately satisfies the zeroth moment

$$\int\frac{d\omega}{2\pi}S(k,\omega)=S(k)$$

as it should. The quantal first moment is given by

$$\int_0^\infty \omega S(k,\omega)\frac{d\omega}{2\pi}=\frac{\hbar k^2}{2m} \tag{8.9}$$

and substituting the model dynamical structure factor into this result we find again the Feynman result (8.4) for $S(k)$.

However, the above argument can now be generalised away from the long wavelength limit if we assume that elementary excitations with a more general dispersion relation exist. That is, we simply generalise the phonon model of the dynamic structure factor to read

$$S(k,\omega)=2\pi S(k)\delta(\omega-\omega_k) \tag{8.10}$$

when we find from the first moment (8.9) that the eigen frequencies of the elementary excitations are given in terms of the static structure factor by

$$\omega_k = \frac{\hbar k^2}{2mS(k)} \tag{8.11}$$

Using $S(k)$ from experiment (see *figure 8.1*) this dispersion curve has the qualitative features of the Landau excitations, and in particular shows a 'roton minimum' near the wave number of the peak in $S(k)$. Agreement with the observed dispersion curve, that we shall discuss in section 8.4, is quantitatively poor, but, as we shall see, the basic idea of elementary excitations in liquid He^4 remains valid provided that it is supplemented by allowance for their interactions leading to a 'multiphonon' component in the spectrum. In particular, it is the latter part of the spectrum which tends to the free particle behaviour

$$\omega_k \rightarrow \frac{\hbar k^2}{2m} \tag{8.12}$$

for $k \rightarrow \infty$, rather than the one-phonon spectrum of eqns (8.10) and (8.11).

8.2 Form of ground-state wave function

Similar methods to those developed earlier for classical liquids are highly relevant in discussing the properties of the ground state of liquid He^4. Ideally, of course, one would like to calculate $S(k)$ or $g(r)$ from first principles, given only the pair interaction $\phi(r)$. However, in the absence of such calculations, insight can be gained into the problem by making direct use of the measured $S(k)$.

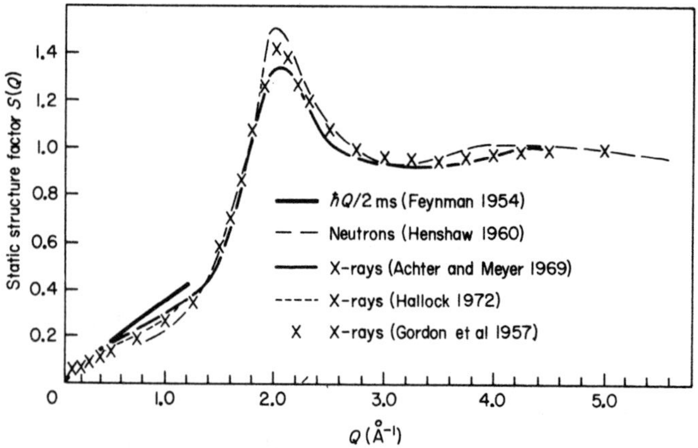

Figure 8.1 Static structure factor of liquid He^4, as determined by various experiments (from Woods and Cowley, 1973). The references shown are given in Woods and Cowley's paper

We can, in fact, obtain rather direct information about the ground-state wave function in liquid He^4, provided we assume for the boson fluid the Jastrow–Bijl form

$$\Psi(\boldsymbol{R}_1, \ldots, \boldsymbol{R}_N) = \prod_{i<j} \exp\left[\tfrac{1}{2}u(R_{ij})\right] \tag{8.13}$$

Forming $\Psi^*\Psi$ and hence obtaining the various distribution functions, it can be seen that there is a complete analogy with classical statistical mechanics. Then the approximate theories of structure that we discussed earlier are immediately applicable, with $-\phi(r)/k_BT$ in the classical theory replaced by $u(r)$. In particular at large r, since $c(r) \sim -\phi(r)/k_BT$ we have that $u(r)$ behaves asymptotically as $c(r)$. But we have

$$\tilde{c}(k) = \frac{S(k)-1}{S(k)} \rightarrow -\frac{1}{S(k)} = -\frac{2mv_s}{\hbar k} \tag{8.14}$$

at small k. Thus, it follows immediately by Fourier transform that

$$c(r) \sim u(r) \sim -\frac{mv_s}{\hbar\pi^2\rho} r^{-2} \tag{8.15}$$

The correlations are seen then, in the Jastrow–Bijl wave function, to be of very long range, as a consequence of the existence of well-defined long-wavelength phonons.

Use of the experimental data for $S(k)$ will, in principle, allow us to obtain $u(r)$ for all r. Thus, using the Abe (HNC) approximation, which reads in this case

$$u(r) - \ln g(r) = h(r) - c(r) \tag{8.16}$$

we can extract $u(r)$ from the experimental data. The result, first obtained by Wu and Feenberg (1961), is shown in *figure 8.2*. The function $u(r)$ is seen to become strongly negative inside the core, as is to be expected. Actually, the above equation can be written explicitly in terms of $S(k)$ as

$$u(r) = \ln g(r) - \frac{1}{8\pi^3\rho} \int d\boldsymbol{r} \exp(-i\boldsymbol{k}\cdot\boldsymbol{r}) \frac{[S(k)-1]^2}{1+\xi[S(k)-1]} \tag{8.17}$$

with $\xi = 1$. The factor ξ was introduced by Wu and Feenberg to facilitate comparison with their numerical solution of the Born–Green equation, also shown in the figure. Unfortunately, however, the choice $\xi \neq 1$ spoils the asymptotic result.

The conclusion is that the approximate theories lead to two-particle correlation functions $u(r)$ in fair agreement with one another. These can then be used to calculate the ground state energy since it is readily verified that with the Jastrow–Bijl function the mean kinetic energy $\langle T \rangle$ is given by

$$\langle T \rangle = \frac{N\rho\hbar^2}{8m} \int d\boldsymbol{r}\, g'(r) u'(r) \tag{8.18}$$

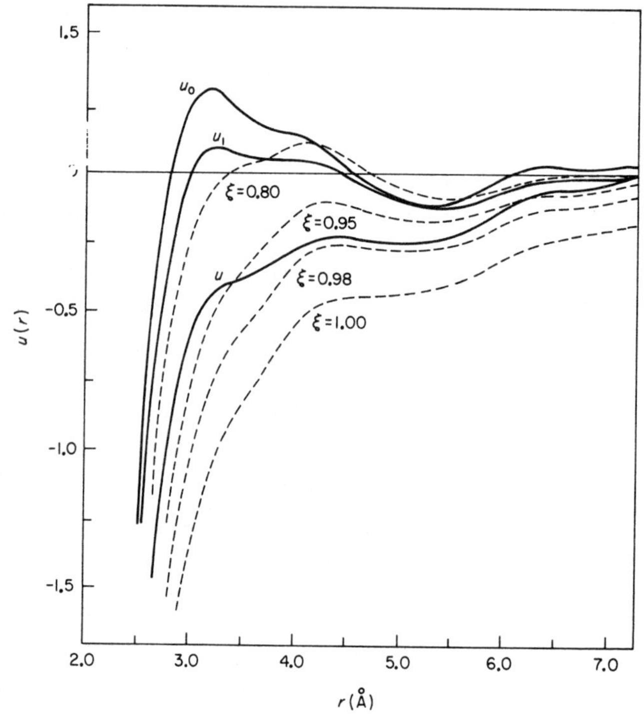

Figure 8.2 Two-body function $u(r)$ as determined by Wu and Feenberg (1961). The full curve, marked u, is based on the superposition approximation, while the broken curve, marked $\xi = 1$, is based on the Abe approximation

and the mean potential energy $\langle\phi\rangle$ is

$$\langle\phi\rangle = \tfrac{1}{2}N\rho \int \mathrm{d}\boldsymbol{r}\, g(r)\phi(r) \tag{8.19}$$

Wu and Feenberg obtained reasonably satisfactory results for the ground-state energy and the conclusion is that a very useful approximation is available to the true ground-state wave function for He^4. For further discussion the reader is referred to the work of Reatto and Chester (1967) and to the book by Feenberg (1969).

8.3 Excitation spectrum of liquid helium-four

Very extensive and detailed experimental information has been obtained on the excitation spectrum of density fluctuations in liquid He^4 through ultrasonic measurements, Brillouin and Raman light scattering, and inelastic neutron scattering. Of course, ultrasonic and Brillouin scattering measurements focus on the properties of the phonon-like excitations, which in both He^4 and He^3 at sufficiently low temperatures have a very long lifetime τ such that $\omega\tau \gg 1$ even at ultrasonic frequencies. This is, of course, the regime of strict validity of the Feynman theory. We shall discuss the properties of collisionless (or 'zero') sound later for the quantal fluid He^3, and refer the reader to the recent article of Woods and Cowley (1973) for a review of the experimental and theoretical work on sound wave excitations in liquid He^4. We shall instead review the main features of the neutron and Raman spectrum of liquid He^4 and their interpretation; more details on these topics can be obtained from the above-mentioned article of Woods and Cowley.

Figure 8.3 reports the distribution of inelastically scattered neutrons for increasing values of the momentum transfer, as measured by Cowley and Woods (1971) and by Woods, Svensson and Martel (1972). The results refer to a temperature of 1.1 K, but are felt to be representative of zero temperature as no significant temperature dependence is observed below 1.5 K. At small wave vectors ($k \lesssim 0.4$ Å^{-1}) the main feature of the spectrum is a very sharp peak, whose width is determined by the experimental resolution. A second, much broader peak at higher energy gradually grows in intensity as k increases, becoming comparable in height to the first peak for $k \sim 2.5$ Å^{-1} and remaining the only outstanding feature of the spectrum for $k \gtrsim 3.5$ Å^{-1}. The position of the first sharp peak as a function of wave number is reported in *figure 8.4* (from Cowley and Woods, 1971), while *figure 8.5*, from the work of the same authors, shows the excitation spectrum in the (ω, k) plane, the hatched area corresponding to the second, broad peak.

It is natural to assume, in view of these observations, that the dynamic structure factor can be written as the sum of a 'single-excitation' spectrum, corresponding to the sharp peak, and of a 'multi-excitation' spectrum, corresponding to the broad peak:

$$S(k, \omega) = 2\pi Z(k)\delta(\omega - \omega_k) + S_{\text{II}}(k, \omega) \tag{8.20}$$

Since the multi-excitation spectrum is expected to behave as k^4 for long wavelengths (Miller, Pines and Nozières, 1962), the strength $Z(k)$ of the single-excitation spectrum coincides in this limit with the static structure factor, namely

$$\lim_{k\to 0} [Z(k) = S(k)] = \frac{\hbar k}{2mv_s} \tag{8.21}$$

according to (8.4). The extent to which this relation is verified in the neutron experiments is illustrated in *figure 8.6*, while *figure 8.7* reports the shape of $Z(k)$ as a function of wave number as determined in the same experiments. The strength of the single-excitation peak becomes very small for $k \gtrsim 2.5$ Å^{-1} and vanishes for $k \gtrsim 3.5$ Å^{-1}; correspondingly, the dispersion curve in *figure 8.4* bends over and flattens out at a value of the energy which is very close to twice

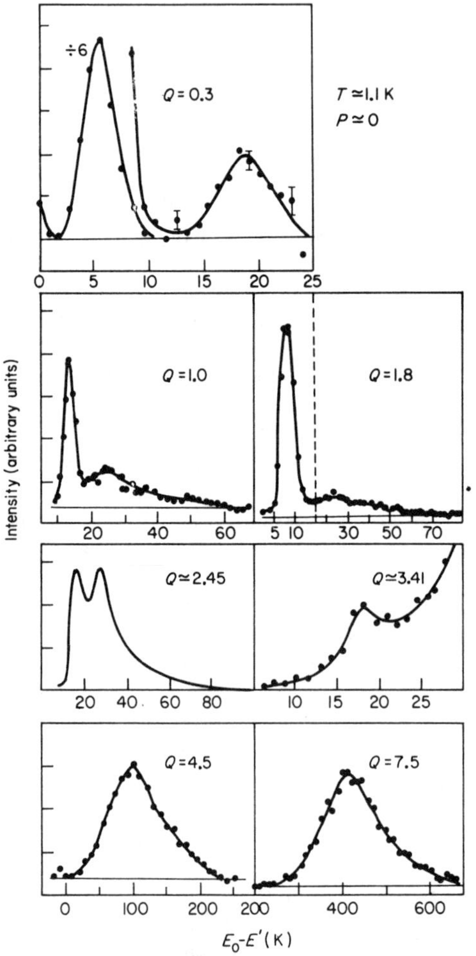

Figure 8.3 Energy distribution of scattered neutrons from liquid He^4 at various momentum transfers (from Woods and Cowley, 1973)

the energy Δ of the roton minimum. This behaviour can be interpreted in accord with the early predictions of Pitaevskii (1959) as due to decay of a single excitation into two excitations. Precisely, such decay may occur either when the group velocity of the excitation begins to exceed the sound velocity, which experimentally occurs at $k = 2.27\ \text{Å}^{-1}$ when the dispersion curve in fact begins to bend over, or when the excitation energy begins to exceed 2Δ, at which point

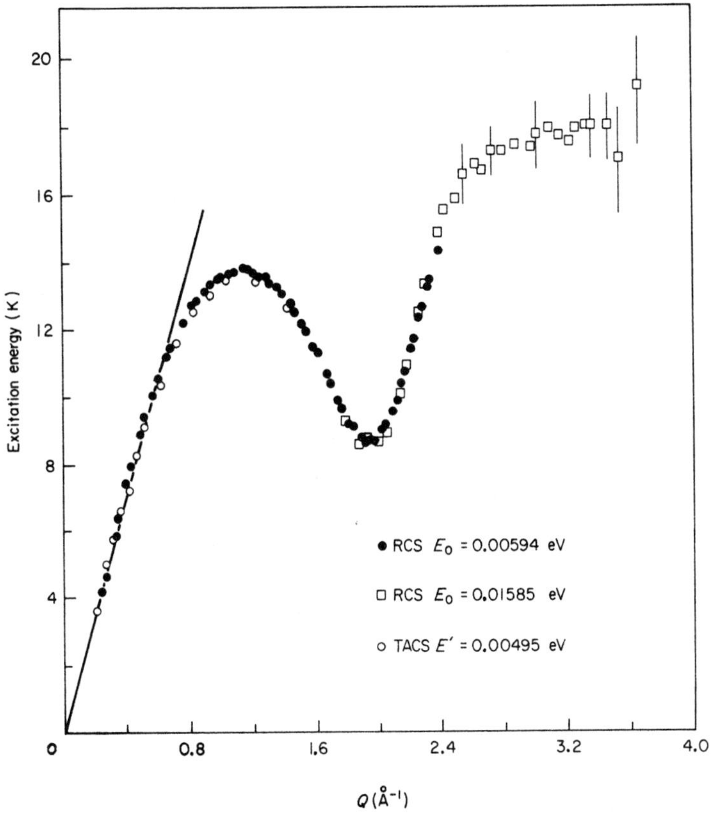

Figure 8.4 Dispersion relation of the single-excitation spectrum in liquid He^4, as determined by neutron inelastic scattering (from Cowley and Woods, 1971)

the single-excitation peak should cease to exist through processes of decay into two rotons.

The multi-excitation spectrum is expected to be dominated by two-excitation processes at long wavelength, while higher order processes should become increasingly important as k increases, as evidenced by the large spread in energy transfers shown by the multi-excitation neutron spectrum in *figure 8.5*. Very detailed information on $S(k, \omega)$ at high frequencies and long wavelengths is available through Raman scattering experiments: as shown in *figure 8.8* (from Greytak et al., 1970), the Raman spectrum peaks at energies slightly lower than 2Δ, which has been interpreted as evidence for the existence of an attractive

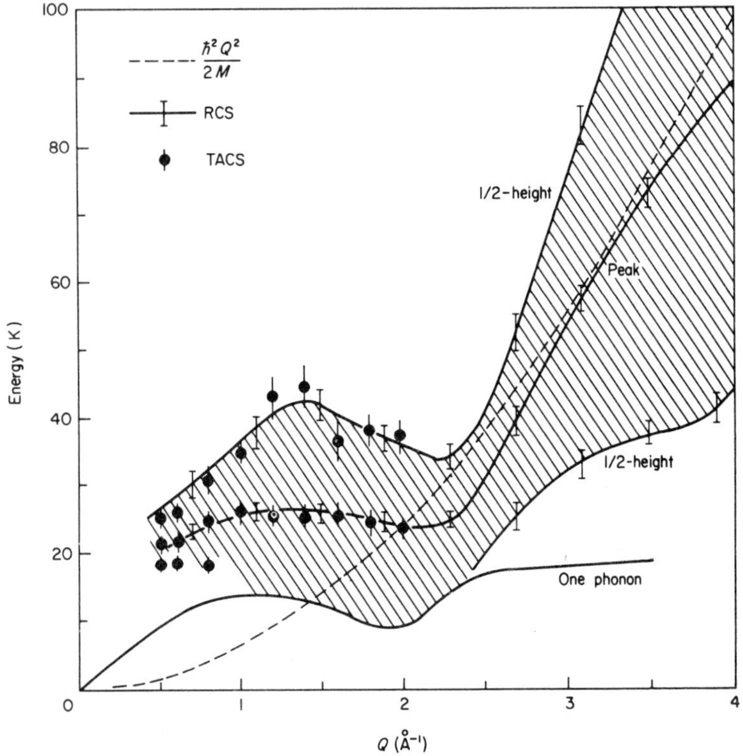

Figure 8.5 Excitation spectrum of liquid He^4 in the (ω, k) plane, as determined by neutron inelastic scattering (from Cowley and Woods, 1971). The hatched area describes the broad 'multi-excitation' peak within its half-width. The broken line is the free-particle dispersion relation, $\omega_k = \hbar k^2/2m$

interaction between two rotons. Approximate theories which include these interactions (see for example Zawadowski, Ruvalds and Solana, 1972) are in satisfactory agreement with the Raman data.

Roton–roton interactions, as well as interference with the single-excitation spectrum, will of course affect the detailed shape of the multi-excitation spectrum in the intermediate range of wave number. At large wave numbers, however, this spectrum should approach an independent-particle regime, with a Doppler-broadened peak at energy $\hbar^2k^2/2m$. This behaviour is indicated in *figure 8.5* by the broken curve, and while the data approximately approach this situation there are deviations in detail in the peak position, as well as oscillations in the peak width, which extend at least up to $k \sim 10$ Å^{-1}. Theoretical analyses of this

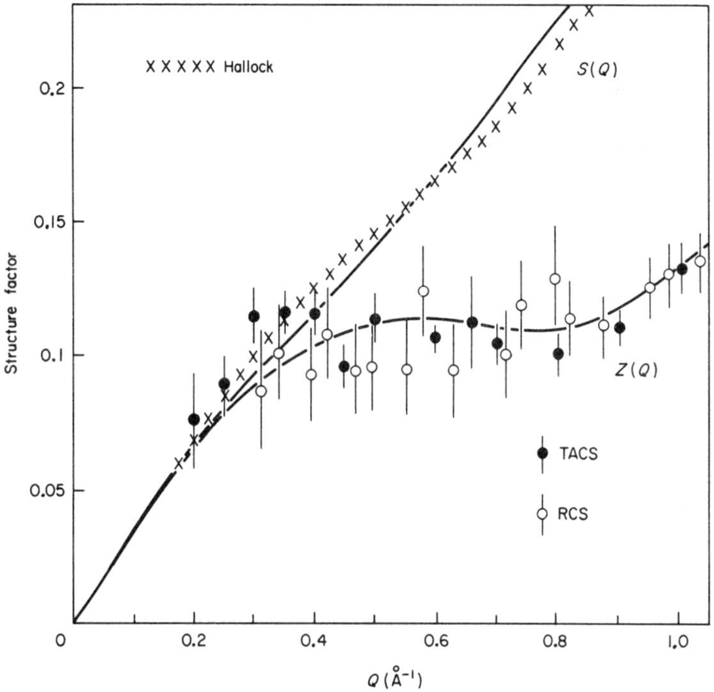

Figure 8.6 Static structure factor and one-phonon strength $Z(k)$ at long wavelengths in liquid He^4 (from Cowley and Woods, 1971)

behaviour have been given, for instance, by Kerr, Pathak and Singwi (1970), on the basis of the mean-field theory of section 5.4.2, suitably extended to account for the quantal character of the single-particle momentum distribution.

Interest in scattering with large momentum transfers from liquid He^4 arises (Hohenberg and Platzman, 1966) from the possibility of observing the presence of a condensate of a macroscopic number of atoms in the zero-momentum state. All the atoms in a perfect Bose gas would be in such state at low enough temperature, but the shape of the single-particle momentum distribution in real liquid He^4 is strongly distorted by the interactions from this ideal behaviour. The presence of a condensate should contribute a sharp peak at energy $\hbar^2k^2/2m$ superposed on the Doppler-broadened peak at the same energy contributed by atoms in $p \neq 0$ states. It appears, however, that such experiment might require somewhat larger momentum transfers than are explored at present. Available evidence on the condensate is still indirect, based on approximate analyses of the scattering also as a function of temperature, which lead to estimates mostly

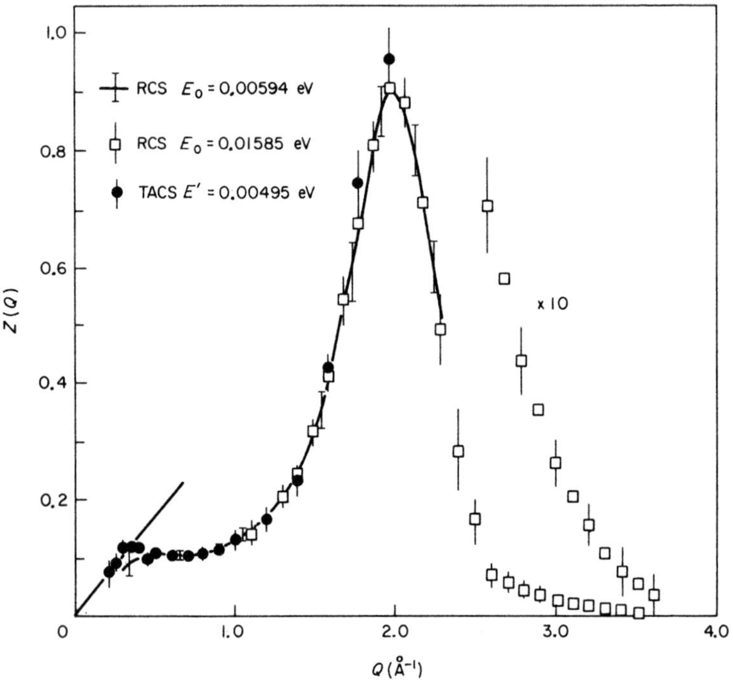

Figure 8.7 Single-excitation strength in liquid He^4 from neutron inelastic scattering (from Cowley and Woods, 1971)

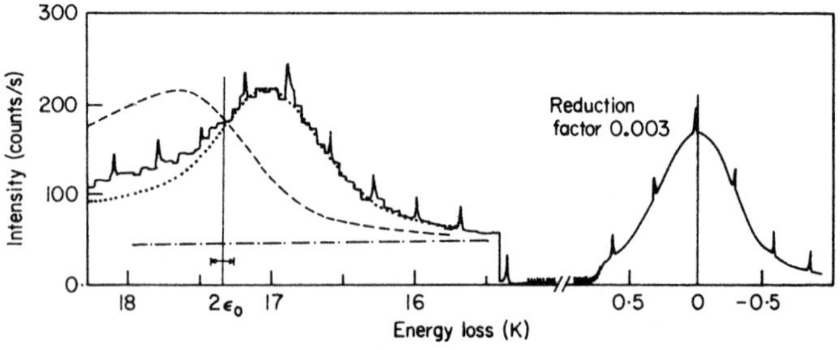

Figure 8.8 Brillouin (right-side peak) and Raman light scattering from liquid He^4 (from Greytak et al., 1970). ε_0 is the energy of the roton minimum. The dotted line represents a theoretical fit which includes the effects of roton–roton interactions (omitted in the broken curve)

lying between 0.06 and 0.17, with large uncertainties, for the fraction of atoms in the zero-momentum state.

Considerable experimental evidence on the pressure and temperature dependence of the excitation spectrum of liquid He^4 is also available, and is reviewed in the article of Woods and Cowley (1973). Of particular interest are the data reported in figure *8.9* (from Woods, 1965), showing the temperature dependence of the single-excitation peak at $k = 0.38$ Å^{-1}. The peak clearly broadens with increasing temperature, but distinctly persists above the λ point and is still visible near the boiling point as a shoulder on the quasi-elastic peak. An interpretation of these data in terms of mean-field theory has been proposed by Pines (1966). Thus, this part of the excitation spectrum is not intimately related to the characteristics of superfluidity, but is rather similar to the collective modes

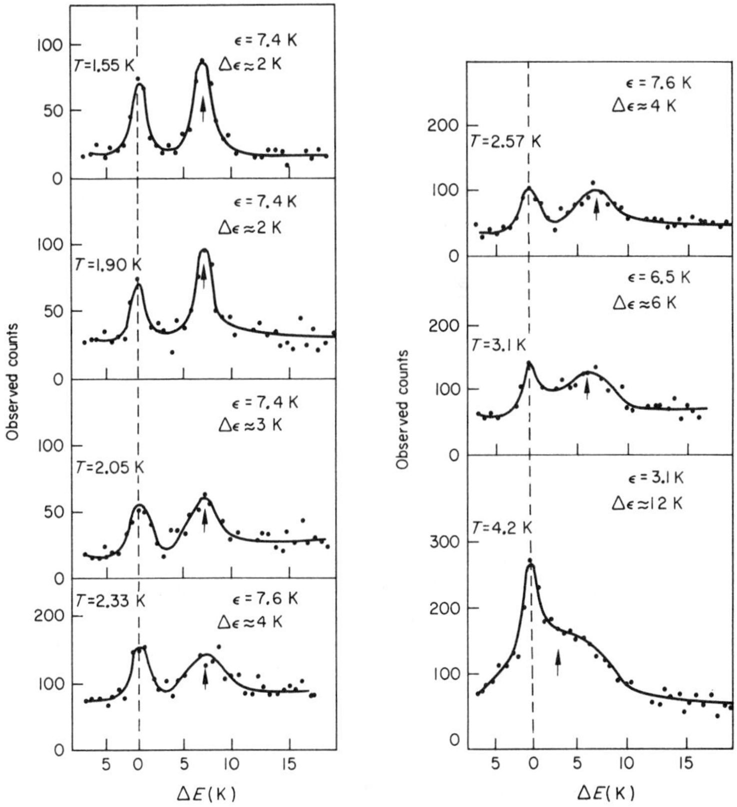

Figure 8.9 Temperature dependence of the one-phonon spectrum of liquid He^4 at $k = 0.38$ Å^{-1} (from Woods, 1965)

observed in the same range of wave vector in classical liquids such as rubidium (see chapter 4). The narrow roton excitation is, instead, related to superfluidity: the width of this excitation is indeed observed to increase sharply as the fluid is heated towards the λ point (Dietrich et al., 1972; see *figure 8.10*).

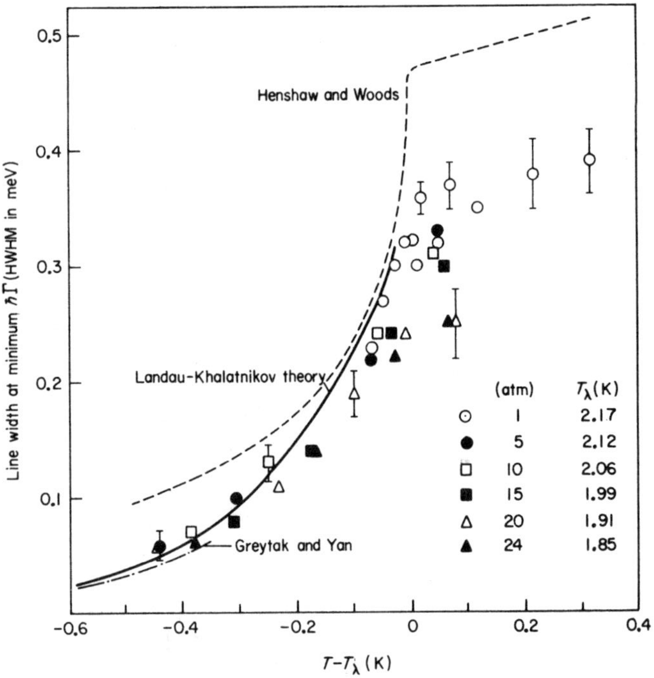

Figure 8.10 Temperature dependence of roton line width (from Dietrich et al., 1972)

8.4 Theories of the van Hove function for liquid helium-four

It was early realised by Feynman and Cohen (1956) that the simple Feynman picture of independent elementary excitations in liquid He^4 had to be extended to account for interactions between the excitations. The wave function of a Feynman excitation is given by the density fluctuation operator ρ_k acting on the ground-state wave function Ψ,

$$\Psi_k = \rho_k \Psi \tag{8.22}$$

except for a normalisation factor which is obviously given by $[NS(k)]^{1/2}$ if Ψ is normalised. The current associated with this state (or, more precisely, with a localised excitation built as a wave packet of such states) is soon shown to be

$\hbar \boldsymbol{k}/m$, whereas the current should be divergence-free for a true stationary state. The missing current resides in the backflow of the fluid around the localised 'bare' excitation.

To see how backflow can be built into the theory, let us consider with Feynman and Cohen the simpler case of a particle tearing through the liquid with velocity v. For weak interactions, we may regard the particle as an external perturbation described by a density

$$\rho_e(\boldsymbol{r}, t) = \delta(\boldsymbol{r} - vt) \tag{8.23}$$

or, in Fourier transform

$$\rho_e(\boldsymbol{k}, \omega) = 2\pi\delta(\omega - \boldsymbol{k} \cdot v) \tag{8.24}$$

the associated current being

$$I_e(\boldsymbol{k}, \omega) = 2\pi v\delta(\omega - \boldsymbol{k} \cdot v) \tag{8.25}$$

The distortion of the liquid around the perturbation is given by linear response theory by

$$\rho(\boldsymbol{k}, \omega) = \chi(k, \omega)\tilde{\phi}(k)\rho_e(\boldsymbol{k}, \omega) \tag{8.26}$$

$\tilde{\phi}(k)$ being the Fourier transform of the pair interaction. By the continuity equation we find the longitudinal current in the liquid to be

$$\begin{aligned} I(\boldsymbol{k}, \omega) &= \frac{\omega}{k^2} \boldsymbol{k}\rho(\boldsymbol{k}, \omega) \\ &= 2\pi\boldsymbol{k} \frac{\boldsymbol{k} \cdot v}{k^2} \chi(k, \omega)\tilde{\phi}(k)\delta(\omega - \boldsymbol{k} \cdot v) \end{aligned} \tag{8.27}$$

For small velocity v we can take $\omega = 0$ in the response function entering (8.27), and for $\boldsymbol{k} \to 0$ we can then use the result (Pines, 1966; see eqn 1.27′ in section 3.3.2)

$$\lim_{k \to 0} \chi(k, 0) = -\frac{N}{m v_s^2} \tag{8.28}$$

which is the equivalent, for a quantal fluid, of the Ornstein–Zernike relation (1.27) for a classical fluid. If we further assume the interaction to be short ranged and thus replace $\tilde{\phi}(k)$ by a constant $\tilde{\phi}$, we then find by Fourier transform of (8.27) that the induced longitudinal current far from the particle has the form of a dipolar backflow,

$$I(\boldsymbol{r}, t) \sim \frac{\tilde{\phi}}{4\pi} \frac{N}{m v_s^2} \nabla(v \cdot \nabla) \left.\frac{1}{R}\right|_{\mathbf{R} = \mathbf{r} - \mathbf{v}t} \tag{8.29}$$

The above derivation is due to Pines and Nozières (1966), who also show that for a *charged* fluid the induced current (8.27) is such as to exactly compensate the current (8.25) in the limit $v \to 0$, $\boldsymbol{k} \to 0$.

To apply the above calculation to a localised Feynman excitation, we can take from (8.1) for small k

$$\tilde{\phi}(k) \sim m v_s{}^2/N \tag{8.30}$$

showing that the total longitudinal current, $I(\boldsymbol{k}, \omega) + \boldsymbol{k}[\boldsymbol{k} \cdot I_e(\boldsymbol{k}, \omega)]/k^2$, vanishes. Feynman and Cohen accordingly write the wave function of an excitation in the form

$$\begin{aligned} \Psi_k &= \left[\sum_i \exp(i\boldsymbol{k} \cdot \boldsymbol{R}_i) \exp\left(i \sum_{j(\neq i)} f(R_{ij}) \right) \right] \Psi \\ &\cong \left\{ \sum_i \exp(i\boldsymbol{k} \cdot \boldsymbol{R}_i) \left[1 + i \sum_{j(\neq i)} f(R_{ij}) \right] \right\} \Psi \end{aligned} \tag{8.31}$$

and choose the function $f(r)$ in the form

$$f(r) = \mathrm{A}\, \boldsymbol{k} \cdot \boldsymbol{r}/r^3 \tag{8.32}$$

as imposed by dipolar flow at large r. Although they determine the constant A variationally, its value is found to differ by only 3% from the value given by the macroscopic calculation sketched above. The result is a marked improvement of the single-excitation dispersion curve, especially for wave numbers in the region of the roton minimum and lower.

Very similar results are obtained by a more conventional treatment of the interactions between Feynman excitations, developed by Jackson and Feenberg (1962); see also the book of Feenberg (1969). We can write the Hamiltonian of the excitations in liquid He^4 in much the same way as the Hamiltonian of an anharmonic solid, in the form

$$H = \sum_k \hbar \omega_k{}^0 a_k^+ a_k + \sum_{kk'} V_{kk'} [a_{k'}^+ a_{k-k'}^+ a_k + a_k^+ a_{k'} a_{k-k'}] + \cdots \tag{8.33}$$

where a_k and a_k^+ are the annihilation and creation operators for an elementary excitation of wave vector $\boldsymbol{k}$. The first term is the Feynman Hamiltonian discussed in section 8.1, and $\omega_k{}^0$ should therefore be identified with the Feynman frequency given in (8.11). The next term allows for processes such as the decay of an excitation into two excitations, via an interaction matrix element $V_{kk'}$. Higher order interactions will be given by successive terms in the expansion,

Jackson and Feenberg evaluate the energy correction due to the truncated Hamiltonian (8.33) by second-order Brillouin–Wigner perturbation theory, with the result

$$\omega_k = \omega_k{}^0 + \tfrac{1}{2} \sum_{k'} \frac{|V_{kk'}|^2}{\omega_k - \omega_{k-k'}^0 - \omega_{k'}^0} \tag{8.34}$$

Both this energy shift and the real decay processes allowed by the Hamiltonian (8.33) are compactly described by writing the self-energy correction for an elementary excitation in the form

$$\sum(k,\omega)=\tfrac{1}{2}\sum_{k'}|V_{kk'}|^2\left\{\frac{1}{\omega-\omega^0_{k-k'}-\omega^0_{k'}}+i\pi\delta(\omega-\omega^0_{k-k'}-\omega^0_{k'})\right\} \tag{8.35}$$

in terms of which the van Hove function is then given by

$$S(k,\omega)=2\,\mathrm{Im}\,\frac{1}{\omega-\omega_k{}^0-\Sigma(k,\omega)} \tag{8.36}$$

The results of the calculation of Jackson and Feenberg, based on estimates of the matrix element $V_{kk'}$ in terms of the measured structure factor $S(k)$, are reported in *figure 8.11*. The dispersion curve is considerably improved over the simple

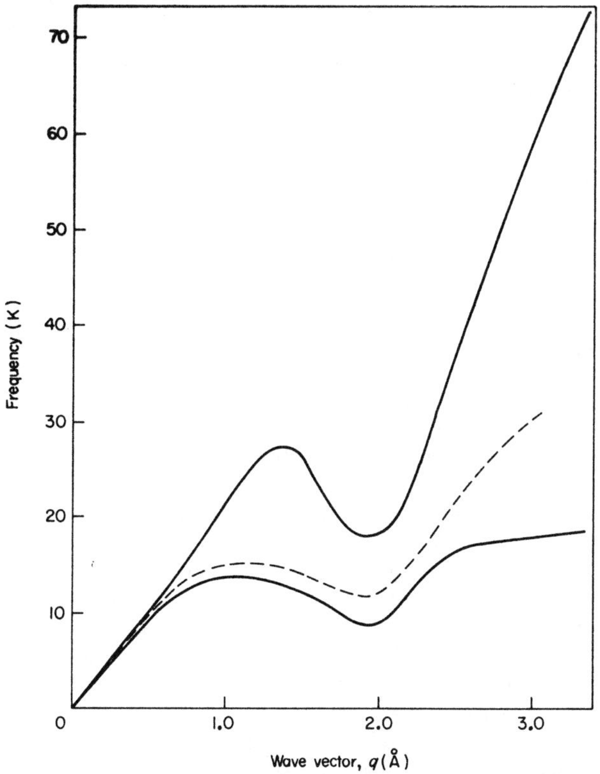

Figure 8.11 Comparison between the experimental dispersion curve for elementary excitations in liquid He^4 and the theoretical results of Feynman (higher continuous curve) and Jackson and Feenberg (broken curve) (from Woods and Cowley, 1973)

calculation of Feynman, although the agreement with experiment is still not very good, especially at large wave vectors where the higher-order processes should be becoming more important.

The new neutron measurements that we have reviewed in the preceding section have renewed great interest in more precise calculations of the van Hove function in liquid He^4, based on extensions of the approach of Jackson and Feenberg. Thus, Zawadowski, Ruvalds and Solana (1972) consider an expression of the self-energy of the type (8.35), in which, however, they replace $\omega_k{}^0$ by the true single-excitation energy so as to make some allowance for higher-order processes, and focus attention on the contributions from the roton part of the single-excitation spectrum. In this simplest approximation, the imaginary part of the self-energy would then describe the decay of a single excitation into two rotons and would obviously jump discontinuously from zero to a finite value at an energy 2Δ. The inclusion of roton–roton attractive interactions leads to a two-roton bound state and thus to a delta-function peak at energy slightly less than 2Δ. Including, then, a finite roton lifetime, this peak is broadened and one ends up with a two-roton continuum with a peak at an energy close to 2Δ and a long tail at higher energies. The features of $S(k, \omega)$, thus consisting of a narrow single-excitation peak shifted to lower energy relative to the earlier calculations and of a broad peak near 2Δ with a long high-frequency tail, result then in qualitative agreement with the measured spectrum in the intermediate range of wave number.

8.5 Quantised vortices in superfluid helium-four

To conclude the discussion of He^4 we shall make some very brief remarks on quantised vortices. That quantised vortex lines and rings can be excited in superfluid He^4 is now amply demonstrated (Hall, 1960; Rayfield and Reif, 1964). These vortices are quantised, the circulation associated with the superfluid flow being an integral multiple of h/m, m being the mass of a helium atom.

Experiment demonstrates that, apart from the vortex core, the flow field associated with such a vortex is similar to that of a classical vortex—irrotational and incompressible. However, the classical theory of vortices contains no unique theory of the core. The data (Hall, 1960) suggest that the core of a quantum vortex in He^4 has a scale of a few interatomic distances.

Feynman (1955) was the first author to show how a wave function could be constructed to (*a*) lead to the velocity field as a classical vortex line and (*b*) take into account the strong interatomic correlations present in He^4. Subsequent work by Chester, Metz and Reatto (1968) has led to the construction of a model wave function for a quantised vortex line and ring. The corresponding variational energies were then calculated using the integral equation derived by Percus and Yevick (see chapter 2) for classical fluids. Prior to this work the only theory of quantised vortices was the order-parameter of Hartree theory of Gross (1961), Pitaevskii (1961) and Fetter (1965). While the energies of Chester,

Metz and Reatto are close to those of this Hartree theory, this later work is giving a quantitative theory of the core of a quantum vortex in He^4. We must refer the reader to the original papers for further details.

8.6 Landau theory of normal helium-three

According to our earlier discussion of the spectrum of the degenerate electron plasma, we expect that a degenerate liquid of *neutral* Fermi particles in the normal state will have a density fluctuation excitation spectrum consisting of (*a*) a single particle-hole continuum similar to that sketched in *figure 7.5* for an ideal Fermi gas, but modified in detail by the interactions; (*b*) a multiple particle-hole continuum essentially extending over the whole (ω, k) plane, with vanishing spectral density as $k \to 0$; and (*c*) a collective phonon-like mode which, under certain conditions for the interactions, will lie outside the single particle-hole continuum and will therefore be essentially undamped in the limit $k \to 0$ (as was the case for the plasmon in the electron liquid). Such a collisionless sound mode will begin to be damped for increasing wave vector through processes of decay into two (or more) particle-hole pairs and will ultimately merge in some manner into the single particle-hole continuum. The situation may thus be expected to be quite similar, qualitatively, to that of a degenerate liquid of *charged* Fermi particles, the only qualitative difference being between the dispersion relation of zero sound in the neutral fluid $(\omega_s(k) = sk + \cdots)$ and that of the plasmon in the charged fluid $(\omega_p(k) = \omega_p + \alpha k^2 + \cdots)$, such difference arising, of course, from the macroscopic electric field which accompanies the collective mode in the charged fluid.

Much less experimental information is available on the excitation spectrum of liquid He^3 than is the case for liquid He^4. The absorption cross-section for neutrons by He^3 nuclei is far bigger than the scattering cross-section, and inelastic neutron scattering experiments are therefore very difficult and have only recently begun. Both theory and experiment have so far explored in detail only the portion of the (ω, k) plane near the origin. In this region, precisely because multi-pair excitations have a negligible spectral weight, the theory developed by Landau (1957) for Fermi liquids in the degenerate, normal state becomes applicable.

The basic assumption of the theory is that, as the interactions are gradually turned on in passing from the perfect gas to the liquid, the role of the gas particles in the classification of the levels is assumed by 'quasi-particles', each of which possesses definite momentum. In particular, if we consider a particle added to the fluid in a state infinitely close to the Fermi surface, then this particle will remain indefinitely in this momentum state. Indeed, because of the Fermi statistics its scattering out of this state would lead to a state containing two particles outside the Fermi surface and a hole inside it, but the initial disposable momentum being infinitesimal, the phase space available for this scattering process is negligible (as stated under (*b*) above). The whole argument obviously assumes the absence of Fermion pairing leading to a BCS gap and to superfluidity, an hypothesis which is now known to be invalid for He^3 at the lowest temperatures (see section 8.7).

Since the theory is restricted to particle (and hole) states very near to the Fermi surface, it can be used to evaluate the thermodynamic properties of the normal Fermi fluid only at low temperatures. In practice, the temperature range involved for liquid He^3 lies between 0.1 K and 0.003 K, the upper limit being estimated by the condition $T \leqslant 0.1\ \mu/k_B$ with $\mu/k_B \sim 1$ K, as is easily estimated from the density and the mass, and the lower limit being fixed by the transition to the superfluid state.

Although the Landau theory for both the thermodynamic and the dynamic properties of He^3 was originally proposed on phenomenological grounds, a sound theoretical proof of its validity for $\hbar\omega \ll \mu$ and $\hbar v_F k \ll \mu$ can be given (see Nozieres, 1963).

8.6.1 Thermodynamic properties

Following Landau we thus assume that the free energy of the Fermi liquid at low temperatures can be expanded into powers of the deviation δn_p of the occupation number of the single-particle state $\boldsymbol{p}$ (where $\boldsymbol{p}$ subsumes both a momentum and a spin index) from its value in the ground state. Namely, we write

$$\delta F = \sum_p (\varepsilon_p - \mu)\delta n_p + \tfrac{1}{2} \sum_{p,p'} f_{pp'} \delta n_p \delta n_{p'} + \cdots \tag{8.37}$$

where $f_{pp'}$ are parameters describing the interactions between quasi-particles. At this stage, both ε_p and $f_{pp'}$ may be viewed as defined as the first and second functional derivative of the free energy with respect to the occupation numbers. In particular, $f_{pp'}$ is then symmetrical in $\boldsymbol{p}$ and $\boldsymbol{p}'$.

In practice, we need to know ε_p only in the vicinity of the Fermi surface, and $f_{pp'}$ only on the Fermi surface. Introducing a series expansion of ε_p near the Fermi surface, the first term

$$\boldsymbol{v}_p = \nabla_p \varepsilon_p \tag{8.38}$$

plays the role of a group velocity of a quasi-particle. On the Fermi surface we can write for an isotropic fluid

$$\left[\boldsymbol{v}_p = \frac{1}{m^*}\boldsymbol{p}\right]_{p=p_F} \tag{8.39}$$

which defines the effective mass m^* of a quasi-particle on the Fermi surface. By using the principle of Galilean relativity that the momentum arriving at a unit volume must be equal to the density of mass flow,

$$\sum_p \boldsymbol{p} n_p = \sum_p m n_p \nabla_p \varepsilon_p \tag{8.40}$$

and taking the variational derivative we find

$$\sum_p \boldsymbol{p}\delta n_p = \sum_p m\left[\delta n_p \nabla_p \varepsilon_p + n_p \sum_{p'} \nabla_p f_{pp'} \delta n_{p'}\right] \tag{8.41}$$

whence, since δn_p is arbitrary,

$$\begin{aligned}\boldsymbol{p}/m &= \nabla_p \varepsilon_p + \sum_{p'} \nabla_{p'} f_{pp'} n_{p'} \\ &= v_p - \sum_{p'} f_{pp'} \frac{\partial n_{p'}}{\partial \varepsilon_{p'}} v_{p'}\end{aligned} \tag{8.42}$$

This relation allows the effective mass m^* to be calculated in terms of the interaction parameters. The factor $\dfrac{\partial n_{p'}}{\partial \varepsilon_{p'}} = -\delta(\varepsilon_{p'} - \mu)$ implies that the integral is, in fact, over the Fermi surface.

For an isotropic fluid in the absence of magnetic field, $f_{pp'}$ on the Fermi surface depends only on the angle between the momenta $\boldsymbol{p}$ and $\boldsymbol{p}'$, that we shall denote by ϑ, and on the relative orientation of the spins σ and σ'. We can then write, in the notation of Pines and Nozières (1966),

$$\begin{aligned} f^{\uparrow\uparrow}_{pp'} &= f^{\mathrm{s}}_{pp'} + f^{\mathrm{a}}_{pp'} \\ f^{\uparrow\downarrow}_{pp'} &= f^{\mathrm{s}}_{pp'} - f^{\mathrm{a}}_{pp'} \end{aligned} \tag{8.43}$$

where f^{s} and f^{a} depend only on the angle ϑ. These functions can be expanded in a series of Legendre polynomials,

$$f^{\mathrm{s(a)}}_{pp'} = \sum_{l=0}^{\infty} f_l^{\mathrm{s(a)}} P_l(\cos\vartheta) \tag{8.44}$$

It is also convenient to separate out a factor ν which is the density of states at the Fermi surface,

$$\nu = \frac{1}{V}\sum_p \delta(\varepsilon_p - \mu) = \frac{m^* p_{\mathrm{F}}}{\pi^2 \hbar^3} \tag{8.45}$$

and write

$$f_l^{\mathrm{s(a)}} = \nu^{-1} F_l^{\mathrm{s(a)}} \tag{8.46}$$

Using then the addition theorem for spherical harmonics, eqn (8.42) leads to

$$m^*/m = 1 + \tfrac{1}{3} F_1{}^{\mathrm{s}} \tag{8.47}$$

The effective mass can be determined from the experimental specific heat, whose leading term derives from the first term on the right-hand side of (8.37) and has the same form as for an ideal Fermi gas,

$$C(T) = \frac{\pi^2}{3} \nu k_{\mathrm{B}}{}^2 T + \cdots \tag{8.48}$$

Experiment demonstrates, however, that CT^{-1} is still increasing noticeably with decreasing temperature below 0.1 K (Wheatley, 1966), the interpretation of this behaviour (Amit, Kane and Wagner, 1968) being the presence of logarithmic terms due to spin fluctuations (Doniach and Engelsberg, 1966; Berk and Schrieffer, 1966).

The compressibility K and the magnetic susceptibility χ of the Fermi liquid can be similarly expressed through the interaction parameters introduced in eqn (8.37). Let us consider in particular the compressibility, which according to the thermodynamic relation $K^{-1}=\rho^2\left(\frac{\partial\mu}{\partial\rho}\right)_T$ can be calculated from the change in chemical potential $d\mu$ needed to accommodate an extra number $dN=Vd\rho$ of particles. If dp_F is the necessary swelling of the (spherical) Fermi surface, we can write from (8.37) and (8.38)

$$\mathrm{d}\mu=v_F\,\mathrm{d}p_F+\sum_{p'} f_{pp'}\delta n_{p'} \tag{8.49}$$

with

$$\delta n_p=-\frac{\partial n_p}{\partial \varepsilon_p} v_p\,\mathrm{d}p_F \tag{8.50}$$

Thus, the integral in (8.49) is again limited to the Fermi surface, and using the Legendre polynomial expansion we find

$$\mathrm{d}\mu=v_F\,\mathrm{d}p_F+F_0{}^s v_F\,\mathrm{d}p_F \tag{8.51}$$

and

$$\frac{\mathrm{d}N}{V}=\frac{1}{V}\sum_p \delta n_p=\nu v_F\,\mathrm{d}p_F \tag{8.52}$$

whence

$$K=\frac{1}{\rho V}\left(\frac{\partial N}{\partial\mu}\right)_{V,T}=\frac{\nu/\rho^2}{1+F_0{}^s} \tag{8.53}$$

A similar calculation for the susceptibility, in which the Fermi surfaces for the two spin orientations are oppositely shifted by a static magnetic field, leads to the expression

$$\chi=\mu_B{}^2\frac{\nu}{1+F_0{}^a} \tag{8.54}$$

where μ_B is the Bohr magneton, $\mu_B=e\hbar/2mc$. Thus, the simple relations holding between the specific heat, the compressibility and the susceptibility in the ideal Fermi gas are no longer true in the interacting Fermi liquid, being modified through the appearance of Landau parameters in (8.53) and (8.54).

Table 8.1 collects the values for these Landau parameters in He^3 as determined experimentally by Wheatley (1966). It may be noted that the interactions are strong and repulsive, and rapidly increasing with pressure, as one would expect for a dense fluid with hard core repulsions; and that the repulsion is slightly weaker for parallel spins since F_0^{a} is negative, as a result of the exchange effects tending to keep parallel spins apart. This tendency to ferromagnetism clearly implies, from (8.54), a strong enhancement of the magnetic susceptibility. We shall see below that the repulsive interaction is a prerequisite for zero sound, and that the tendency to ferromagnetism plays a crucial role in the superfluid phases.

Table 8.1 Effective mass and Landau parameters for He^3 (from Wheatley, 1966)

Pressure	m^*/m	F_0^{s}	F_0^{a}	F_1^{s}	F_1^{a}
0.28	3.1	10.77	−0.67	6.25	−0.72
27.0	5.8	75.63	−0.72	14.35	−0.66

A similar discussion can, of course, be developed for a degenerate Fermi plasma. The theoretical results for the electron liquid at metallic densities indicate that the Landau interaction effects are essentially negligible for the mass and 20–40% enhancement on the susceptibility, depending on the density (Pizzimenti, Tosi and Villari, 1971), while the effect on the compressibility is a major one, with the effective interaction being attractive and leading to an instability for $r_s \sim 5$ (see for example Vashishta and Singwi, 1972). For electrons in metals, on the other hand, strong effective electron–electron interactions can arise, depending on the system, through the electron–ion interaction (Rice, 1968).

8.6.2 Transport equation and zero sound

To apply the Landau theory to the dynamics of the normal Fermi liquid at small k and ω, we shall have to consider a deviation $\delta n_p(\boldsymbol{r}, t)$ from the ground state distribution, defined for $\boldsymbol{p}$ on the Fermi surface. We shall, of course, consider small deviations of the simple form

$$\delta n_p(\boldsymbol{r}, t) = \delta n_p(\boldsymbol{k}, \omega) \exp\left[i(\boldsymbol{k} \cdot \boldsymbol{r} - \omega t)\right] \tag{8.55}$$

By analogy with the Boltzmann equation derived in appendix 5.1, we can write a linearised transport equation for quasi-particles in the form

$$\frac{\partial \delta n_p(\boldsymbol{r}, t)}{\partial t} + \boldsymbol{v}_p \cdot \nabla_r \delta n_p(\boldsymbol{r}, t) + \boldsymbol{F}_p(\boldsymbol{r}, t) \cdot \nabla_p n_p{}^0 = \left.\frac{\mathrm{d}n_p(\boldsymbol{r}, t)}{\mathrm{d}t}\right|_{\mathrm{coll}} \tag{8.56}$$

where $n_p{}^0$ denotes the unperturbed distribution, $\boldsymbol{F}_p$ is the force acting on a quasi-particle in the state $\boldsymbol{p}$, and the right-hand side gives the dissipative effect due to collisions between excited quasi-particles. In accord with the discussion of the previous section, we shall write the force $\boldsymbol{F}_p(\boldsymbol{r}, t)$ as the sum of an external force $\boldsymbol{F}_p{}^{\mathrm{e}}(\boldsymbol{r}, t)$ and of a force arising from the interactions between the quasi-particles,

$$\boldsymbol{F}_p(\boldsymbol{r}, t) = \boldsymbol{F}_p{}^{\mathrm{e}}(\boldsymbol{r}, t) - \sum_{p'} f_{pp'} \nabla_r \delta n_{p'}(\boldsymbol{r}, t) \tag{8.57}$$

For a more detailed derivation of the Landau transport equation and an evaluation of the collision term, as well as for details on the subsequent discussion, the reader should refer to the book of Pines and Nozieres (1966).

As usual, we characterise the effects of collisions through a relaxation time τ. Then, for frequencies such that $\omega\tau \ll 1$, the collision integral plays a dominant role in forcing hydrodynamic behaviour on the fluid, characterised by ordinary sound waves with a speed of sound determined by the compressibility K and by transport coefficients such as viscosity η and thermal conductivity κ. The calculation of these properties by the Landau theory leads to $\eta \propto T^{-2}$ and $\kappa \propto T^{-1}$ (Abrikosov and Khalatnikov, 1959; see also Emery, 1964). We are here interested, on the other hand, in the opposite regime $\omega\tau \gg 1$, where the collision integral can be omitted from eqn (8.56). We thus write the collisionless transport equation, in the absence of external forces, in the form

$$(\boldsymbol{k} \cdot \boldsymbol{v}_p - \omega)\delta n_p(\boldsymbol{k}, \omega) - \boldsymbol{k} \cdot \boldsymbol{v}_p \frac{\partial n_p{}^0}{\partial \varepsilon_p} \sum_{p'} f_{pp'}\, \delta n_{p'}(\boldsymbol{k}, \omega) = 0 \tag{8.58}$$

Clearly, non-trivial solutions of this equation describe collective modes of motion of the fluid, with an eigen frequency which varies linearly with the wave number, since the secular equation depends only on the ratio k/ω.

In principle, many such modes are possible, determined by the dependence of $\delta n_p(\boldsymbol{k}, \omega)$ on spin and on the angular coordinates ϑ and ϕ which define the direction of the momentum vector $\boldsymbol{p}$ on the Fermi surface. The modes can be classified by noting that, given the symmetry properties of the Landau parameters that we discussed in the preceding section, modes in which the two spin populations move in phase are decoupled from those in which they move out of phase, as are modes with different m in an expansion of $\delta n_p(\boldsymbol{k}, \omega)$ in spherical harmonics $Y_{lm}(\vartheta, \phi)$. Let us then consider a spin-symmetric mode with $m = 0$, in which the Fermi surface is distorted as in a *longitudinal* compressional wave. This type of distortion (analogous to a c_{11}-type wave in a solid) should be contrasted with a spherically symmetrical distortion of the Fermi surface, as in the calculation of the compressibility given in the preceding section. This mode is in essence the so-called zero-sound mode.

To obtain an explicit analytic solution for the zero-sound mode, Pines and Nozières (1966) adopt a model in which all the components $F_l{}^s$ of the Landau parameters are made to vanish except $F_0{}^s$. This is a useful simplification, because all the components of the distortion with different l are coupled in a complicated manner in the equation of motion. It is then easy to show that the equation of motion (8.58) admits the non-trivial solution

$$\delta n_p(k, \omega) \sim \delta(\varepsilon_p - \mu) \frac{\cos\vartheta}{\lambda - \cos\vartheta} \tag{8.59}$$

where $\lambda = \omega / k v_F$, provided that λ is given by (Khalatnikov and Abrikosov, 1958)

$$\tfrac{1}{2}\lambda \ln \frac{\lambda+1}{\lambda-1} = 1 + \frac{1}{F_0{}^s} \tag{8.60}$$

This admits real solutions (i.e., an undamped collective mode) provided that $F_0{}^s$ is positive (i.e., in the case of repulsive interactions). The velocity of zero-sound in this model, as a function of the strength of the interactions, is reported in *figure 8.12* and compared with the analogous velocity of ordinary sound, which from (8.53) is

$$\lambda_1 = \left[\tfrac{1}{3}\frac{m^*}{m}(1+F_0{}^s)\right]^{1/2} \tag{8.61}$$

in units of the Fermi velocity v_F.

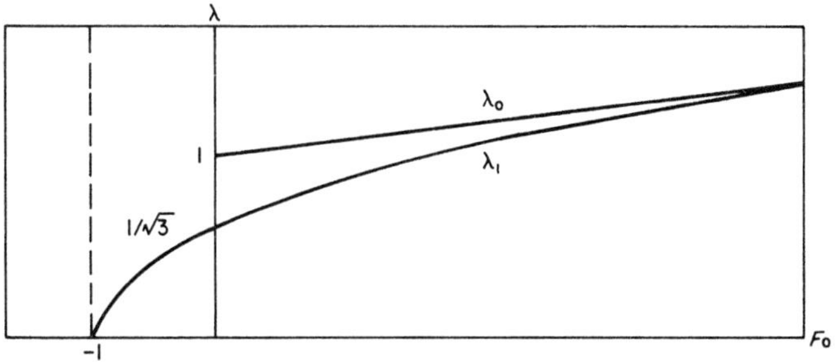

Figure 8.12 Speed of zero sound (λ_0) and of first sound (λ_1), in units of the Fermi velocity, as functions of the Landau parameter $F_0{}^s$ for a model of He^3 having $F_l{}^s = 0$ for $l \geqslant 1$ (from Pines and Nozières, 1966)

Experimental data on the propagation and attenuation of sound waves in liquid He^3 as functions of temperature are reported in *figure 8.13* (from Abel, Anderson and Wheatley, 1966). The observed behaviour is in accord with the theory. Starting on the high-temperature side, the fluid is in the ordinary hydrodynamic regime, where it supports ordinary sound waves damped by ordinary viscosity, which increases with decreasing temperature. As the temperature becomes sufficiently low, however, the liquid enters the regime $\omega\tau > 1$, where the attenuation is now decreasing with decreasing temperature as the liquid approaches the ideal collisionless region. The increase of the speed of sound in the transition from ordinary to zero sound is only a few per cent, in qualitative accord with the results of *figure 8.12* and with the fact that the effective interactions are very strong in liquid He^3.

8.7 Low-temperature phases of helium-three

8.7.1 Phase diagrams

After the original discovery (Osheroff, Richardson and Lee, 1972) of pressure

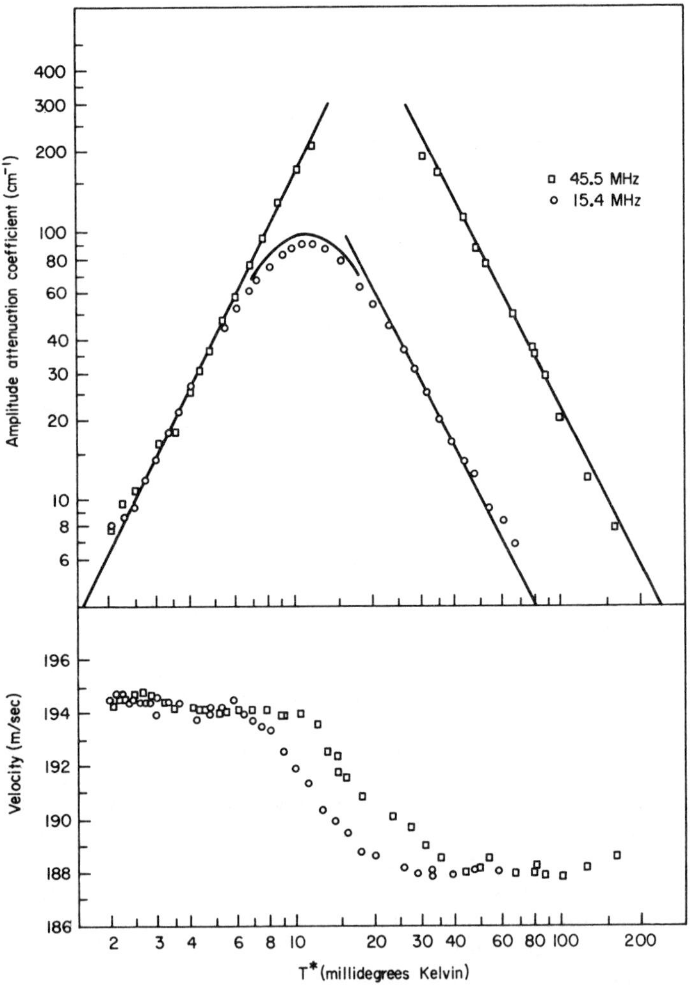

Figure 8.13 Experimental attenuation coefficient and velocity of sound waves as functions of temperature in liquid He^3 at 0.32 atm, for two values of the frequency (from Abel, Anderson and Wheatley, 1966). The straight lines through the attenuation points are based on the Landau theory

anomalies in liquid He^3 below 3 mK, a considerable amount of experimental and theoretical work has been done on the new low-temperature phases. The phase diagram in the p–T plane in zero magnetic field is schematically reproduced in *figure 8.14*. Considering, in particular, the behaviour of the liquid as it is cooled along the melting curve, a transition from the normal Fermi liquid (N) phase to a new (A) phase is encountered at 2.7 mK, the transition having been identified also in experiments of nuclear magnetic susceptibility (Osheroff et al., 1972), heat capacity (Webb et al., 1973; Dundon, Stolfa and Goodkind, 1973), sound attenuation (Lawson et al., 1973; Paulson, Johnson and Wheatley, 1973), viscous damping of a vibrating wire (Alvesalo et al., 1973), and heat flow (Greytak et al., 1973). As the liquid is cooled to still lower temperatures, a second transition is observed at 2 mK, to a second (B) phase. The transition from phase A to phase B is now known to be a first-order transition, having a small latent heat (~20 ergs/mole) associated with it, while no latent heat is apparently associated with the transition from phase N to phase A.

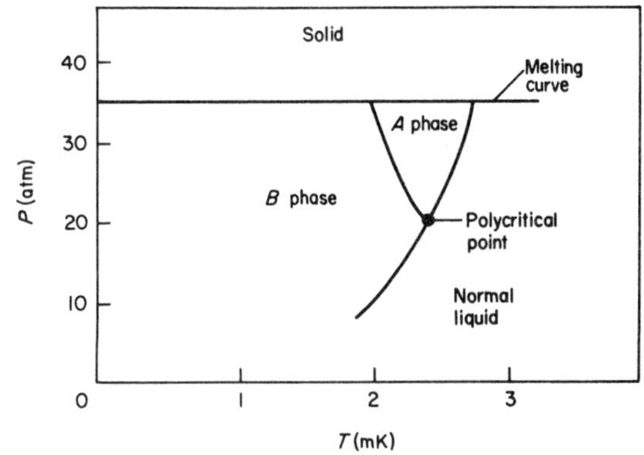

Figure 8.14 Schematic p–T phase diagram for liquid He^3 at low temperatures, in zero magnetic field

The phase diagram in the H–T plane, at a pressure on the melting curve, is sketched in *figure 8.15*. As the magnetic field is increased, the N–A transition broadens and then divides into two distinct branches (Osheroff, Richardson and Lee, 1972; Lawson et al., 1973), with a temperature separation between the two phase boundaries which increases *linearly* with the magnetic field and becomes only of the order of 0.1 mK for fields of the order of 10 kiloOersted. On the contrary, the phase boundary between the A and B phases shifts to lower temperatures *quadratically* with the magnetic field, with an estimated critical field of 5.5 kOe at zero temperature.

8.7.2 Experimental information on the low-temperature phases

A summary of the experimental information available on the low-temperature phases of He^3, which is particularly illuminating for the suggested interpretation of the nature of the new phases, has been presented by Richardson and Lee

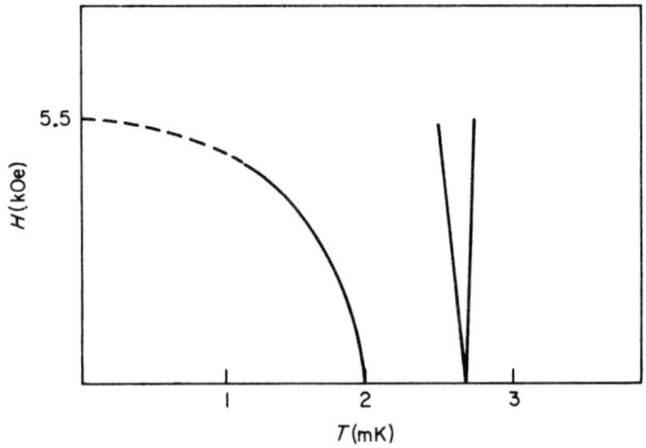

Figure 8.15 Schematic H–T phase diagram for liquid He^3 at low temperatures.

(1973) and by Mermin and Ambegaokar (1973). As already indicated by the phase diagrams, the two phases have different behaviours in several respects. Specifically (see also schematic representation in *figure 8.16*):

(1) The heat capacity shows a strong anomaly at the transition from the N phase to the A phase, the anomaly being of the kind usually drawn to illustrate

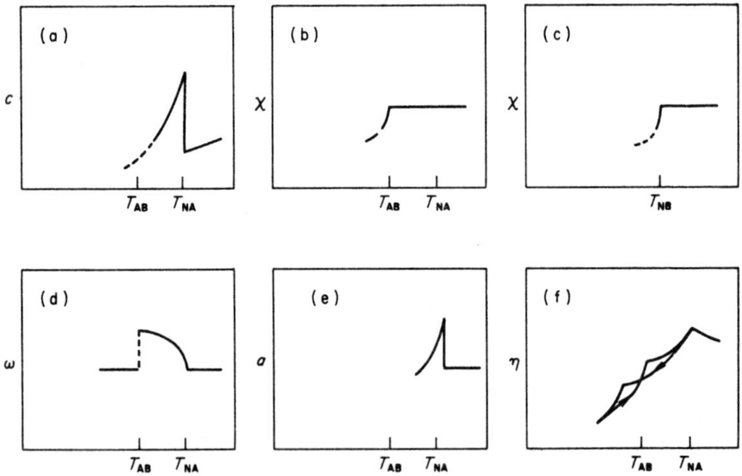

Figure 8.16 Schematic behaviour of physical properties of liquid He^3 at low temperatures (redrawn from Richardson and Lee, 1973). The various properties are: heat capacity (*a*); magnetic susceptibility at pressures above (*b*) and below (*c*) the polycritical point; frequency of nmr transverse resonance (*d*); sound attenuation (*e*); and viscosity (*f*)

an Ehrenfest second-order phase transition, with $\Delta c/c_n \sim 1.6$–1.9, to be compared with the case of the superconducting transition where $\Delta c/c_n = 1.42$. In contrast, there is no strong evidence for a specific heat anomaly across the *A–B* line.

(2) The magnetic susceptibility in the *A* phase appears to be unaltered from the normal state susceptibility, and is therefore independent of temperature as for a Pauli paramagnet. In contrast, the susceptibility decreases sharply on entering the *B* phase, both above and below the polycritical point.

(3) The nmr signal in a transverse configuration (namely, in an rf field perpendicular to the static field), which in the *N* phase lies at the Larmor value $\omega_L = \gamma H$ as for a gas of spins, shifts away from the Larmor value on entering the *A* phase. The shift increases with decreasing temperature, but the signal returns to the Larmor value on entering the *B* phase.

(4) For the normal phase the viscosity behaves as T^{-2}, as appropriate for a normal Fermi liquid, and also the ultrasonic attenuation behaves as in a normal Fermi liquid. On entering the *A* phase, the viscosity drops sharply but continuously, while the ultrasonic attenuation shows a sharp narrow anomaly at the *N–A* transition. In the *B* phase the viscosity drops even more sharply, with some hysteresis being observed at the *A–B* transition.

(5) The thermal conduction has also been measured, and whereas it is proportional to T^{-1} for a Fermi liquid, the *A* phase shows a non-linear, history-dependent behaviour, with the thermal conductivity increasing slowly with falling temperature. The *B* phase exhibits a linear reproducible behaviour, with a discontinuous increase at T_{BN}.

(6) The above evidence, especially that on heat capacity and on transport, points to a superfluid behaviour for the low-temperature phases. Actually the most direct way to prove superfluidity is to search for a fourth sound mode, that is, propagating pressure waves. The experiments, carried out by Kojima, Paulson and Wheatley (1974), show the existence of fourth sound waves, and one can derive from these data the superfluid density which comes out to be rather small for $T > 0.8\,T_c$, with little or no observable change at the *A–B* transition. The latter finding is in apparent contrast with the viscosity measurements of the Helsinki group, referred to under (4) above.

8.7.3 Theoretical developments

A number of authors, very early, proposed that He^3 could undergo a BCS-type transition to a superfluid state (Pitaevskii, 1960; Brueckner et al., 1960). Between He^3 atoms, there is of course van der Waals attraction and it is therefore possible to form some sort of condensed state like the superconducting state of metals. The following discussion is based on short reviews of Mermin and Ambegaokar (1973) and of Leggett (1973*b*).

In the Bardeen–Cooper–Schrieffer (1957) theory of superconductivity, the crucial role is played by Cooper pairs with zero angular momentum ($L = 0$) and zero spin ($S = 0$), the spin part of the pair wave function having thus the form

$1/\sqrt{2}(\uparrow\downarrow - \downarrow\uparrow)$. This theory leads, in particular, to a specific heat which, as remarked above, jumps by a factor of 1.42 at the critical temperature, and to a spin susceptibility which drops off sharply below the critical temperature. In theory, $\chi \to 0$ as $T \to 0$ because the pairs have $S = S_z = 0$ and therefore cannot be polarised by a magnetic field.

We must now consider how to pass from a situation in which the Cooper pairs are constructed from electrons, and the basic interaction is between electrons, mediated by the phonons, to the situation of liquid He^3 where the Cooper pairs must be built from He^3 neutral atoms, and the forces are hard core repulsion plus van der Waals attraction. It was early realised that Cooper pairs in He^3 will preferably form in states with $L \neq 0$, since in *s* states we would have overlapping hard cores. In states of higher orbital angular momentum, the pairs can take advantage of the van der Waals attraction and avoid the hard core repulsion. This implies, however, that the gap parameter will generally be anisotropic.

Assuming, therefore that the *A* and *B* phases are the result of Cooper pairing with $L \neq 0$, the experimental evidence provides important clues on the structure of the pairs. Pairs with even *L* have, necessarily, a spin wave function of the form $1/\sqrt{2}(\uparrow\downarrow - \downarrow\uparrow)$ with $S = S_z = 0$ (singlet state), for which the theory predicts a quadratic dependence of the transition temperature on magnetic field and a vanishing zero-field susceptibility as $T \to 0$. This behaviour is obviously not in accord with the observations for the *A* phase. Instead, one can explain the splitting of the *N–A* transition in magnetic field, and the unchanged value of the zero-field susceptibility, by postulating an equal-spin-pairing state, where pairs are in odd-*L* states (possibly, but not necessarily, $L = 1$) with spin wave functions of the types $\uparrow\uparrow$ and $\downarrow\downarrow$ with $S = 1$ and $S_z = \pm 1$. In this case the application of a magnetic field simply produces opposite linear shifts of the Fermi energies for the two spin orientations, leading to a linear splitting of the transition.

While singlet pairing is definitely excluded by the experimental evidence for the *A* phase, the evidence described above is not so definite for the *B* phase. In principle, one could hypothesise a singlet-pairing state, or a triplet-pairing state of the type proposed by Balian and Werthamer (1963) in which there is a component with $S_z = 0$, having spin wave function $1/\sqrt{2}(\uparrow\downarrow + \downarrow\uparrow)$, in addition to the components with $S_z = \pm 1$. In both cases, the suppression of the transition temperature would be quadratic in the field, but in the latter case the susceptibility would tend to a finite value for $T \to 0$, in contrast to the vanishing value predicted for the singlet state.

The predictions made on the basis of the various pairings are collected in *Table 8.2*. This includes also the predictions made with regard to nmr† experiments, both in the transverse (rf field perpendicular to static field) and in the longitudinal (rf field parallel to static field) configuration. It may be noted that the assumption of an equal-spin-pairing state for the *A* phase is consistent with

†The microscopic theory of nmr in anisotropic superfluid He^3A has been given by Leggett (1973).

Table 8.2. Properties of Fermion pairings

	l even, S=0		*l odd, S=1*	
	$l=0$	$l\neq 0$	EPS	BW
	$\frac{1}{\sqrt{2}}(\uparrow\downarrow-\downarrow\uparrow)$	$\frac{1}{\sqrt{2}}(\uparrow\downarrow-\downarrow\uparrow)$	$\uparrow\uparrow, \downarrow\downarrow$ $(S_z=\pm 1)$	$\uparrow\uparrow, \downarrow\downarrow, \frac{1}{\sqrt{2}}(\uparrow\downarrow+\downarrow\uparrow)$ $(S_z=1, 0, -1)$
gap	isotropic	anisotropic	anisotropic	isotropic, $l=1$ anisotropic, $l\neq 1$
$\chi\,(T\rightarrow 0)$	0	0	χ_n	$\frac{2}{3}\chi_n$
nmr shift (transv.)	No	No	Yes	No
nmr shift (long.)	No	No	Yes	Yes

the nmr experiments in the transverse configuration discussed in the preceding section, and that an nmr experiment in the longitudinal configuration would discriminate between the singlet pairing and the Balian–Werthamer pairing for the *B* phase. In fact, longitudinal resonance experiments have now been carried out (Osheroff and Brinkman, 1974; Bolzer et al., 1974; Webb, Kleinberg and Wheatley, 1974) and, together with other evidence on the behaviour of the phases in a magnetic field (Paulson, Kojima and Wheatley, 1974), they are again consistent with the above identification of the *A* phase, and tend to support the identification of the *B* phase, as triplet states (see also Osheroff, 1974)

Thus, the experimental evidence available at the time of writing is compatible with the hypotheses of an equal-spin-pairing state for the *A* phase and of a Balian–Werthamer-type state for the *B* phase—although the *B* phase may really be more complicated than is now supposed. Of course, to derive magnetic superfluid phases it is necessary to invoke a significant role of the exchange interaction between the nuclear spins, leading to parallel spins and a tendency to ferromagnetism. As we noted in the discussion of the lowest Landau parameters for the normal phase of He^3, there is indeed a slight attraction between parallel spins relative to anti-parallel spins.

Why, then, do two cases occur? We give here a brief introduction to the mechanism proposed by Anderson and Brinkman (1973). According to their calculation, the Balian–Werthamer triplet state does indeed arise because of an attractive contribution to the effective interactions arising from virtual exchange of spin fluctuations (Doniach and Engelsberg, 1966; Berk and Schrieffer, 1966) but one should take into account how the introduction of a BCS gap feeds back on this mechanism. The changes in susceptibility brought about by pairing are such as to reduce the effective interactions in the Balian–Werthamer state, while this effect is not operative for some of the anisotropic states which may therefore be stabilised. Among the indications provided by the theory is that the critical temperature should decrease, and the *B* and *A* transitions converge towards each other with decreasing pressure, as spin fluctuations become less important at lower pressure, and that the specific heat anomaly should be larger than in the BCS theory, both points being in qualitative agreement with experiment.

8.8 Solutions of helium-three in liquid helium-four

The experiments of Edwards et al. (1965) and Anderson et al. (1966) show that He^3 atoms in dilute solutions behave like a normal Fermi liquid, as predicted by Landau and Pomeranchuk (1948); see also Pomeranchuk (1949). Many of the low temperature properties of He^3 in solution are qualitatively similar to those of pure liquid He^3. The experiments also indicate that there is a weak and mainly attractive interaction between He^3 atoms in dilute solutions in He^4. The interest in these solutions thus arises because:

(1) here is a Fermi liquid system in which the density and degeneracy temperature is a variable;

(2) the effective interaction between the He^3 quasi-particles is sufficiently weak to apply perturbation theory to treat the effective potential, in calculating thermodynamic quantities and transport coefficients.

From later theoretical work (for example Emery, 1966; Bardeen, Baym and Pines, 1966; Baym and Saam, 1968) a qualitative understanding of various properties has been obtained, but the determination of the parameters of the Fermi liquid remains on a phenomenological footing. More microscopy studies were carried out by later workers (see for example Davison and Feenberg, 1969). Even more recently, Ostgaard (1970) has applied the method of the reaction matrix, obtained from Brueckner's many-body theory, to estimate various low-temperature properties of the solutions.

In Ostgaard's work, the system is viewed as a low-density Fermi liquid with He^3 quasi-particles formed by He^3 atoms in superfluid He^4. The single-particle spectrum is determined by knowledge of an effective mass, and an effective interaction between the He^3 quasi-particles is derived. The Landau f function is calculated from the reaction matrix and the coefficients of its expansion in Legendre polynomials determined, in quite good agreement with experimental results. The various properties, such as the compressibility, the quasi-particle effective mass or specific-heat ratio and the magnetic susceptibility, can be obtained as functions of the He^3 concentration in the solution and the maximum solubility of He^3 in liquid He^4 is estimated, in good agreement with the measured value. Ostgaard has also calculated transport coefficients (viscosity, thermal conduction and spin diffusion) and has found, again, remarkably good agreement with the experimental results.

Rather than go into all these aspects in detail we shall restrict ourselves here to a brief survey of the work of Bardeen, Baym and Pines (1967). These authors determine the approximate form of the effective interaction between He^3 atoms in superfluid He^4 from experimental data, the input information being data on spin diffusion and phase separation in dilute mixtures. The interaction is found to be weak, and attractive at long wavelengths. The Fermi liquid parameters for the normal state of He^3 in solution lead to results for the effective mass and spin susceptibility in agreement with experiment.

Pines (1963*a*) and van Leeuwen and Cohen (1962) had earlier suggested that there should be an attractive interaction between two He^3 atoms in solution, arising from the exchange of a He^4 phonon. In the long wavelength limit, this part of the effective interaction can be determined exactly with the aid of thermodynamic or deformation-potential arguments, and it is given by

$$\begin{aligned} V_0^{\mathrm{ph}} &= (\partial\mu_{3\uparrow}/\partial\rho_4)_{\rho_3}(\partial\rho_4/\partial\rho_{3\downarrow})_{\mathrm{p}} \\ &= -(1+\alpha)^2 m_4 s^2/\rho_4 \end{aligned} \tag{8.62}$$

where $\mu_{3\uparrow}$ is the chemical potential of a spin-up He^3, ρ_4 is the He^4 particle density, $\rho_{3\downarrow}$ is the density of He^3 with spin-down and $\alpha \simeq 0.28$ is the relative increase in effective volume of the mixture resulting from the replacement of a He^4 atom by a He^3 atom. The quantity $m_4 s^2/\rho_4$, where m_4 is the atomic mass of He^4 and s is the sound velocity in pure He^4, determines essentially the scale of interaction energies.

However, there is an additional 'direct' part of the interaction, which is determined for very dilute systems by thermodynamics as

$$V_0^{\mathrm{dir}} = \left(\frac{\partial\mu_{3\uparrow}}{\partial\rho_{3\downarrow}}\right)_{\rho_4} = (1+2\alpha)m_4 s^2/\rho_4 \tag{8.63}$$

The resultant long wavelength limit result is then,

$$V_0 = -\alpha^2 m_4 s^2/\rho_4 \tag{8.64}$$

in excellent agreement with the empirically determined interaction.

The reason for the major cancellation between the direct and the phonon-induced contributions to the effective interaction $[\alpha^2/(1+\alpha)^2 \cong 1/20]$ is due to the fact that He^3 is an isotopic impurity. The interactions of He^3 and He^4 atoms are identical, but the He^3 atoms have a smaller mass and occupy a slightly larger volume in the liquid than the He^4 atoms. For a dilute system, the change in the total free energy resulting from the replacement of two He^4 atoms by two He^3 atoms is essentially that due to the interaction of two holes in the liquid of relative size α.

8.8.1 Thermodynamics and the effective interaction

As He^3 atoms move through the liquid they displace the He^4 in their way. The He^4 atoms, driven by local changes in their chemical potential μ_4, move in such a way as to keep μ_4, or equivalently the local pressure, constant. Thus, the effective interaction between two He^3 atoms at long wavelengths is simply related to the net change in energy of the system on substituting two He^3 atoms for two He^4 atoms at constant pressure.

The He^3 chemical potential μ_3 is simply the free energy required to add one He^3 atom to the system at constant pressure. Thus, we write

$$\mu_3 = \mu_f + \mu' \tag{8.65}$$

where μ_f is the chemical potential of a non-interacting Fermi gas of effective mass m and density ρ_3.

The effective interaction $V(\boldsymbol{r})$ is assumed to have a Fourier transform $V(\boldsymbol{k})$ through

$$V(\boldsymbol{r}) = \int \frac{\mathrm{d}\boldsymbol{k}}{(2\pi)^3} V_k \exp(i\boldsymbol{k}\cdot\boldsymbol{r}) \tag{8.66}$$

and this describes the amplitude of the scattering of two quasi-particles of opposite spin†. Thus V_0 is the derivative of μ' for a He^3 atom of one spin (↑ say) with respect to a uniform change, at constant μ_4, of the density of He^3 with ↓ spin:

$$V_0 = (\partial\mu'_\uparrow/\partial\rho_{3\downarrow})_{\mu_4} \tag{8.67}$$

The indentification of V_0 is analogous to the Landau definition of the effective interaction between quasi-particles in a pure Fermi liquid. Since $\mu_{f\uparrow}$ is independent of $\rho_{3\downarrow}$, we can write

$$V_0 = (\partial\mu_3{\uparrow}/\partial\rho_3{\downarrow})_{\mu_4} \tag{8.68}$$

If we regard $\mu_{3\uparrow}$ as a function of $\rho_{3\downarrow}$ and ρ_4, we can write alternatively

$$V_0 = \left(\frac{\partial\mu_{3\uparrow}}{\partial\rho_{3\downarrow}}\right)_{\rho_4} + \left(\frac{\partial\mu_{3\uparrow}}{\partial\rho_4}\right)_{\rho_3}\left(\frac{\partial\rho_4}{\partial\rho_{3\downarrow}}\right)_{\mu_4} \tag{8.69}$$

The first part of V_0 in the right-hand side is clearly the contribution to the effective interaction when *no* variation in background density of He^4 occurs. The second term is the phonon-induced interaction between He^3 atoms. $(\partial\rho_4/\partial\rho_{3\downarrow})_{\mu_4}$, equal to $(\partial\rho_4/\partial\rho_3)_p$ in the limit of zero concentration, is the density fluctuation induced in the He^4 background by a unit change in ρ_3 at constant pressure. This can be written as $-(1+\alpha)$. The quantity $(\partial\mu_{3\uparrow}/\partial\rho_4)_{\rho_3}$ measures the response of a second He^3 atom to that density variation and is given by $-(1+\alpha)m_4 s^2/\rho_4$.

We can now calculate the term $(\partial\mu_{3\uparrow}/\partial\rho_{3\downarrow})_{\rho_4}$ in V_0, which tells us the shift in the chemical potential of a spin-up He^3 atom due to unit change of the density of ↓ spin He^3 atoms, the background density of He^4 atoms remaining constant. To evaluate this term, we have to consider the changes in the physical properties of the solution resulting from the substitution of He^3 atoms for He^4 atoms. To see

†For particles of the same spin the scattering amplitude includes a further exchange term.

this, we write

$$\left(\frac{\partial\mu_{3\uparrow}}{\partial\rho_{3\downarrow}}\right)_{\rho_4}=\left(\frac{\partial\mu_4}{\partial\rho_4}\right)_{\rho_3}+\left[\frac{\partial}{\partial\rho_4}(\mu_{3\downarrow}-\mu_4)\right]_{\rho_3}+\left[\frac{\partial}{\partial\rho_{3\downarrow}}(\mu_{3\uparrow}-\mu_4)\right]_{\rho_4} \tag{8.70}$$

Here we have added and subtracted $(\partial\mu_4/\partial\rho_4)_{\rho_3}$ and $(\partial\mu_{3\downarrow}/\partial\rho_4)_{\rho_3}$, and have used the identity

$$\left(\frac{\partial\mu_4}{\partial\rho_{3\downarrow}}\right)_{\rho_4}=\left(\frac{\partial\mu_{3\downarrow}}{\partial\rho_4}\right)_{\rho_3} \tag{8.71}$$

The physical interpretation of the direct interaction is now clear. The first term on the right-hand side is the interaction energy of two He^4 atoms, before being displaced by two He^3 atoms. This is

$$\left(\frac{\partial\mu_4}{\partial\rho_4}\right)_{\rho_3}=m_4s^2/\rho_4 \tag{8.72}$$

The second term is the energy change, beyond $\mu_{3\downarrow}-\mu_4$ evaluated at equilibrium, in replacing one of the He^4 atoms by the ↓ spin He^3 atom, in the presence of the other He^4 atom. This energy is

$$\left[\frac{\partial}{\partial\rho_4}(\mu_{3\downarrow}-\mu_4)\right]_{\rho_3}=\alpha m_4s^2/\rho_4 \tag{8.73}$$

The third term is the additional energy (beyond $\mu_{3\uparrow}-\mu_4$), required to replace the other He^4 by the ↑ spin He^3. Bardeen et al. show that this is the same as the second term. Hence

$$V_0^{\text{dir}}=(1+2\alpha)m_4s^2/\rho_4 \tag{8.74}$$

The resultant long-wavelength interaction is clearly

$$V_0=-\alpha^2m_4s^2/\rho_4 \tag{8.75}$$

This is an exact description of the physical processes considered, namely one-phonon exchange and associated direct term of the same order. Two-phonon exchange is expected to correct this result by terms of higher order in α^2. Taking the experimental value $\alpha=0.28$, one gets

$$V_0=-0.078m_4s^2/\rho_4 \tag{8.76}$$

in good agreement with the empirically determined value

$$V_0=-0.075m_4s^2/\rho_4 \tag{8.77}$$

8.8.2 Microscopic considerations

We shall briefly summarise the microscopic approach to the ground-state energy ε of the He^3–He^4 mixture.

The exact Hamiltonian of the mixture may be written as

$$\mathrm{H}=\sum_{\mathrm{He}^4}\frac{p_i^{\,2}}{2m_4}+\sum_{\mathrm{He}^3}\frac{p_i^{\,2}}{2m_3}+\tfrac{1}{2}\sum_{i,J}\phi(|\boldsymbol{R}_i-\boldsymbol{R}_j|) \tag{8.78}$$

ϕ being the bare interatomic potential. Since $m_3=3m_4/4$, H can also be expressed as

$$\mathrm{H}=\sum_{i}\frac{p_i^{\,2}}{2m_4}+\tfrac{1}{2}\sum_{i,j}\phi(|\boldsymbol{R}_i-\boldsymbol{R}_j|)+\sum_{\mathrm{He}_3}\frac{p_i^{\,2}}{6m_4} \tag{8.79}$$

The sum in the first term is now over *both* the He^3 and He^4 atoms. The first two terms are formally the Hamiltonian for $N=N_3+N_4$ He^4 atoms, and the last term clearly represents the additional kinetic energy of the He^3 atoms due to their lighter mass. This last term, the additional zero point energy of the He^3 atoms, is essentially the perturbation that leads to an effective interaction.

Now the ground-state energy can be estimated by taking as trial wave function the *true* ground-state wave function of pure He^4 at the *same* particle density ρ as in the mixture. Such a function obviously does not take account of the exclusion principle for He^3 atoms of the same spin. This is *not* important for calculating α or the interaction between opposite-spin particles, for to calculate these two quantities we need consider only one or two He^3 atoms in the matrix He^4 liquid. In this low concentration limit, the trial wave function would be the *exact* ground state if the masses of He^3 and He^4 were identical. With this variational wave function we can evidently write

$$\varepsilon(\rho)=\varepsilon_0(\rho)+\rho_3\varepsilon_1(\rho) \tag{8.80}$$

where $\varepsilon_0(\rho)$ is the ground-state energy per unit volume of pure He^4 at density ρ. Also $\varepsilon_1(\rho)=\langle p^2/6m_4\rangle$ is one-third of the average kinetic energy per particle in pure He^4 at density ρ and $T=0$.

The linearity in ρ_3 is a consequence of the identity of the interatomic forces between isotopes. (Also, correction terms are expected from exchange effects; these are *not* important for our present purposes, as discussed above). Prigogine, Bingen and Cohen (1963)—see also Prigogine (1958)—have proposed an 'interpolation formula'

$$\varepsilon(\rho)=(1-x)\varepsilon_0(\rho)+x\varepsilon_3(\rho) \tag{8.81}$$

at concentration x, which is evidently equivalent to

$$\varepsilon(\rho)=\varepsilon_0(\rho)+\rho_3[\varepsilon_3(\rho)-\varepsilon_0(\rho)]/\rho \tag{8.82}$$

where $\varepsilon_3(\rho)$ is the energy density of pure He^3 at density ρ. However, the coefficient of ρ_3 in this equation is not sufficiently accurate at low concentrations of He^3, as the molar volume data indicate.

We can now calculate the chemical potentials μ_4 and μ_3 as

$$\mu_4 = \left(\frac{\partial \varepsilon}{\partial \rho_4}\right)_{\rho_3} = \frac{\partial \varepsilon_0}{\partial \rho} + \rho_3 \frac{\partial \varepsilon_1}{\partial \rho} \tag{8.83}$$

and

$$\mu_3 = \left(\frac{\partial \varepsilon}{\partial \rho_3}\right)_{\rho_4} = \mu_4 + \varepsilon_1(\rho) \tag{8.84}$$

The assumed trial function thus yields the result that the difference in the chemical potentials $\mu_3 - \mu_4$ is simply the additional zero-point energy $\varepsilon_1(\rho)$ of a He^3 atom. This quantity, furthermore, depends only on the density of the system. The derivative $\left[\frac{\partial}{\partial \rho_{3\downarrow}}(\mu_{3\uparrow} - \mu_4)\right]_{\rho, \rho_{3\uparrow}}$ therefore vanishes, and we recover $V_0 = -\alpha^2 m_4 s^2/\rho_4$.

8.8.3 Explicit variational form for atomic volume change α

The chemical potential μ_4 has a term

$$\frac{\partial \varepsilon_1}{\partial \rho} = \frac{\partial}{\partial \rho}\left\langle \frac{p^2}{6m_4} \right\rangle \tag{8.85}$$

which can be regarded as a Landau 'Fermi liquid' effect. It is the modification of the extra zero-point energy of the spin-down He^3 already present, caused by the addition of an extra He^4 or spin-up He^3.

An explicit expression for α is readily obtained, since

$$\alpha = \frac{\rho_4}{m_4 s^2} \frac{\partial}{\partial \rho_4}(\mu_3 - \mu_4) \tag{8.86}$$

the derivative being evaluated at constant ρ_3 in the limit as the concentration vanishes. Using the above variational result for the difference in chemical potential, we get

$$\alpha = \frac{\rho_4}{m_4 s^2} \frac{\partial}{\partial \rho_4}\left\langle \frac{p^2}{6m_4} \right\rangle \tag{8.87}$$

Thus in the present calculation α is determined directly in terms of the properties of the ground state of pure He^4, and in particular by the density dependence of its average kinetic energy per particle. McMillan (1965) has calculated this density dependence and with his value one finds $\alpha = 0.30$.

However, Bardeen, Baym and Pines (1966) give a rough estimate of α which is sufficiently illuminating to summarise here.

Let us regard each He^4 atom as a hard sphere of effective diameter d moving in a fixed impenetrable shell formed by the atom's neighbours. If a is an average near-neighbour distance, the hard sphere moves as a point particle in a shell of radius a–d. Hence, from elementary considerations, the ground state energy is

$$\left\langle \frac{p^2}{2m_4} \right\rangle = \frac{\pi^2\hbar^2}{2m_4(a-d)^2} \tag{8.88}$$

If we assume $a \sim \rho_4^{-1/3}$ and d independent of ρ_4 then

$$\rho_4 \frac{\partial\langle p^2\rangle}{\partial\rho_4} = \tfrac{2}{3}\pi^2\hbar^2 a/(a-d)^3 \tag{8.89}$$

and hence from the above formula for α

$$\alpha \simeq \left(\frac{\pi\hbar}{3m_4 sa}\right)^2 \left[1 - \frac{d}{a}\right]^{-3} \tag{8.90}$$

De Boer (1957) has used the above ground-state energy to calculate the zero-point energy of solid He^4 and finds good numerical agreement if he chooses the empirical value $d = 2.0$ Å and takes $a = 2^{1/6}\rho_4^{-1/3}$ appropriate to an fcc lattice. Using this value of d, and $a = 3.8$ Å, an average near-neighbour distance taken from neutron scattering experiments, then $\alpha = 0.31$, quite close to the experimental value 0.28.

The satisfactory agreement between the variational calculation of α and the experimental value is good confirmation that the origin of the effective interaction is the difference in zero-point motion of the He^3 and He^4 atoms.

CHAPTER 9

CRITICAL PHENOMENA

So far we have been mainly dealing, explicitly or implicitly, with liquids in the region of the triple point, where liquid and solid densities are comparable. However, in this chapter we shall discuss the region near the critical point, where liquid and gas densities become comparable. Although the arguments we presented earlier relating the direct correlation function $c(r)$ to the pair potential $\phi(r)$ break down here, since the required inequality $h^2 < |c|$ is violated, the direct correlation function still plays a significant role. We shall first discuss a phenomenological approach, then thermodynamic scaling, and also we shall deal with the nature of the correlation functions as the critical point is approached.

9.1 Phenomenology

The most obvious definition of the critical point is that point at which the isotherm has a point of inflexion satisfying

$$\left(\frac{\partial p}{\partial \rho}\right)_{\mathrm{T}}=0, \quad \left(\frac{\partial^2 p}{\partial \rho^2}\right)_{\mathrm{T}}=0 \tag{9.1}$$

Another definition which is sometimes adopted involves the specific heat at constant volume c_{V}, for from thermodynamics we can show that

$$-\frac{1}{c_{\mathrm{V}}}\left(\frac{\partial p}{\partial V}\right)_{\mathrm{T}} \geqslant 0 \tag{9.2}$$

the equality being satisfied at the critical point. We shall concern ourselves briefly later with the behaviour of c_{V} near T_{c}, the critical temperature.

From eqn (9.1), it follows immediately that the isothermal compressibility K_{T} and therefore $S(0)$, the long wavelength limit of the structure factor, diverge at the critical point. Usually the form of K_{T} is taken as (see below)

$$K_{\mathrm{T}}=(T-T_{\mathrm{c}})^{-\gamma} \tag{9.3}$$

measured at the critical density ρ_{c} and for insulating liquids $\gamma \sim 1.1$. As we shall see below, this is near to the value for a van der Waals fluid.

Also, we have that the difference between liquid and gas densities is given by

$$\rho_{\mathrm{liq}}-\rho_{\mathrm{gas}}=(T-T_{\mathrm{c}})^{\beta} \tag{9.4}$$

and β takes the values 0.35 ± 0.02 for insulating fluids, from available experimental data (see, for example, Domb and Green 1972).

9.1.1 van der Waals equation of state

To understand how forms like (9.3) and (9.4) can arise let us start from the undoubtedly oversimplified model of the van der Waals equation of state. In fact, we shall see that the predictions of this equation would be valid whenever a Taylor expansion of the equation of state around the critical point can be carried out.

The considerations based on the van der Waals equation are elementary and we shall only note that, rewriting

$$\frac{p}{k_B T} = \frac{\rho}{1 - b\rho} - \frac{a\rho^2}{k_B T} \tag{9.5}$$

in terms of reduced variables $p^* = p/p_c$, etc., we find, using the condition (9.1) to determine p_c, ρ_c and T_c,

$$\frac{p^*}{T^*} = \frac{8\rho^*}{3 - \rho^*} - \frac{3\rho^{*2}}{T^*} \tag{9.6}$$

The critical exponents appearing in eqns (9.3) and (9.4) can now be estimated by expanding about the critical point, and then, with $\Delta p = p - p_c$, etc., we find

$$\Delta p = a_1 \Delta T + b_1 \Delta T \Delta\rho + d(\Delta\rho)^3 + \cdots \tag{9.7}$$

Although the constants a_1, b_1 and d can be determined, of course, from the van der Waals equation of state, the exponents γ and β in eqns (9.3) and (9.4) do not depend on them, but only on the gross (but, we shall see later, inaccurate!) assumption that we can Taylor expand the equation of state about the critical point.

Now it follows from eqn (9.7) that

$$b_1 \Delta T + d(\Delta\rho)^2 \sim \frac{\Delta p}{\Delta\rho} - a_1 \frac{\Delta T}{\Delta\rho} \tag{9.8}$$

and both quantities on the right-hand side $\to 0$ as $T \to T_c$. Hence it follows that $\Delta\rho \propto (\Delta T)^{1/2}$ and hence, by comparison with eqn (9.4), we find $\beta = \frac{1}{2}$ in this theory. This is in conflict with the experimental value $\beta = 0.35 \pm 0.02$ quoted above for insulating liquids. It is clear then that the above Taylor expansion does not work for insulating fluids.

We can immediately calculate $(\partial p/\partial\rho)_T$ from eqn (9.7) and we find

$$\left(\frac{\partial p}{\partial\rho}\right)_T \sim b_1 \Delta T + 3d(\Delta\rho)^2 \tag{9.9}$$

But since we have shown that $b_1 \Delta T + d(\Delta\rho)^2 \to 0$ as $T \to T_c$ we can write at the critical density ρ_0

$$K_{\mathrm{T}}=\left[\frac{1}{\rho}\left(\frac{\partial p}{\partial \rho}\right)_{\mathrm{T}}\right]^{-1}=\frac{\text{const}}{\Delta T} \tag{9.10}$$

giving the exponent $\gamma=1$, in reasonable agreement with experiment.

Before turning to a more fundamental approach, it will be a useful preliminary to enquire how the pair correlation function $g(r)$ behaves in the neighbourhood of the critical point from a classical theory.

9.1.2 Ornstein–Zernike theory and correlation length

The original argument of Ornstein and Zernike (1918) for the form of $g(r)^1 1 \equiv h(r)$ near the critical point, started out from eqn(2.40) for the direct correlation function $c(r)$. They made essentially two assumptions:

(1) That $c(r)$ is short ranged compared with $h(r)$. This is only true, of course, near the critical point, from the arguments of chapter 2.

(2) That $h(r')$ in the integral in eqn (2.40) can be Taylor expanded around the point r.

From (2), the first term $h(r)$ in this expansion gives a contribution $h(r)\rho \int c(r)\, \mathrm{d}\boldsymbol{r}$ to the convolution, the term grad h integrates to zero, while the term proportional to $|\boldsymbol{r}'-\boldsymbol{r}'|^2$ evidently contributes

$$\text{constant } \nabla^2 h(r) \int cr^2\, \mathrm{d}\boldsymbol{r} \tag{9.11}$$

There is no reason why $\int cr^2\, \mathrm{d}\boldsymbol{r}$ should vanish and therefore, using (q) above, we find the following differential equation for $h(r)$:

$$\nabla^2 h = \text{constant}\left[1-\rho \int c(\boldsymbol{r})\, \mathrm{d}\boldsymbol{r}\right] h$$

$$= \text{constant } [1-\tilde{c}(0)]h$$

$$= \kappa^2 h \tag{9.12}$$

Since $S(0)\rightarrow\infty$ at T_c from eqn (9.1), and $S(0)=[1-\tilde{c}(0)]^{-1}$ from the Fourier transform of eqn (2.40), it follows that κ in eqn (9.12)$\rightarrow 0$ as $T\rightarrow T_c$. The solution of (9.12), which decays to zero at infinity, is evidently

$$h = \text{constant } \frac{\exp(-\kappa r)}{r} \tag{9.13}$$

This equation leads us to an important characterisation of critical behaviour through a correlation length $\xi=\kappa^{-1}$, ξ becoming infinite as $T\rightarrow T_c$.

In terms of a conventional critical exponent ν, we write

$$\xi \propto |T-T_c|^{-\nu} \tag{9.14}$$

Equation (9.13) then leads at $T=T_c$ to

$$h(r) \sim \frac{\text{const}}{r}: \qquad S(k) \sim \frac{\text{const}}{k^2} \tag{9.15}$$

While these results, (9.15), are useful first approximations, we shall give arguments below that, in practice, there is a further critical exponent η associated with the total correlation function h. In fact we define η by generalising (9.15) to read, at $T=T_c$

$$h(r) \sim \frac{\text{const}}{r^{1+\eta}}: \qquad S(k) \sim \frac{\text{const}}{k^{2-\eta}} \tag{9.16}$$

Experimental evidence points to the fact that η is small (~ 0.1). Theories based on the ideas of the renormalisation group (see, for example, Hubbard and Schofield, 1972), which we shall briefly sketch below (section 9.4), give us a definite procedure for getting rather precise numerical estimates of γ, β and η.

But before doing this, the above discussion of $h(r)$ near T_c prompts us to record how some thermodynamic equations allow one to relate 3- and 4-particle correlation functions to the pair function $g(r)$ near T_c.

9.2 Asymptotic form of three-particle correlation function near the critical point.

From eqns (1.15), (1.11) and (1.27) for the pressure p, the energy E and the compressibility, it is straightforward to calculate, in the terms of the pair potential $\phi(r)$ and $g(r)$, the first and second isothermal density derivatives of p and E. Next, one notes the properties of the critical point summarised in eqn (9.1), plus the fact that both derivatives of E remain finite at T_c. It will also be assumed in what follows (see Malomuzh, Oleinik and Fisher, 1973) that $g(r)$, $\partial g(r)/\partial\rho$ and $\partial^2 g(r)/\partial\rho^2$, regarded as functions of r, are continuous at small distances and integrable in all states of the assembly, including the critical point.

Then the force equation, (2.13), can be utilised to gain information about the three-particle correlation function g_3 in the vicinity of T_c. In particular, if we write $r=|\mathbf{r}_2-\mathbf{r}_1|$, $R=|\mathbf{r}_3-\mathbf{r}_1|$, and deal with the limiting case $R \gg r$, then (see Fisher, 1972)

$$g_3(\mathbf{rR}) = g(r) + \sum_l A_l(rR) P_l(\cos\vartheta) \tag{9.17}$$

As remarked in chapter 2, only the $l=1$ term actually enters the force equation. Using eqn (2.13), the expression for p, and the fact that $S(0) \to \infty$ at T_c, it can be shown that

$$A_1(rR) = (\tfrac{1}{2})\, r \left[2g(r) + \rho \frac{\partial g(r)}{\partial \rho} \right] g'(R) + \text{terms falling off faster than } R^{-3}. \tag{9.18}$$

Similarly from eqn (2.20), it can be shown that the $l=0$ term $A_0(rR)$ has the asymptotic form

$$A_0(rR)=\left[2g(r)+\rho\frac{\partial g(r)}{\partial\rho}\right][g(R)-1]+\text{terms falling off faster than } R^{-3}. \quad (9.19)$$

Finally, the $l=2$ term is given by

$$A_2(rR)=\tfrac{1}{8}r^2\left[2g(r)+\rho\frac{\partial g(r)}{\partial\rho}\right][g''(R)-g'(R)/R]$$
$$+\text{terms falling off faster than } R^{-3} \quad (9.20)$$

while $A_l(rR)$, $l\geqslant 3$ falls off faster than R^{-3}.

The above results, with the assumptions stated, follow from the distribution function theory of liquids, near T_c. If, however, we are willing to combine these results for A_0, A_1 and A_2 with the bold assumption of Polyakov (1970), based on arguments of conformal invariance (see also Fisher (1973)), that

$$\langle\rho(\boldsymbol{r}_1)\rho(\boldsymbol{r}_2)\rho(\boldsymbol{r}_3)\rangle\approx\frac{\text{constant}}{|\boldsymbol{r}_1-\boldsymbol{r}_2|^{\beta/\nu}|\boldsymbol{r}_2-\boldsymbol{r}_3|^{\beta/\nu}|\boldsymbol{r}_3-\boldsymbol{r}_1|^{\beta/\nu}} \quad (9.21)$$

then it follows that

$$\frac{\beta}{\nu}=\frac{(1+\eta)}{2} \quad (9.22)$$

This, as discussed further below, is one of the so-called scaling relations between critical exponents. We need to emphasise here that while the form (9.21) is *consistent* with the microscopic theory derived above, it does *not* follow in any obvious general way from microscopic considerations. A parallel discussion of four-particle correlations to that based on eqn (9.17) for g_3 has been given by Malomuzh, Oleinik and Fisher (1973).

This is the point at which to note that a study of critical phenomena in a wider context than the liquid–vapour critical point has led, both experimentally and theoretically, to the conclusion that there are *universal* features associated with critical behaviour which transcend individual examples, such as magnetic critical points, liquid–vapour point, etc. In particular, for a very wide class of problems, the critical indices seem to be the same for different systems. This cannot, of course, be seen from any discussion restricted to liquid state equations. We have reached the point, therefore, at which we must turn to more general arguments, based on scaling properties. Although we pose the following considerations mainly in the language of fluids, we shall not hesitate to use examples from magnetism when they prove helpful in this wider context.

9.3 Scaling exponents for fluids

Let us consider a plot of the coexistence curve on the E–ρ plane. It is as shown in *figure 9.1*, where the tangent to the coexistence curve is also drawn. The 'tie' lines

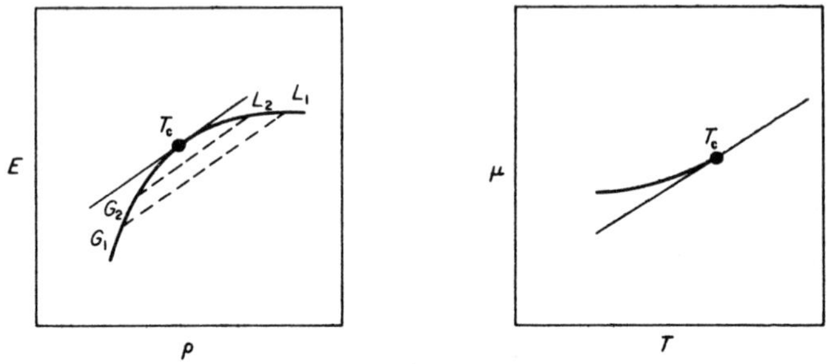

Figure 9.1 Schematic drawing of the liquid–vapour coexistence curve in the E–ρ plane and in the μ–T plane

shown relate liquid states L_1 and L_2 to gaseous states G_1 and G_2 and are parallel to the tangent at T_c. Then one of the important extensive variables in the liquid–gas critical point is

$$\Delta E \rightarrow [E - E_c] - \text{const}\,[\rho - \rho_c] \tag{9.23}$$

where the constant is given by the slope of the tangent to the coexistence curve at the critical point.

The 'order' parameter in the liquid–gas transition is $\rho - \rho_c$, while its intensive conjugate variable is $(\mu - \mu_c) - c(T - T_c)$. The intensive conjugate variable to ΔE is $\Delta T = T - T_c$. The μ–T plane, with the coexistence curve ending at the critical point T_c, is also shown in *figure 9.1.*

Now all of the variables discussed are related to the correlation length ξ, and we introduce therefore scaling exponents through

$$\text{order parameter} \sim \xi^{-x} \tag{9.24}$$

$$\Delta E \sim \xi^{-x_e} \tag{9.25}$$

and the corresponding intensive conjugate variables are given by

$$(\mu - \mu_c) - c(T - T_c) \sim \xi^{-y} \tag{9.26}$$

and

$$\Delta T = T - T_c \sim \xi^{-y_e} \tag{9.27}$$

It is a consequence of thermodynamics that

$$\begin{aligned} x + y &= d \\ x_e + y_e &= d \end{aligned} \tag{9.28}$$

d being the dimensionality of the system.

9.3.1 Thermodynamic scaling

We now introduce the hypothesis (which is actually contained within the renormalisation group—see below) that the singular part of the free energy is a generalised homogeneous function (see Widom, 1965; Domb and Hunter, 1965)

$$F(\lambda^{y_e}\Delta T, \lambda^y H) = \lambda^d F(\Delta T, H) \tag{9.29}$$

Differentiating with respect to H to obtain the equation of state (we use the magnetic analogy, where the magnetisation M is the order parameter and H is the conjugate variable) yields

$$M(\lambda^{y_e}\Delta T, \lambda^y H) = \lambda^{d-y} M(\Delta T, H) \tag{9.30}$$

or choosing $\lambda = |\Delta T|^{-1/y_e}$

$$M(\Delta T, H) = |\Delta T)^{\beta} M(\pm 1, H/|\Delta T|^{\Delta}) \tag{9.31}$$

where $\Delta = y/y_e$ and $\beta = (d/y_e) - \Delta$. The ± 1 sign arises from the ratio of ΔT to its absolute value; there are two possibilities, $T > T_c$ or $T < T_c$.

Ho and Lister (1969) have verified such thermodynamic scaling by experiments on the ferromagnet $CrBr_3$ (for further experimental discussions, see Heller, 1967). They plot $h = |H|/|\Delta T|^{\gamma+\beta}$ versus $m = |M|/|\Delta T|^{\beta}$, where $\gamma + \beta = \beta\delta = \Delta$. For fluids, (see, e.g. Domb and Green, (1974)),

$$H = \pm |M|^{\delta} h(\Delta T/|M|^{1/\beta}), \quad \delta = \Delta/\beta \tag{9.32}$$

and thermodynamic scaling has again been verified experimentally at the critical point of He^4.

Since all exponents depend on two, say y and y_e, there must be relations between exponents. These (see Fisher, 1967; Stanley, 1971; Jones and March, 1973) include

$$2 - \alpha = \beta(\delta + 1) \tag{9.33}$$

and

$$\gamma = \beta(\delta - 1) \tag{9.34}$$

For He^4 it is found that $\delta = 4.45$ and $\beta = 0.359$ from experiment. Here α is the exponent connected with specific heat through

$$C = A(\Delta T)^{-\alpha} \tag{9.35}$$

while we have already defined β and γ for a fluid, the corresponding equations for a ferromagnet being

$$\chi = \Gamma(\Delta T)^{-\gamma} \tag{9.36}$$

and

$$M = B(\Delta T)^{\beta}, \qquad |H| = D|M|^{\delta} \tag{9.37}$$

Classically, as we saw above, $\beta = \frac{1}{2}$ but in practice it is much nearer to $\frac{1}{3}$.

Thermodynamic scaling predictions can only be true in a very restricted region of the $(\Delta T, H)$ plane. Indeed we are concerned only with the study of the most singular part of the free energy as both ΔT and H tend to zero.

9.3.2 Further remarks on length scaling

We turn from thermodynamic to length scaling (see Brout, 1974 for a review) and, of course, as we discussed above, a crucial feature of critical phenomena is the divergence of the correlation length ξ at T_c. This led Kadanoff (1966)—see also Kadanoff et al. (1967)—to construct a phenomenological theory using ξ as the unique divergent element.

In spin language, we shall define the correlation length as follows. Let μ_i be a spin variable at site i. The correlation function is then $S(\boldsymbol{R}_i - \boldsymbol{R}_j) = \langle\mu_i\mu_j\rangle - \langle\mu_i\rangle\langle\mu_j\rangle$, $\boldsymbol{R}_i$ being the position of the ith site. Translational symmetry requires that, as indicated, S is a function only of the vector distance between the sites. The correlation length, apart from a possible multiplying constant, is then given by

$$\sum_{\mathbf{R}} R^2 S(R) \Big/ \sum_{\mathbf{R}} S(R) \equiv \xi^2 \equiv \kappa^{-2} \tag{9.38}$$

where $\kappa \to 0$ as $T \to T_c$. From our earlier considerations $\kappa \propto t^{\nu}$ $(t = |T - T_c|/T_c > 0,$ $H = 0)$.

Kadanoff now achieves the homogeneity form of the free energy F by dividing the system into cells of linear dimension $L(L \ll \xi)$ and using the collection of cells to define a new Ising model, characterised by an effective moment per cell and an effective interaction, or alternatively by an effective field H and an effective reduced temperature t. He assumes these to scale with a power of L, via

$$H_{\text{eff}} = L^{1/y} H, \qquad t_{\text{eff}} = L^{1/x} t \tag{9.39}$$

Since there are N/L^d cells, d being the dimensionality, and the cell partition function is assumed the same as that of the original assembly, one gets the condition

$$L^{-d} F(L^{1/y} H, L^{1/x} t) = F(H, t) \tag{9.40}$$

This is of the form required by homogeneity of the free energy provided

$$2 - \alpha = xd, \quad \Delta = x/y \tag{9.41}$$

Further useful relations can be obtained by requiring that at $H = 0$ the correlation function of the cell model scales according to

$$S(t, R) = C_L S(t_{\text{eff}}, R/L) = C_L S(L^{1/x} t, R/L) \tag{9.42}$$

where C_L has to be used to (re)normalise $S(R)$. The general solution, then, is of the form

$$S(t, R) = C(t)\tilde{S}(t^x R) \tag{9.43}$$

From (9.42) and the relation $\kappa \propto t^{\nu}$, we then obtain $x=\nu$ and hence

$$2-\alpha=\nu d \tag{9.44}$$

This relation is exact for the Onsager (1944) solution of the 2-dimensional Ising model. Deviations from numerical calculations in three dimensions appear at the time of writing to be 2 or 3%; ν being 0.64 for the Ising model (and $\alpha=\frac{1}{8}$).

9.3.2.1 Scaling of correlation function
Evidently from (9.43), length scaling gives

$$S(R)=C(t)\tilde{S}(\kappa R) \tag{9.45}$$

since $\kappa \propto t^{\nu}$ and $x=\nu$. $C(t)$ is to be determined from

$$\chi=\frac{\partial^2 F}{\partial H^2}=\int S(R)\,\mathrm{d}^{\mathrm{d}}\boldsymbol{R} \propto t^{-\gamma} \tag{9.46}$$

and hence

$$\begin{aligned} C(t)\int \tilde{S}(\kappa R)\,\mathrm{d}^{\mathrm{d}}\boldsymbol{R} &\sim C(t)\kappa^{-\mathrm{d}} \\ &\propto C(t)t^{-\nu\mathrm{d}} \\ &=t^{-\gamma} \end{aligned} \tag{9.47}$$

or

$$C(t)\sim t^{-\gamma+\nu\mathrm{d}} \tag{9.48}$$

Hence the Fourier transform of S is

$$S(k)\propto t^{-\gamma}\tilde{\tilde{S}}(k/\kappa) \tag{9.49}$$

At the critical point, as discussed earlier in eqn (9.16), one writes

$$S(R)\sim R^{-\mathrm{d}+2-\eta}, \qquad S(k)\sim k^{-2+\eta} \tag{9.50}$$

But for k finite, eqns (9.49) and (9.50) indicate that $\lim_{t\to 0} S(k)$ exists. Therefore the $t^{-\gamma}$ singularity in (9.49) must cancel and hence

$$\tilde{\tilde{S}}(x)\sim x^{-\gamma/\nu} \tag{9.51}$$

and

$$\lim_{t\to 0} S(k)\sim k^{-\gamma/\nu} \tag{9.52}$$

Comparison with (9.50) then yields

$$\gamma=(2-\eta)\nu \tag{9.53}$$

a relation confirmed by the available numerical calculations (for the 3-dimensional Ising model, $\gamma=\frac{5}{4}, \nu=0.64$, $\eta=0.055$, in agreement with (9.53)).

It is worth remarking here, as Jasnow, Moore and Wortes (1969) have pointed out, that a weaker condition than (9.49) is sufficient for (9.53) to hold. This is *not* true on the other hand for (9.44).

Though Kadanoff's ideas are therefore very important and *almost* true, at the time of writing anomalies appear to exist which prevent this approach being fully quantitative.

9.4 Introduction to renormalisation: the Gell–Mann–Low argument

The Ornstein–Zernike discussion of section 9.1.2 shows that, at T_c, the structure factor $S(k)$ behaves as k^{-2}, in at least the crudest form of the theory. Also the reciprocal correlation length κ is zero, as we saw at T_c. The theory of renormalisation (with $\kappa=0$) starts from a result of Gell–Mann and Low (1951) that the Green function contains an arbitrariness of scale. This arbitrariness can be transferred to a choice of reference momentum λ, with $\lambda \ll \Lambda$, Λ being some cut-off momentum. The structure factor S (propagator) is referred to this momentum. The first step then picks out the anomalous part by writing

$$S(k)=k^{-2}s(k/\Lambda) \tag{9.54}$$

The renormalisation approach, as we shall see below, then relates $s(k/\Lambda)$ to $s(\lambda/\Lambda)$. Since s is dimensionless, its k dependence can only involve the remaining length of the problem, the cut-off Λ^{-1}.

From work in field theory it is known that the characteristic of the self-energy of a renormalisable field theory (see for instance Bogoliubov and Shirkov, 1959) is that it has a quadratic divergence in Λ^2 and a logarithmically divergent derivative (the limit is $\Lambda\to\infty$, k fixed). The relation between the self-energy Σ and S is explicitly

$$S^{-1}(k)=k^2+\Sigma(0)-\Sigma(k^2) \tag{9.55}$$

where this form already includes the statement that the Ornstein–Zernike length κ^{-1} tends to infinity. By Taylor expansion around $k=0$, the logarithmic divergence of the derivative of Σ, say Σ', implies that S^{-1} has logarithmically divergent terms. The achievement of renormalisation is that these terms are removed. Since a straightforward expansion in k^2 is not possible (e.g. in lowest order Σ' contains $\ln(k/\Lambda)$), renormalisation must be carried out about some finite momentum λ, finite implying $\lambda\neq 0$ but $\lambda/\Lambda\to 0$.

The renormalisation amplitude $Z(\lambda)$ is introduced through

$$S_R(\lambda)=Z^{-1}(\lambda)S(\lambda)=\lambda^{-2} \tag{9.56}$$

Therefore at $k=\lambda$, rewriting (9.55) in the identical form

$$S^{-1}(\lambda)=[1-\Sigma'(\lambda)]\left[\lambda^2+\frac{\Sigma(0)-\Sigma(\lambda)+\lambda^2\Sigma'(\lambda)}{1-\Sigma'(\lambda)}\right] \tag{9.57}$$

the second term in the second square bracket involves at most a second derivative divided by a first derivative and has a vanishing limit as $\Lambda\to\infty$. Eqns (9.56) and (9.57) then yield

$$Z(\lambda)=[1-\Sigma'(\lambda)]^{-1} \tag{9.58}$$

By the same reasoning it can be shown that

$$\lim_{\Lambda\to\infty}[\Sigma(k^2)-\Sigma(0)]/[1-\Sigma'(\lambda)]$$

exists. This then establishes that $S_{R\lambda}(k)\equiv Z^{-1}(\lambda)S(k)$ has a finite limit as $\Lambda\to\infty$. Hence we find

$$S_{R\lambda}(\lambda)=Z^{-1}(\lambda/\Lambda)S(\lambda)=\lambda^{-2} \tag{9.59}$$

and

$$\begin{aligned} S_{R\lambda}(k)&=Z^{-1}(\lambda/\Lambda)S(k)\\ &=Z^{-1}(\lambda/\Lambda)k^{-2}s(k/\Lambda)\\ &=k^{-2}S_{R\lambda}(k/\lambda) \end{aligned} \tag{9.60}$$

These equations yield

$$S_{R\lambda}(k/\lambda)=Z^{-1}(\lambda/\Lambda)s(k/\Lambda), \qquad S_{R\lambda}(1)=1 \tag{9.61}$$

which has the unique solution of a power law,

$$S_{R\lambda}=(k/\lambda)^\eta, \qquad Z=A(\lambda/\Lambda)^\eta, \qquad S=A(k/\Lambda)^\eta \tag{9.62}$$

Unitarity requires $\eta>0$ and hence we conclude that $Z\to0$ as $\lambda/\Lambda\to0$. Thus, we regain the generalised Ornstein–Zernike form

$$S(k)\propto k^{-2}(k/\Lambda)^\eta \tag{9.63}$$

A generalisation of this argument can be effected, for $T\neq T_c$, i.e. $\kappa\neq0$. The point to be watched is that $\Sigma'(0)$ is infinite. The renormalisation formalism handles this by writing

$$\begin{aligned} S_R^{-1}&=ZS^{-1}=k^2+\kappa^2-Z[\Sigma(k^2)-\Sigma(0)-k^2\Sigma'(0)]\\ &=k^2+\kappa^2-\Sigma_R(k^2) \end{aligned} \tag{9.64}$$

where Σ_R is the renormalised self-energy. This evidently satisfies $\Sigma_R(0)=0$. Though we omit all details, the calculation of the self-energy graphs turns out to be expressed in terms of S_R itself. In this way, κ and Z are determined self-consistently.

In appendix 9.1, one way by which, at least in principle, a calculation of the renormalisation amplitude Z, along with an appropriate Green's function, can yield explicit forms for the critical exponents is outlined. This is referred to as asymptotic scaling, the basic relation being the Callan–Symanzik equation (A9.1.17). An alternative route which permits, at least in principle, the direct calculation of critical exponents from the renormalisation group has been given by Di Castro (1972).

9.5 Critical exponents by expansion methods

According to present knowledge of critical phenomena in second-order phase transitions, the critical exponents appear to be universal functions of (*a*) the dimensionality d of the system, and (*b*) the number n of components of the order parameter, unless the forces are long range. In this latter case an additional parameter comes in.

Although the present range of interest lies in the domain $0 \leqslant n \leqslant 3$, $0 \leqslant d \leqslant 3$, it proves of considerable interest to examine:

(1) The line $d=4$, and its neighbourhood where $\varepsilon = 4-d$ is small (Wilson, 1972).

(2) The limit $n \to \infty$, and the vicinity where n^{-1} is small (Ma, 1973; Brezin and Wallace, 1973).

It is possible, then, by generalising to the (n, d) plane, to make predictions on trends of variations of the critical temperature for a given system and of the critical exponents as functions of n and d. Interpolation methods then allow useful predictions on the critical exponents of systems not yet studied experimentally.

The boundaries are often useful starting points for expansion methods, namely the $\varepsilon = (4-d)$ and $1/n$ expansions. Some attention has also been given to the cases $n = -2$ (d arbitrary) and $d = 1$ and 0 (n arbitrary) (see Toulouse, 1973).

9.5.1 The ε expansion

One of the main achievements in the application of field theory to critical phenomena has been to make clear that the line $d=4$ separates the mechanisms involved. Thus, $d>4$ obeys the molecular field theory of section 9.1 and $d<4$ the scaling type theories discussed in sections 9.3 and 9.4.

The physically interesting case is $\varepsilon > 0$: the three-dimensional case corresponding to $\varepsilon = 1$. Not only is ε good for qualitative considerations, but it seems useful as, at very least, a semi-quantitative tool.

The results of the ε expansion (Wilson, 1972; Wilson and Fisher, 1972) are:

$$\gamma = 1 + \frac{1}{2}\frac{n+2}{n+8}\varepsilon + \frac{1}{4}\frac{(n+2)(n^2+22n+52)}{(n+8)^3}\varepsilon^2 + O(\varepsilon^3) \tag{9.65}$$

and

$$\eta = \frac{1}{2}\frac{n+2}{(n+8)^2}\varepsilon^2 + \frac{1}{2}\frac{n+2}{(n+8)^2}\left[6\frac{3n+14}{(n+8)^2} - \frac{1}{4}\right]\varepsilon^3 + O(\varepsilon^4) \tag{9.66}$$

from which all other indices can be calculated by means of the scaling relations given earlier in eqns (9.33), (9.34), (9.44) and (9.53).

The value of γ obtained for the Ising model is 1.244, compared with the best calculation of 5/4: remarkable agreement. η is much less well given, often in error by 50–100%.

9.5.2 Remarks on Lagrangians and Hamiltonians

To conclude these remarks on renormalisation, we want to comment that Landau, long ago, on phenomenological grounds, wrote down a Lagrangian L of the form

$$L=\frac{1}{2}\int \mathrm{d}^d\boldsymbol{r}\left[|\nabla M|^2+\lambda_2 M^2+\sum_{n=2}^{\infty}\{\lambda_{2n}/(2n)!\}M^{2n}\right] \tag{9.67}$$

which can be used to calculate critical fluctuations of wave number k such that $ka \ll 1$, a being the interatomic spacing.

Problems which appear to fall into this class are (*a*) the self-avoiding walks ($n=0$); (*b*) the Ising model ($n=1$); (*c*) the λ point of He^4 ($n=2$); (*d*) the Heisenberg model ($n=2$); and (*e*) the spherical model ($n=\infty$).

It appears that in the liquid–vapour critical phenomena, the Lagrangian would have odd terms in the order parameter†. Brout (1974) conjectures that whereas $\delta=5$ seems consistent with (*a*)–(*e*) above, the liquid–vapour critical point has $\delta<5$, this being connected with the odd terms in the Lagrangian.

The final concept we should at least refer to here is Wilson's idea of a fixed point in the space of the Hamiltonians. One has reached a fixed point when one has renormalised the Hamiltonian to a form which is invariant to the form and behaviour of the starting parameters, always giving the *same* final Hamiltonian, with perhaps different values of the coefficients.

9.6 Critical exponents determined from light scattering

9.6.1 Critical opalescence

In an earlier chapter (section 4.3.5) we explicitly related the van Hove function $S(k, \omega)$ to the intensity of scattered light. Our interest here is in critical opalescence, and in particular we shall consider the example of light scattered in the critical region by carbon dioxide, according to the calculations of Mountain (1966). This system has $T_c=31$ C, $p_c=73$ atm, $\rho_c=236$ amagat (for CO_2, 1 amagat$=0.001\,98$ g/cm^3).

Figure 9.2 shows the width of the central line, while the shift of the Brillouin lines is shown in *figure 9.3*. The fraction of the light contained in the central line $(1-c_V/c_p)$ is shown in *figure 9.4*.

In *figure 9.5*, the quantity $\sigma(k, \omega)=S(k, \omega)/S(k)$ is plotted against ω for CO_2 at $T=32.1$ C, $\rho=230$ amagat and $k=10^5$ cm^{-1}. Near the critical point the central line of $\sigma(k, \omega)$ is dominant, due to the large value of c_p/c_V. As the scattering angle is decreased, the width of the central line is strongly reduced, since $k \propto \sin\frac{1}{2}\vartheta$, ϑ being the scattering angle.

†See, however, Hubbard and Schofield (1972) who apply the Wilson theory to the liquid–vapour critical point. A fixed point, relating to that of the Ising model and giving the same critical indices, is located.

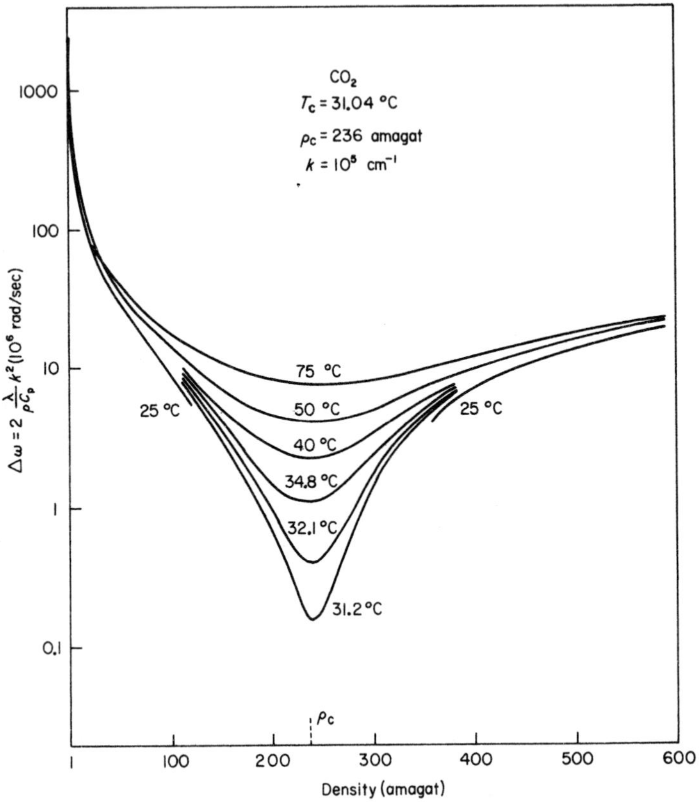

Figure 9.2 Calculated half-width of the Rayleigh peak (for $k = 10^5$ cm^{-1}) in carbon dioxide as a function of density and temperature, approaching the critical point (from Mountain, 1966)

Finally, in *figure 9.6*, $\sigma(k, \omega)$ is shown for the same density and k value, but for a higher temperature $T = 75$ C. The width of the Brillouin lines was estimated by taking the bulk and shear viscosity to be equal.

9.6.2 Light scattering from xenon

Giglio and Benedek (1969) have measured using a helium–neon laser source the angular dependence of the intensity of light scattered from xenon near its critical point. Ornstein and Zernike (1914; 1918) predicted that near the critical point of a single-component fluid, spatial correlations in the density fluctuations would produce angular anisotropy in the intensity of the scattered light. Until

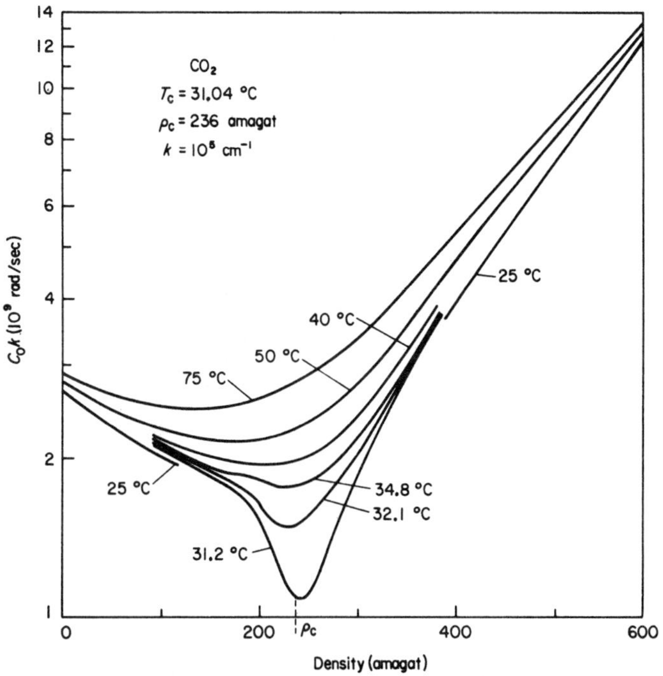

Figure 9.3 Calculated shift of the Brillouin lines (for $k = 10^5$ cm^{-1}) in carbon dioxide as a function of density and temperature, approaching the critical point (from Mountain, 1966)

the work of Giglio and Benedek, there does not seem to have been any direct experimental confirmation of their prediction.

They find, at all temperatures and densities they have studied, that the reciprocal of the scattering intensity varies linearly with $\sin^2(\frac{1}{2}\vartheta)$, ϑ being the scattering angle, as Ornstein and Zernike predicted. Their data provide an accurate measurement of the correlation length ξ and the isothermal compressibility K_T. The critical exponent ν determining the behaviour of the correlation length is found to have the value $\nu = 0.57 \pm 0.05$.

9.7 Dynamics in critical phenomena

The oldest theory of critical dynamics goes back to van Hove (1954*b*). To explain this in general terms, consider the usual time-dependent correlation function

$$\langle \delta Q(\boldsymbol{r}, t) \delta Q(0, 0) \rangle$$

for a variable Q, δQ being the fluctuation of this variable about its average value. As usual, we work with its Fourier transform, $Q(\boldsymbol{k}, \omega)$.

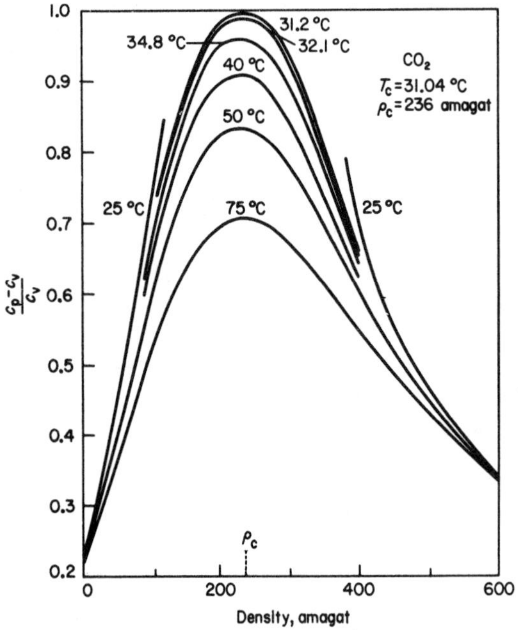

Figure 9.4 Calculated fraction of scattered light contained in the central line for carbon dioxide near the critical point (from Mountain, 1966)

Consider the situation in which $Q(\boldsymbol{r}, t)$ relaxes to its equilibrium value via a single, non-propagating mode. If Q is chosen such that it is conserved by the equations of motion of the system (i.e. $\int \mathrm{d}\boldsymbol{r}\, Q(\boldsymbol{r}, t)$ is independent of time), then, at any temperature $T \neq T_c$, the relaxation of Q as $k \rightarrow 0$ is given by a diffusion equation. Then the characteristic frequency associated with the relaxation rate of Q can be written as

$$\bar{\omega} = \lambda_Q k^2 / \chi_Q \tag{9.68}$$

where λ_Q is the transport coefficient for the variable Q, while χ_Q is the susceptibility describing the linear response of the system to an applied field conjugate to Q. In the case of entropy fluctuations, for example, λ_E is the thermal conductivity while χ_E is the heat capacity per unit volume at constant pressure, ρc_p, and hence

$$\bar{\omega}_E = \lambda_E k^2 / \rho c_p \tag{9.69}$$

If the variable Q is not conserved, on the other hand, then the relaxation to equilibrium should take place at a finite rate even at $k = 0$ and hence

$$\bar{\omega} = \Lambda_Q / \chi_Q \tag{9.70}$$

where Λ_Q is the 'kinetic coefficient' for Q.

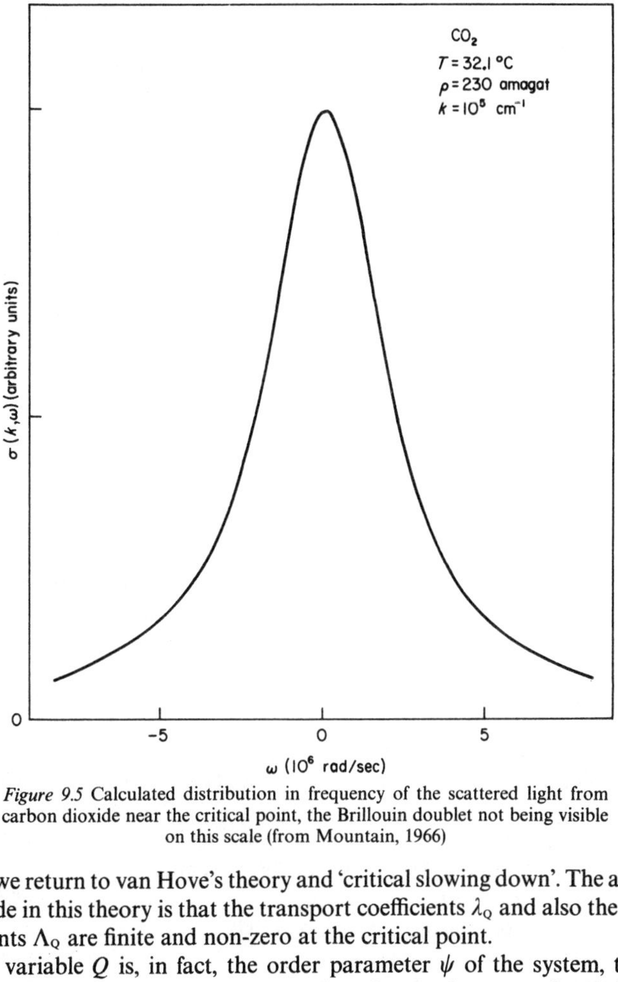

Figure 9.5 Calculated distribution in frequency of the scattered light from carbon dioxide near the critical point, the Brillouin doublet not being visible on this scale (from Mountain, 1966)

Now we return to van Hove's theory and 'critical slowing down'. The assumption made in this theory is that the transport coefficients λ_Q and also the kinetic coefficients Λ_Q are finite and non-zero at the critical point.

If the variable Q is, in fact, the order parameter ψ of the system, then χ_Q diverges near T_c as the inverse coherence length κ, to the power $(2-\eta)$, i.e.

$$\chi_Q \propto \kappa^{-2+\eta} \tag{9.71}$$

where η is a small positive quantity. If the relaxation of ψ is described by a single non-propagating mode, then the van Hove theory predicts a critical slowing down of the order parameter fluctuations as $k \to 0$, near T_c, with

$$\varpi \propto \kappa^{2-\eta} k^2 \qquad \text{if } \psi \text{ is conserved} \tag{9.72}$$

$$\varpi \propto \kappa^{2-\eta} \qquad \text{if } \psi \text{ is not conserved} \tag{9.73}$$

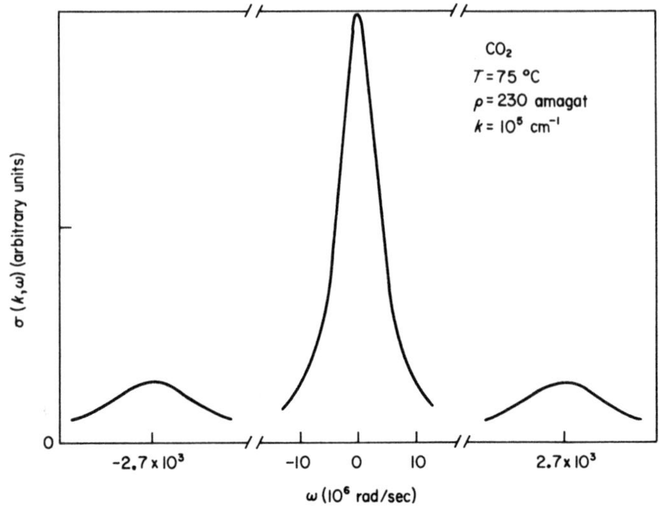

Figure 9.6 Calculated distribution in frequency of the scattered light from carbon dioxide away from the critical point (from Mountain, 1966)

Unfortunately, there are a variety of situations in which both theory and experiment are *not* in agreement with the predictions of this theory.

We are specifically interested here in the liquid–gas critical point, as well as in the critical point for phase separation (see below). In both these cases, the order parameter is conserved and is dominated by a non-propagating diffusive mode at long wavelength, both above and below the transition. For instance, in the liquid–gas transition the order parameter may be taken to be $\rho - \rho_c$, and long wavelength density fluctuations near T_c are dominated by entropy fluctuations at constant pressure, whose relaxation is described by the thermal conductivity formula (9.69). Nevertheless, a mode–mode analysis (see for example Kawasaki, 1970; Stanley, 1971) shows that the conventional theory is not correct It is also of interest to mention here the work of Swift and Kadanoff (1968) on transport coefficients near the λ-transition of He^4.

9.7.1 Comments on universality

Motivating the discussion is the notion of 'universality' for dynamic critical phenomena.

As in the case of static critical phenomena, universality does *not* mean that all systems have the same critical exponents, but that there exist large classes of systems with the *same* critical behaviour. As an example, all Ising models of a given lattice dimensionality *d*, with short-range forces, are supposed to have the *same* static critical behaviour, this being independent of the details of the underlying lattice.

In the language of the renormalisation group, details of the lattice are irrelevant variables, which disappear from the problem near T_c.

Renormalisation group ideas applied to dynamic critical phenomena clearly imply that most of the details of the dynamics are also irrelevant near T_c and that large classes of systems will show the same dynamic critical behaviour. It is clear from various examples, however, that a class of systems showing the same static critical behaviour may be divided into several classes of dynamic behaviour. One familiar example is the classical isotropic Heisenberg ferromagnets and antiferromagnets: they have identical equilibrium properties but *very* different time dependences, both in the hydrodynamic and in the critical regime. It should be said here, however, that the dynamical characteristics of the liquid–vapour critical phenomena may have somewhat complex features. For a discussion of these difficulties, the reader may refer to Kadanoff (1970).

9.7.2 Dynamic scaling

Following Halperin and Hohenberg (1967)—see also Halperin (1973)—we shall say that the correlation function $Q(\mathbf{k}, \omega)$ obeys dynamic scaling if:

(1) At $T=T_c$, for sufficiently small $\mathbf{k}$, $Q(\mathbf{k}, \omega)$ has the form

$$\left.\begin{aligned} Q(\mathbf{k}, \omega) &\sim 2\pi\bar{\omega}_{\mathbf{k}}^{-1} Q(\mathbf{k}) f_Q(\omega/\bar{\omega}_{\mathbf{k}}) \\ \bar{\omega}_k &= \text{const}\, k^z \end{aligned}\right\} \tag{9.74}$$

where $Q(\mathbf{k})$ is the Fourier transform of the equal-time correlation function, z is an as yet unspecified exponent and f_Q is a function normalised to unity and depending only on $\omega/\bar{\omega}_{\mathbf{k}}$. The dynamic scaling hypothesis supposes also that at T_c, f_Q is a well-behaved function of its argument, with a characteristic width of order unity. Thus the spectrum of $Q(\mathbf{k}, \omega)$, at fixed $\mathbf{k}$, has a single relaxation frequency, $\bar{\omega}_{\mathbf{k}}$.

(2) For temperatures near T_c, the dynamic scaling hypothesis requires that the function $Q(\mathbf{k}, \omega)$ has a form similar to (9.74), but now the functions depend in an essential way on the ratio k/κ, where κ, the inverse coherence length, goes to zero at T_c. (Along the coexistence curve or the 'critical isochore', κ varies as $(T-T_c)^\nu$). For example, away from the critical point, $\bar{\omega}_k$ is assumed to take the form

$$\bar{\omega}_{\mathbf{k}} = const\; k^z \Omega(k/\kappa) \tag{9.75}$$

where Ω depends on k only through the ratio k/κ. In the limit $k/\kappa \to 0$, the characteristic frequency will have the form

$$\bar{\omega} \sim k^z (k/\kappa)^x \tag{9.76}$$

where x is determined by macroscopic or hydrodynamic considerations. Usually x will be 0, 1 or 2, depending on the variable Q under consideration.

As discussed in detail by Halperin (1973), the exponent z depends not only on whether or not the order parameter ψ is conserved, but also on whether or not the energy is conserved. When ψ is not conserved, conventional theory would yield $z=2-\eta$, whereas when ψ is conserved, $z=4-\eta$.

When neither ψ nor E are conserved, the renormalisation methods have been applied to some simple models by both the ε expansion and the $1/n$ expansion. The exponent z in this case is found to be greater than the conventional results by terms of $0(\varepsilon^2)$ or $0(1/n)$.

When ψ is conserved but not E, evidence points to the applicability of conventional theory to all orders in ε. When ψ is not conserved but E is, the conventional theory again appears to fail, as discussed more fully by Halperin, Hohenberg and Ma (1972) – see also Halperin (1973).

9.7.3 Measurement of kinematic viscosity by light scattering

To conclude this discussion we should mention that Zolleweg, Hawkins and Benedek (1971) have measured the effective kinematic viscosity of xenon along its coexistence curve in the temperature range $0.070\,K < T_c - T < 5\,K$ using light scattering. Their finding is that the effective kinematic viscosity is slowly varying along the coexistence curve and is given explicitly by

$$\eta_k = [448 + 12(T_c - T)] \pm 15 \text{ mS} \tag{9.77}$$

9.8 Phase separation in alloys and critical point of mixing

We have discussed earlier the relation for the mean square fluctuation in the concentration Δc, defined by

$$\Delta c = N^{-1}[(1-c)\Delta N_1 - c\Delta N_2] \tag{9.78}$$

where N is the total number of atoms in the alloy, N_1 being the number of type 1 and N_2 of type 2. In a conformal solution we saw that

$$N\langle(\Delta c)^2\rangle = S_{cc}(0) = \frac{c(1-c)}{1+2c(1-c)w/k_BT} \tag{9.79}$$

where w is the interchange energy

It can be seen that if w is negative then the solution has a critical point of mixing, the critical temperature T_c being given by

$$T_c = \tfrac{1}{2}|w|/k_B \tag{9.80}$$

the maximum in $S_{cc}(0)$ always occurring at $c=\frac{1}{2}$ in a conformal solution.

However, we should caution the reader that the above argument is oversimplified. Thus, the expression for the free energy of mixing, namely

$$\Delta G_m \simeq Nk_BT[c \ln c + (1-c)\ln(1-c)] - Nc(1-c)w \tag{9.81}$$

from which (9.79) results is too simple to adequately describe partially miscible liquids and the thermodynamic theory of fluctuations breaks down near the critical point. The main point of the argument is that $\langle(\Delta c)^2\rangle$ diverges at the critical temperature of phase separation.

9.8.1 Inelastic neutron scattering in nearly critical metal alloys

In the following discussion on phase separation in alloys we shall focus on some neutron scattering results obtained for liquid metal mixtures (Brumberger, Alexandropoulos and Claffey, 1967; Wignall and Egelstaff, 1968).

Earlier studies of the critical scattering of light and X-rays from binary solutions were made on complex molecular systems and have been reviewed by Egelstaff and Ring (1968). There are many binary metal systems exhibiting a phase separation region (see Mott, 1957). These systems have the advantage over the earlier studies of being monatomic. However, only neutron or X-ray experiments are feasible because of the absorption of light by metals.

As discussed in chapter 6, if a mixture is irradiated by monochromatic radiation of wavelength λ, the intensity of coherent scattered radiation at an angle ϑ is given by (see for example Lomer and Low, 1965)

$$I(k)=\left\langle \sum_{\mathrm{ij}} a_{\mathrm{i}}a_{\mathrm{j}} \exp\left[i\boldsymbol{k}\cdot(\boldsymbol{R}_{\mathrm{i}}-\boldsymbol{R}_{\mathrm{j}})\right]\right\rangle \tag{9.82}$$

where $\boldsymbol{k}=\boldsymbol{k}_0-\boldsymbol{k}'$ (or $k\simeq 2k_0 \sin(\vartheta/2)$, if $k_0\simeq k'$), $\boldsymbol{k}_0$ and $\boldsymbol{k}'$ are the wave vectors of the incident and scattered neutrons, respectively, a_{i} and $\boldsymbol{R}_{\mathrm{i}}$ are the scattering length and position vector of the ith nucleus.

The expression derived by Wignall and Egelstaff (1968) for $I(0)$ can be written

$$I(0)=(c_1a_1+c_2a_2)^2\langle(\Delta N)^2\rangle + \frac{\rho N^2}{c_1}\{c_1{a_1}^2v_2+c_2{a_2}^2v_1-(c_1a_1+c_2a_2)^2\rho v_1v_2\}\langle(\Delta c_1)^2\rangle \tag{9.83}$$

Here v_{i} are the partial molar volumes of the components and $c_{\mathrm{i}}=N_{\mathrm{i}}/N$, where $N=N_1+N_2$. As emphasised above, $\langle(\Delta c)^2\rangle$ diverges at the temperature T_{c} for phase separation. However, $\langle(\Delta N)^2\rangle$ remains finite and thus the scattering near T_{c} is dominantly from the concentration fluctuations.

Wignall and Egelstaff analyse their data on Bi–Zn, Pb–Ga and Bi–Ga, using 5.3 Å neutrons, for small scattering angles, in terms of the result (see for example Fisher, 1964)

$$I(k)/F\simeq(c_1a_1+c_2a_2)^2\langle(\Delta N)^2\rangle+\frac{k_{\mathrm{B}}T}{c}\left(\frac{\partial c_1}{\partial\mu}\right)_{\mathrm{T,p}}\frac{[A(T)]^{2-\eta}}{[\{A(T)\}^2+k^2]^{1-(1/2)\eta}} \tag{9.84}$$

where $F=c_1{a_1}^2v_2+c_2{a_2}^2v_1-\rho v_1v_2(c_1a_1+c_2a_2)^2$. Here $A(T)$ is the reciprocal of the mean size of a fluctuation in concentration. At large scattering angles the oscillations in the liquid structure can be observed and then the intensity can no longer be approximated by the above equation.

The estimated critical compositions for the three systems are Bi 31%–Ga 69% at 262 C; Ga 58%–Pb 42% at 606 C; and Bi 17%–Zn 83% at 605 C. Analysing the experimental data using (9.84), with the usual exponents γ and ν defined through

$$(\partial c_1/\partial\mu)_{T,p} \propto |T-T_c|^{-\gamma}$$
$$A(t) \propto |T-T_c|^{\nu} \tag{9.85}$$

the value $\gamma=1.0\pm0.2$ was obtained from the data for Bi–Ga. For the same mixture, ν was estimated as 0.5 ± 0.15.

In further experiments on inelastic scattering of slow neutrons, Egelstaff and Wignall (1970) have studied the temperature dependence of critical opalescence in the Bi–Ga system for five different concentrations. These measurements allow estimates of the critical indices η and δ to be made. The results are $\eta<0.2$, $\delta=4.5\pm0.5$, consistent with the relation (9.53) namely

$$\eta=2-3\frac{\delta-1}{\delta+1} \tag{9.86}$$

9.8.2 Transport coefficients

The order parameter for the binary liquid case may be taken to be $x-x_c$, where x is the relative concentration of the two components. The characteristic frequency determining the rate of relaxation of the concentration fluctuations is given by

$$\varpi=\lambda_x k^2/\chi_x \tag{9.87}$$

where λ_x is the transport coefficient for relative diffusion of the two components, while χ_x is the susceptibility for concentration changes. This may be defined as the derivative with respect to x of the difference in chemical potentials of the species.

The susceptibility χ_x diverges as $\kappa^{-2+\eta}$ near T_c (as also in the liquid–gas transition). The mode–mode approach similarly predicts that, as with the liquid–gas transition, the transport coefficient diverges as κ^{-1}.

The divergence of the transport coefficient in the binary liquid can be understood by an argument due to Arcovito *et al.* (1969). Let us divide the liquid into regions the size of the coherence length, each having a volume κ^{-3}, and let $\delta\bar{x}$ be the deviation of the concentration in a given volume κ^{-3} from the average value in the sample as a whole.

The fluctuations in $\delta\bar{x}$ are given by the equipartition law as

$$\langle(\delta\bar{x})^2\rangle \simeq T\chi_x\kappa^3 \tag{9.88}$$

Now apply a field $\boldsymbol{F}$ which exerts a force proportional to the deviation in relative concentration $\delta\bar{x}$. Then the liquid in any given volume κ^{-3} will tend to drift in the field at a velocity v such that the applied force $f_{\text{app}}=\delta\bar{x}\kappa^{-3}\boldsymbol{F}$ is balanced by the viscous drag on the volume, which is given by Stokes' law as

$$f_{\text{vis}} \simeq -\eta^*\kappa^{-3}\kappa^2 v \tag{9.89}$$

where η^* is an effective viscous coefficient. It is then a simple matter to get v, add up the contributions of all the volumes κ^{-3} to get the transport coefficient

$$\lambda_x \simeq T\chi_x\kappa^3/\eta^*\kappa^2 \sim \kappa^{-1+\eta}/\eta^* \tag{9.90}$$

It follows that either λ_x diverges or η^* diverges or both coefficients diverge. In fact, detailed mode–mode coupling shows that both the true viscosity and the effective viscosity coefficient are *at most* weakly divergent. Thus, formula (9.90) essentially establishes the divergence of λ_x.

Finally, we note that Kawasaki (1970) has obtained an expression for the mutual diffusion coefficient D_{12} of a binary mixture in the neighbourhood of the critical temperature,

$$D_{12} = k_B T/(6\pi\eta\xi) \tag{9.91}$$

where η is the viscosity and ξ the correlation length. This is immediately seen to be analogous to the Einstein–Stokes equation $D = k_B T/(6\pi\eta a)$ for the mass diffusion coefficient for particles of radius a.

Berge and Dubois (1971) have combined measured data on D with that on viscosity in the case of a cyclohexane–aniline critical mixture. Their work verifies closely Kawasaki's formula for this system, near the critical temperature T_c. Its applicability over a wide range of temperatures above T_c is determined experimentally. It also allows a direct determination of the correlation length ξ in a domain where other methods are not very suitable.

CHAPTER 10

THE LIQUID SURFACE

10.1 Introduction

It will occasion no surprise that, when we come to apply molecular distribution function theory to surfaces, we must generalise some of the concepts. Most obviously, instead of the homogeneity of the one-body density in the bulk liquid, if we have a planar surface perpendicular to the z axis, then one of the basic quantities characterising the liquid surface is the density $\rho(z)$. This, of course, is a particular case of a general local density $\rho(\boldsymbol{r})$, giving the number of molecules per unit volume at position $\boldsymbol{r}$. Similarly, we define a two-body function $\rho^{(2)}(\boldsymbol{r}_1, \boldsymbol{r}_2)$ as the average density of pairs of molecules such that one molecule is situated in the element of volume $d\boldsymbol{r}_1$ and the other in $d\boldsymbol{r}_2$.

If $\phi(R)$ is the usual pair potential between two molecules with separation R, it can be shown (see appendix 10.1) that the pressure $p^t(z)$, normal to an element of area parallel to the z axis, is given by

$$p^t(z_1) = k_B T\rho(z_1) - \frac{1}{2}\int d\boldsymbol{R}\,\frac{d\phi}{dR}\,\frac{(x_2-x_1)^2}{R}\,\rho^{(2)}(z_1, \boldsymbol{R}) \tag{10.1}$$

a result due to Kirkwood and Buff (1949). Here the x axis is parallel to the surface of the liquid and perpendicular to the element of area across which the pressure p^t is exerted. It is clear that the first term in the pressure is a 'kinetic' contribution, while the second arises from the pair potential acting between molecules.

Away from the surface $p^t(z)$ becomes equal to p_0, the equilibrium pressure in the system. We can now argue that the pressure deficit $[p_0 - p^t(z)]$ in the surface layer manifests itself macroscopically as a tension exerted by the fluid on the walls of the container. The magnitude of this tension per unit length, which is evidently the surface tension σ, is given by

$$\sigma = \int_{-z_l}^{z_g} [p_0 - p^t(z)]\,dz \tag{10.2}$$

But since $p^t(z)$ differs from p_0 only in the surface region, the limits can be replaced by $-\infty$ and $+\infty$.

We can now use this equation to evaluate σ in terms of the molecular distribution functions. Thus we write

$$\sigma=\int_0^{\infty}[p_0-p^{t}(z)]\,dz+\int_{-\infty}^{0}[p_0-p^{t}(z)]\,dz \tag{10.3}$$

and, if we introduce the expression for $p^{t}(z)$ in terms of $\rho^{(2)}$, we find

$$\sigma=k_BT\int_0^{\infty}[\rho_g-\rho(z)]\,dz+k_BT\int_{-\infty}^{0}[\rho_l-\rho(z)]\,dz$$
$$+\tfrac{1}{2}\int_0^{\infty}dz\int d\mathbf{R}\,\frac{d\phi}{dR}\frac{(x_2-x_1)^2}{R}[\rho^{(2)}(z,\mathbf{R})-\rho_g{}^{(2)}(R)]$$
$$+\tfrac{1}{2}\int_{-\infty}^{0}dz\int d\mathbf{R}\,\frac{d\phi}{dR}\frac{(x_2-x_1)^2}{R}[\rho^{(2)}(z,\mathbf{R})-\rho_l{}^{(2)}(R)] \tag{10.4}$$

But we can choose the Gibbs surface as that for which the adsorption is zero, and then the sum of the first two integrals is zero. Thus we have for the surface tension the result

$$\sigma=\tfrac{1}{2}\int_0^{\infty}dz\int d\mathbf{R}\,\frac{d\phi}{dR}\frac{(x_2-x_1)^2}{R}[\rho^{(2)}(z,\mathbf{R})-\rho_g{}^{(2)}(R)]$$
$$+\tfrac{1}{2}\int_{-\infty}^{0}dz\int d\mathbf{R}\,\frac{d\phi}{dR}\frac{(x_2-x_1)^2}{R}[\rho^{(2)}(z,\mathbf{R})-\rho_l{}^{(2)}(R)] \tag{10.5}$$

This is an exact expression for a one-component fluid with pair interactions.

Needless to say, it is very difficult to calculate $\rho^{(2)}(z,\mathbf{R})$ and therefore all the work based on this method has had to relate $\rho^{(2)}$ to the bulk pair function plus the one-body density $\rho(z)$.

The most simple (though drastic) assumptions are: (*a*) to assume the gas is of negligible density, and (*b*) to take the liquid homogeneous right up to the dividing surface. Under these conditions, the above expression for σ becomes

$$\sigma=\frac{\pi}{8}\int_0^{\infty}\rho_l{}^{(2)}(R)\frac{d\phi}{dR}\quad R^4\,dR \tag{10.6}$$

$\rho_l{}^{(2)}(R)$ being now the bulk pair function in the liquid. We shall return to the numerical consequences of this formula below. However, we have used this introduction to an approach to calculate the surface tension σ to motivate a discussion of the way we can hope to calculate the density and the pair function in the presence of the surface. Most of the rest of this chapter is devoted to this problem.

10.2 Molecular distribution functions near a liquid surface

Let us work with distribution functions $\rho^{(n)}$ giving us probabilities of finding

n atoms simultaneously in $\mathrm{d}\boldsymbol{r}_1$ around $\boldsymbol{r}_1$, $\mathrm{d}\boldsymbol{r}_2$ around $\boldsymbol{r}_2 \ldots, \mathrm{d}\boldsymbol{r}_n$ around $\boldsymbol{r}_n$. Here we concern ourselves explicitly with the singlet distribution function $\rho(z)$, and with the pair function $\rho^{(2)} = \rho^2 g(\boldsymbol{r}_1, \boldsymbol{r}_2)$, ρ being the average density in the bulk liquid.

Before discussing the equation satisfied by $\rho(z)$ (involving, inevitably, $\rho^{(2)}$), let us note that for a planar surface, not only is $\rho(\boldsymbol{r}) \equiv \rho(z)$ but $g(\boldsymbol{r}_1, \boldsymbol{r}_2)$ does *not* depend on all six components of the vectors $\boldsymbol{r}_1$ and $\boldsymbol{r}_2$, but only on $|\boldsymbol{r}_1 - \boldsymbol{r}_2|$, z_1 and z_2. Needless to say, as we go deep into either the bulk liquid, or the gaseous phase, these correlation functions must go over into the correlation functions characteristic of the single phase.

When we generalise the hierarchical equations to allow for the inhomogeneity of the liquid surface, we obtain an equation, as remarked above, relating $\rho(z)$ and $\rho^{(2)}$, namely

$$k_B T \nabla \rho(\boldsymbol{r}_1) + \int \mathrm{d}\boldsymbol{r}_2 \nabla \phi(|\boldsymbol{r}_1 - \boldsymbol{r}_2|) \rho^{(2)}(z_1, z_2, |\boldsymbol{r}_1 - \boldsymbol{r}_2|) = 0 \tag{10.7}$$

For a detailed derivation of this equation, see Fisher and Bokut (1956); Fisher (1964). However, the essential physics contained in it is that the mean density in the surface of the liquid must adjust itself so that the mean pressure across every plane parallel to the surface is the same. Although equation (10.7) is exact as it stands, progress can only be made with it by inserting some approximation for $\rho^{(2)}$. It is clear on physical grounds that $\rho^{(2)}$ must be extremely complex in its dependence on the proximity of the pair of particles considered, to the surface. Also, $\rho^{(2)}$ must evidently go over to the bulk liquid pair function when both particles are deep in the bulk liquid. Though it does not contain, by any means, the full physics of this situation, some progress has been made with the approximate form

$$\rho^{(2)}(z_1, z_2, r_{12}) = \rho(z_1)\rho(z_2) g(r_{12}) \tag{10.8}$$

where g is the bulk pair function. Such a representation seems first to have been proposed by Green (1960) and has been used by Berry and Resnek (1971). Some further discussion of numerical calculations on the pair function near the liquid surface is given below, in section 10.5.

10.2.1 Approximate theory of surface tension in terms of one-atom density $\rho(z)$

Let us go straight to the determination of the surface tension σ, utilising the above form for $\rho^{(2)}$. Writing $\boldsymbol{R} = \boldsymbol{r}_2 - \boldsymbol{r}_1$ we then find

$$\sigma = \frac{\pi}{2} \int_0^\infty \mathrm{d}R \frac{\mathrm{d}\phi(R)}{\mathrm{d}R} g(R) \left[-\int_{-R}^{R} \mathrm{d}z (R^2 z - z^3) \int_{-\infty}^{\infty} \mathrm{d}z_1 \frac{\mathrm{d}\rho(z_1)}{\mathrm{d}z_1} \rho(z_1 + z) \right] \tag{10.9}$$

The most direct procedure to evaluate σ is now to assume a smooth variation of density between liquid and vapour. Berry, Durrans and Evans (1972) have calculated the surface tension of various insulating liquids by adopting the single parameter form

$$\rho(z)=\begin{cases}\rho[1-\frac{1}{2}\exp(z/L)] & (z<0)\\ \frac{1}{2}\rho\exp(-z/L) & (z>0)\end{cases} \tag{10.10}$$

Inserting this expression into (10.9) yields

$$\sigma=\frac{\pi}{8}\rho^2L^4\int_0^\infty \mathrm{d}R\,\frac{\mathrm{d}\phi(R)}{\mathrm{d}R}\,g(R)\left[\left(\frac{R}{L}\right)^4-8\left(\frac{R}{L}\right)^2+72\right.$$
$$\left.-\exp(-R/L)\left\{4\left(\frac{R}{L}\right)^3+28\left(\frac{R}{L}\right)^2+72\frac{R}{L}+72\right\}\right] \tag{10.11}$$

If we allow the 'surface thickness' L to tend to zero, then this is readily shown to reduce to eqn (10.6) given above, namely

$$\lim_{L\to 0}\sigma=\frac{\pi}{8}\rho^2\int_0^\infty \mathrm{d}R\,\frac{\mathrm{d}\phi(R)}{\mathrm{d}R}\,R^4g(R) \tag{10.12}$$

which is a formula first derived by Fowler (1937).

10.2.2 Results for Lennard–Jones potential

There are now various procedures which can be adopted for calculating σ from this formula. Thus, for neon say, one could make use of the measured bulk pair function $g(r)$. Then it is certainly true that the answer is sensitive to the way $\phi(r)$ is constructed and various authors have discussed this problem. In numerous papers, $\phi(r)$ is calculated to fit experimental quantities but the procedure leaves something to be desired.

At least satisfactory internal consistency can be achieved between $\phi(r)$ and $g(r)$ if use is made of computer experiments. Of course, no proof exists of the consistency of the choice of the exponential density profile, nor of the Green decoupling, so that any conclusions drawn are not complete.

But for neon, using molecular dynamical calculations, it is *not* possible to get agreement with experiment for an abrupt surface. *Table 10.1*, taken from the

Table 10.1. Surface tension for Ne, *with exponential density profile, as function of* L (from McDonald and Freeman, 1973)

L(Å)	0	0.5	1.0	1.5	2.0
σ(erg cm^{-2})	6.8	6.5	5.9	5.1	4.5

work of McDonald and Freeman (1973), shows that the surface thickness L for Ne at 20K must be chosen around 1 Å, although the results are not dramatically sensitive to the choice of L. For example, Fowler's formula, with the same potential, would yield $\sigma=6.8$ erg cm^{-2} while the experimental value is 5.7 erg cm^{-2}. The introduction of $L\neq 0$ reduces the Fowler value, which is satisfactory, but in view of the other uncertainties in the theory, it would be unwise to attribute fully quantitative significance to the value of L thus obtained, although it is physically very reasonable.

10.3 Fluctuation theory and surface tension

The Kirkwood–Buff theory, in terms of the pair function in the presence of the surface, affords an approach which is formally exact. When a decoupling is made, following Green, sensible results follow as we have seen.

However, it is of interest that an alternative (approximate) approach is available via fluctuation theory. It seems that the principal result of the theory was due to Yvon, and was derived independently by Buff. To our knowledge, the first time the derivation is recorded in the literature is through the work of Triezenberg and Zwanzig (1972)—referred to as TZ.

The theory to which this fluctuation approach leads is in terms *both* of the density profile at a liquid–vapour interface and the Ornstein–Zernike direct correlation function near the interface. Contact can also be established with the approximate theories of Fisk and Widom (1969) and of Felderhof (1970).

The argument of TZ goes as follows. They consider a fluid in a vertical gravitational field. The magnitude of the field is assumed to be sufficiently weak so that the only effect is to produce a liquid–vapour interface. The vertical direction is the z axis of a coordinate system and a vector in the xy plane will be denoted by $\boldsymbol{r}$. The Gibbs equimolar surface is used to define the position of the interface and the origin is then chosen so that the equilibrium Gibbs surface is the plane $z = z_0$.

A density fluctuation is now assumed to occur so that the density change is given by

$$\delta\rho(\boldsymbol{r}, z) = \rho(\boldsymbol{r}, z) - \rho(z) \tag{10.13}$$

As the density fluctuates, the location of the Gibbs surface also fluctuates. The instantaneous position of this surface, say $z_0(\boldsymbol{r})$, is given by

$$z_0(\boldsymbol{r}) = \int \mathrm{d}z\, \delta\rho(\boldsymbol{r}, z)/\Delta\rho \tag{10.14}$$

where $\Delta\rho$ is the density difference between the bulk liquid and the gaseous phase. The change in area of the Gibbs surface, δA say, can also be expressed as

$$\delta A = \tfrac{1}{2} \int \mathrm{d}^2\boldsymbol{r} |\nabla_r z_0(\boldsymbol{r})|^2 \tag{10.15}$$

where the integration is over the original Gibbs surface, A_0 say.

In Fourier space,

$$\delta\rho(\boldsymbol{r}, z) = \sum_q \exp(i\boldsymbol{q}\cdot\boldsymbol{r})\, n(q, z) \tag{10.16}$$

where $\boldsymbol{q}$ (like $\boldsymbol{r}$) is a vector in the xy plane. Analogously the position of the Gibbs surface can be Fourier expanded as

$$z_0(\boldsymbol{r})=\sum_q z_0(q)\exp(i\boldsymbol{q}\cdot\boldsymbol{r}) \tag{10.17}$$

where

$$z_0(\boldsymbol{q})=\int \mathrm{d}z\, n(\boldsymbol{q}, z)/\Delta\rho \tag{10.18}$$

Evidently δA can then be expressed in the alternative form

$$\delta A=\tfrac{1}{2}A_0\sum_q q^2|z_0(q)|^2 \tag{10.19}$$

10.3.1 Free energy associated with density fluctuations

From standard fluctuation theory, the free energy change δF is quadratic in the fluctuations and can be written as

$$\delta F=\tfrac{1}{2}k_{\mathrm{B}}T\int \mathrm{d}^2\boldsymbol{r}_1\,\mathrm{d}z_1\,\mathrm{d}^2\boldsymbol{r}_2\,\mathrm{d}z_2\delta\rho(\boldsymbol{r}_1, z_1)K(\boldsymbol{r}_1, z_1; \boldsymbol{r}_2, z_2)\delta\rho(\boldsymbol{r}_2, z_2) \tag{10.20}$$

where the inverse of the kernel K is related to the equilibrium second moments of the density fluctuations by

$$K^{-1}(\boldsymbol{r}_1, z_1; \boldsymbol{r}_2, z_2)=\langle\delta\rho(\boldsymbol{r}_1, z_1)\delta\rho(\boldsymbol{r}_2, z_2)\rangle \tag{10.21}$$

From the statistical mechanics of inhomogeneous systems, this kernel can be expressed in terms of $\rho(z)$ and the direct correlation function $c(\boldsymbol{r}_1, z_1; \boldsymbol{r}_2, z_2)$ (see for example Percus, 1964), by

$$K(\boldsymbol{r}_1, z_1; \boldsymbol{r}_2, z_2)=\frac{\delta(\boldsymbol{r}_1-\boldsymbol{r}_2)\delta(z_1-z_2)}{\rho(z_1)}-c(\boldsymbol{r}_1, z_1; \boldsymbol{r}_2, z_2) \tag{10.22}$$

In Fourier expansion, the free energy change is given by

$$\delta F=\tfrac{1}{2}A_0k_{\mathrm{B}}T\int\int \mathrm{d}z_1\,\mathrm{d}z_2\sum_q n^*(q, z_1)\tilde{K}(q, z_1, z_2)n(q, z_2) \tag{10.23}$$

where

$$\tilde{K}(q, z_1, z_2)=\frac{\delta(z_1-z_2)}{\rho(z_1)}-\tilde{c}(q, z_1, z_2) \tag{10.24}$$

The definition of $\tilde{c}$ is

$$\tilde{c}(q, z_1, z_2)=\int \mathrm{d}^2\boldsymbol{r}\, c(0, z_1; \boldsymbol{r}, z_2)\exp(i\boldsymbol{q}\cdot\boldsymbol{r}) \tag{10.25}$$

At this point all attention is focused on density fluctuations which alter the surface area. The changes in free energy and area associated with these fluctuations having been calculated, the surface tension σ then follows from the thermodynamic definition

$$\delta F=\sigma\delta A \tag{10.26}$$

To proceed, let us consider a spontaneous long wavelength fluctuation in the position $z_0(q)$ of the Gibbs surface. The amplitude of the fluctuation is assumed to be small, so that the surface deviates only by a small amount from planar form. It will now be assumed that this fluctuation is associated with a vertical shift in the equilibrium density profile, namely

$$\rho(z)\rightarrow\rho[z-z_0(q)] \tag{10.27}$$

the shape of the profile remaining unchanged. Thus the change in density can be got by Taylor expanding as

$$n(q, z)\simeq -z_0(q)\frac{\mathrm{d}\rho(z)}{\mathrm{d}z} \tag{10.28}$$

This can now be inserted into the equation for the free energy change, to yield

$$\delta F=\tfrac{1}{2}A_0k_BT\,|z_0(q)|^2\int\int \mathrm{d}z_1\,\mathrm{d}z_2\frac{\mathrm{d}\rho(z_1)}{\mathrm{d}z_1}\tilde{K}(q, z_1, z_2)\frac{\mathrm{d}\rho(z_2)}{\mathrm{d}z_2} \tag{10.29}$$

The next assumption is that the kernel $\tilde{K}$ can be expanded in powers of q, such that

$$\tilde{K}(q, z_1, z_2)= K_0(z_1, z_2)+q^2 K_2(z_1, z_2)+\cdots \tag{10.30}$$

and, in particular, the second-order coefficient is related to the Ornstein–Zernike function by

$$K_2(z_1, z_2)=\tfrac{1}{4}\int \mathrm{d}z_1\,\mathrm{d}z_2\,\mathrm{d}^2\mathbf{r}\,r^2\,c(0, z_1; \mathbf{r}, z_2) \tag{10.31}$$

In the long wavelength limit $q\rightarrow 0$, there is no change in surface area and hence δF is zero. This indicates that $\mathrm{d}\rho(z)/\mathrm{d}z$ is an eigenfunction of the kernel $K_0(z_1, z_2)$ with zero eigenvalue, that is

$$\int \mathrm{d}z_2\,K_0(z_1, z_2)\frac{\mathrm{d}\rho(z_2)}{\mathrm{d}z_2}=0 \tag{10.32}$$

If the direct correlation function is known, this equation can be solved for the equilibrium profile $\rho(z)$.

The remaining term is quadratic in q, and comparing it with the corresponding change in surface area the surface tension turns out to be

$$\sigma=k_BT\int\int \mathrm{d}z_1\,\mathrm{d}z_2\frac{\mathrm{d}\rho(z_1)}{\mathrm{d}z_1}K_2(z_1, z_2)\frac{\mathrm{d}\rho(z_2)}{\mathrm{d}z_2} \tag{10.33}$$

As remarked above, this formula was first derived by Yvon. A different, but equivalent, form has been given by Buff and Lovett (1968).

10.3.2 Approximate forms of direct correlation function

If a local approximation is now made for the direct correlation function by writing

$$\tilde{c}(q, z_1, z_2) \simeq c_0(z_1)\delta(z_1 - z_2) - \tfrac{1}{6} l^2 \left(q^2 - \frac{d^2}{dz_1{}^2} \right) \delta(z_1 - z_2) \tag{10.34}$$

then the approximate theories of Fisk and Widom (1969) and of Felderhof (1970) are regained. Here, the first term is related to the local compressibility of the fluid at density $\rho(z)$, or to the local Helmholtz free energy per particle, $f(\rho)$ at z by

$$k_B T \left(\frac{1}{\rho} - c_0 \right) = \rho \frac{\partial p}{\partial \rho} = \frac{\partial^2}{\partial \rho^2} [\rho f(\rho)] \tag{10.35}$$

The quantity l is the range of the direct correlation function. In this approximation, the density profile is determined by the differential equation

$$\frac{\partial^2 [\rho f(\rho)]}{\partial \rho^2} \frac{d\rho}{dz} + \tfrac{1}{6} k_B T l^2 \frac{d^3 \rho}{dz^3} = 0 \tag{10.36}$$

Integrating the equation leads to the result of Fisk and Widom

$$\frac{\partial}{\partial \rho} [\rho f(\rho)] + \tfrac{1}{6} k_B T l^2 \frac{d^2 \rho}{dz^2} = \text{const} \tag{10.37}$$

In the same approximation, the second-order term K_2 is given by

$$K_2(z_1, z_2) = \tfrac{1}{6} l^2 \delta(z_1 - z_2) \tag{10.38}$$

and if we substitute this into the result (10.33) for σ, we find

$$\sigma = \tfrac{1}{6} k_B T l^2 \int dz \left[\frac{d\rho(z)}{dz} \right]^2 \tag{10.39}$$

10.3.3 Comparison of the Fisk–Widom theory with experiment near critical point

Fisk and Widom have developed the above expression for σ in the region of the critical point. This is of interest because experiments have been carried out for xenon by Zolleweg, Hawkins and Benedek (1971) in which the spectrum of light scattered inelastically from thermal excitations in the surface has been measured. The experimental results can be summarised by the empirical formula

$$\sigma = (62.9 \pm 1.8) \left[1 - \frac{T}{T_c} \right]^{1,302 \pm 0,006} \text{dyn/cm} \tag{10.40}$$

in the temperature range $0.070\ \text{K} < T_c - T < 5\ \text{K}$.

Their results provide a test of the consequences of the Fisk–Widom theory, as follows. The Fisk–Widom relation predicts that the 'interfacial thickness' L defined by

$$L = \frac{1}{c} \frac{1}{\beta^2} \frac{\sigma \rho^2 K_T}{(\rho_l - \rho_v)^2} \tag{10.41}$$

may be identified with the correlation range ξ for temperatures near to T_c. Here ρ_l and ρ_v are the densities of the liquid and vapour phases, while β is the usual critical exponent which describes the shape of the coexistence curve. The constant c is a parameter of order unity, its value being determined by the form of the chemical potential as a function of density in the two-phase region $\rho_v < \rho < \rho_l$.

It should be noted that an immediate consequence of the assumption $L = \xi$ is the Widom (1965) equality

$$\mu + v = \gamma' + 2\beta \tag{10.42}$$

where μ, v and γ' are exponents characterising the temperature variation of the surface tension σ, the correlation range ξ and the quantity $\left.\frac{\partial \rho}{\partial \mu}\right|_T = \rho^2 K_T$. Using the value $\mu = 1.302 \pm 0.006$ and taking from experiment the values $v = 0.57 \pm 0.05$, $\gamma' = 1.21 \pm 0.03$ and $\beta = 0.345 \pm 0.01$ they find $\gamma' + 2\beta - v - \mu = 0.03 \pm 0.06$. Thus the temperature dependence of L is consistent with that of ξ within experimental error.

Zolleweg et al. estimate that $c = 0.83 \pm 0.15$, which is somewhat lower than the estimates of Fisk and Widom.

10.3.4 Surface tension of a hard-sphere fluid

Reiss, Frisch and Lebowitz (1959) have derived an expression for the surface tension of a hard sphere fluid by a simple fluctuation argument which is sufficiently illuminating to be summarised here.

The idea is to relate the work required to create a cavity in the liquid composed of hard spheres of diameter a, to the work needed to insert a solute hard sphere of diameter $(2\lambda - 1)a$. Clearly, around the centre of the solute there is a spherical region of radius λa from which the centres of the solvent atoms are excluded. Thus, the presence of the solute sphere is equivalent to the existence of a spherical cavity of radius at least λa in the liquid.

Reiss et al. prove first that the reversible work $W(\lambda)$ expended in creating such a cavity is related to the value of the solute–solvent pair correlation function evaluated at contact, $G(\lambda)$. The argument involves the consideration of the probability $P(\lambda)$ of observing a fluctuation in which the cavity forms, which is given in terms of the work $W(\lambda)$ as

$$P(\lambda) = \exp\left[-W(\lambda)/k_B T\right] \tag{10.43}$$

We can write for $P(\lambda)$ the integral equation

$$P(\lambda) = 1 - \int_0^{\lambda a} P(\lambda') 4\pi\rho\lambda'^2 a^2 G(\lambda')\, \mathrm{d}(\lambda' a) \tag{10.44}$$

where the integral gives the probability of having at least one particle within distance λa, calculated as the integral of the probability (per unit radial distance) of finding the centre of the nearest particle at a distance $\lambda' a$ away. The latter probability clearly is the product of the probability $P(\lambda')$ of having no particle within distance $\lambda' a$, times the probability of having a particle at distance $\lambda' a$. From (10.43) and (10.44) there follows

$$G(\lambda) = -\frac{1}{4\pi\rho\lambda^2 a^3}\frac{\partial \ln P(\lambda)}{\partial\lambda}$$

$$= \frac{1}{4\pi\rho\lambda^2 a^3 k_B T}\frac{\partial W(\lambda)}{\partial\lambda} \tag{10.45}$$

which is the desired relation between the work of cavity formation and the solute–solvent correlation at contact. The latter quantity can also be interpreted as the density of solvent particles at the surface of the cavity.

To actually relate the above exact properties of the hard-sphere fluid to its surface tension, Reiss et al. note that for λ sufficiently large one can write the macroscopic formula

$$W(\lambda) = \frac{4\pi}{3}\rho a^3\lambda^3 p(\eta) + 4\pi a^2\lambda^2\sigma(\eta)\left[1 - \frac{2\delta(\eta)}{\lambda}\right] \tag{10.46}$$

where $p(\eta)$ and $\sigma(\eta)$ are the pressure and the surface tension of the hard-sphere liquid with packing fraction $\eta = (\pi/6)\rho a^3$, and the last term corrects the surface contribution for finite curvature of the surface through a parameter $\delta(\eta)$. They also note that $P(\lambda)$ can be calculated by elementary arguments for $\lambda \leqslant \frac{1}{2}$ (that is, for zero, or negative radius of the solute!) as

$$P(\lambda) = 1 - \frac{4\pi}{3}\rho a^3\lambda^3 \quad (\lambda \leqslant \tfrac{1}{2}) \tag{10.47}$$

Indeed, a spherical region of radius λa for $\lambda \leqslant \frac{1}{2}$ can contain at most the centre of only *one* particle of diameter a, so that the probability that it be occupied is $(4\pi/3)\rho\lambda^3 a^3$ and the probability that it be empty is as in (10.47). If we now assume that the macroscopic expression (10.46) for $W(\lambda)$ holds down to $\lambda = \frac{1}{2}$, we can calculate the first two derivatives of $W(\lambda)$ at that point from the expressions (10.43) and (10.47), and thus determine $\sigma(\eta)$ and $\delta(\eta)$ in terms of $p(\eta)$. Using the Percus–Yevick expression for the pressure of the hard-sphere fluid†,

$$p(\eta) = \rho k_B T(1 + \eta + \eta^2)(1 - \eta)^{-3} \tag{10.48}$$

†The expression (10.48) for the pressure was actually derived by Reiss, Frisch and Lebowitz (1959) prior to the exact solution of the Percus–Yevick equation for the hard-sphere fluid, by using the relation (10.45) and a relation between the pressure and the pair correlation function at contact in the pure liquid, that is $G(\lambda = 1)$ (see for example Hill, 1955; see also appendix 4.2).

(see eqn (2.79)) the result is

$$\sigma(\eta)=\frac{9k_BT}{2\pi a^2}\frac{\eta^2(1+\eta)}{(1-\eta)^3} \tag{10.49}$$

$$\delta(\eta)=\tfrac{1}{2}\frac{\eta}{1+\eta} \tag{10.50}$$

The *negative* value of $\sigma(\eta)$ is a consequence of the hard-sphere model. Although this aspect of the model is unphysical (and will, of course, be corrected by the addition of an attractive tail in the pair potential) the whole argument illustrates the deep-lying relations between surface properties and correlation functions.

10.4 Calculation of singlet distribution function

Having discussed routes to the calculation of σ, we wish briefly to come back to the question of obtaining $\rho(z)$ from a given pair potential $\phi(r)$. We shall assume in the present discussion that $\phi(r)$ is little affected by passing through the interface between liquid and vapour. This is a good assumption for liquid argon (for example, the van der Waals constant is estimated to vary by 8 or 9% as we go from a density appropriate to the triple point of the bulk liquid to the pair interaction in free space—see Robinson and March, 1972), but is certainly *not* appropriate in a liquid metal.

If we use the Green decoupling, then the form

$$k_BT\frac{\mathrm{d}\rho(z_1)}{\mathrm{d}z_1}=\int \mathrm{d}\boldsymbol{R}\,\frac{z}{R}\frac{\mathrm{d}\phi}{\mathrm{d}R}\rho^{(2)}(z,\boldsymbol{R}) \tag{10.51}$$

becomes, if we neglect the dependence of $g(R)$ on the density along the isotherms near the coexistence curve,

$$K_BT\frac{\mathrm{d}\ln\rho(z_1)}{\mathrm{d}z_1}=2\pi\int_{-\infty}^{\infty} z\,\mathrm{d}\,z\rho(z_1+z)\int_{|z|}^{\infty}\mathrm{d}R\phi'(R)g(R) \tag{10.52}$$

It is convenient to write the kernel as

$$h(z)=\frac{2\pi z}{k_BT}\int_{|z|}^{\infty}\mathrm{d}R\phi'(R)g(R) \tag{10.53}$$

The even function $k(z)$ is now defined as

$$k(z)=-\int_{z}^{\infty}h(z')\,\mathrm{d}z' \tag{10.54}$$

and in terms of this the above equation may be integrated to yield

$$\frac{1}{\rho(z_1)}\exp\left\{-\int_{-\infty}^{\infty}\mathrm{d}zk(z)\rho(z_1+z)\right\}=\mathrm{C} \tag{10.55}$$

where C is a constant independent of z.

The basic relation between liquid and vapour densities will involve the constant

$$K_l(T) = -\int_{-\infty}^{\infty} dz k(z)$$
$$= \frac{4\pi}{3k_BT}\int_0^{\infty} dRR^3\phi'(R)g(R) \tag{10.56}$$

K_l is positive because the integral is dominated by the attractive region in which $\phi'(R)$ is positive. We can now equate the asymptotic forms of the right-hand side of the above equations, when we find

$$C = \frac{1}{\rho_l}\exp(K_l\rho_l) = \frac{1}{\rho_g}\exp(K_l\rho_g) \tag{10.57}$$

Hence, as discussed by Berry and Resnek (1971), one can plot $\exp[K_l(T)\rho]/\rho$ against ρ (see *figure 10.1*), choose a liquid density ρ_l (greater than $1/K_l(T)$) and can then determine from the figure the corresponding vapour density ρ_g. Once these limiting densities are known, the density profile can be obtained by iteration, starting from a zero'th order approximation in which the density changes suddenly from ρ_l to ρ_g at $z=0$. The first iteration is evidently

$$\rho(z) = \rho_l \exp\left\{\int_{-\infty}^{z} dz' k(z')[\rho_l - \rho_g]\right\} \tag{10.58}$$

This theory predicts that the thickness of the transition region between liquid and vapour is comparable with the range of the kernel $k(z)$. This latter range is essentially the range of the interatomic forces.

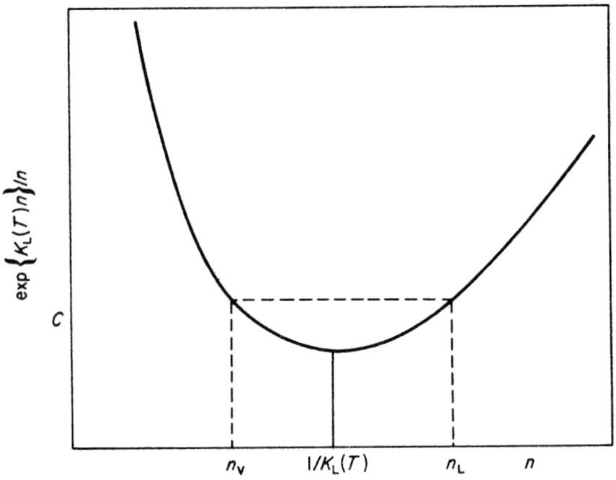

Figure 10.1 Construction relating liquid and vapour densities at a given temperature (from Berry and Resnek, 1971)

Using the empirical law (see Guggenheim, 1945) that

$$\frac{\rho_l-\rho_g}{\rho_c}=\frac{7}{2}\left[1-\frac{T}{T_c}\right]^{1/3} \tag{10.59}$$

where ρ_c and T_c refer to the critical point, the single ρ_l for which liquid and vapour can coexist at a given temperature can be found. We must refer the reader to the paper by Berry and Resnek for further details.

10.5 Pair distribution function near liquid surface

We shall conclude this discussion of the surface of a one-component classical fluid by reporting briefly the work of Croxton and Ferrier (1971*a*). They obtain the pair function in the presence of a planar surface by means of a variational calculation using a trial function of the form

$$\rho^{(2)}(z, \boldsymbol{r}_{12})=\rho_l^2\{1+h(r)[1+\exp(-\alpha(z_1)/r)]\} \times \sum_{l\geqslant}^{\infty}\sum_{m=0}^{\infty} A_{lm}(z,T)P_l^m(\cos\vartheta)\Phi(m\phi) \tag{10.60}$$

The angular dependence is thereby expressed in terms of the spherical harmonics with temperature-dependent coefficients A_{lm}. The function $\alpha(z_1)$ suppresses oscillations in the radial distribution function, but gives the correct limiting forms at $r=0$ and $r=\infty$. $h(r)$ is, as usual, the 'hole' function $g(r)-1$.

Symmetry considerations show that there is no dependence on many of the A_{lm}'s and the assumption is therefore made that a good approximation can be obtained by keeping only the s and p_z terms in the harmonic expansion. Then the form adopted by Croxton and Ferrier (1971*a*) is

$$\rho^{(2)}(z_1\boldsymbol{r}_{12})=\rho_l\{1+h(r)[1+\exp(-\alpha(z_1)/r)]\} \times A_{00}^2(z)[P_0{}^0(\cos\vartheta)+\lambda(z,T)P_1{}^0(\cos\vartheta)]^2 \tag{10.61}$$

Here λ is a mixing coefficient between spherical modes of the bulk liquid and the asymmetric modes of the surface and is such that it vanishes for large values of $|z|$ and also as $T\to T_c$, because the transition range is extensive and there is little anisotropy.

The procedure is then, in principle, to minimise the free energy with respect to this trial function. In practice the internal energy is minimised, assuming that $\rho^{(2)}$ is temperature independent. Suitable boundary conditions on α, A_{00} and A_{10} are taken from the properties of the bulk phases.

The local expectation value of the energy is minimised and leads to

$$A_{00}^2(z_1)=\frac{\rho(z_1)\displaystyle\int\rho(z_2)g(r)\,\mathrm{d}\boldsymbol{r}}{2\rho^2\{1+\frac{1}{3}\lambda^2(z_1)\}\displaystyle\int_0^\infty g(r)r^2\,\mathrm{d}r} \tag{10.62}$$

A_{10} is the product of A_{00} and λ. λ is plotted in *figure 10.2* for three different temperatures. It can be seen that the anisotropy is greater at low temperatures, but falls off rapidly away from the interfacial region.

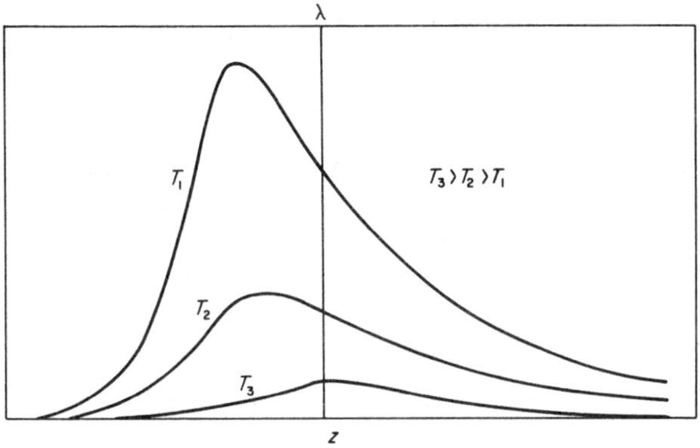

Figure 10.2 Qualitative variation of the mixing coefficient $\lambda(z)$ with temperature (from Croxton and Ferrier, 1971*a*)

On this model the surface tension σ and surface energy E_s are

$$\sigma = -\pi\rho_l^2 \int_0^\infty g(r)\frac{d\phi}{dr} r^3 \, dr \int_{-\infty}^{\infty} \lambda(z)[1+\tfrac{4}{15}\lambda(z)] \, dz \qquad (10.63)$$

and

$$E_s = 2\pi\rho_l^2 \int_0^\infty \phi(r)g(r)r^2 \, dr \left[\int_{-\infty}^{0} \{A_{00}(z)(1+\tfrac{1}{3}\lambda^2(z))-1\} \, dz + \int_0^\infty A_{00}^2(z)(1+\tfrac{1}{3}\lambda^2(z)) \, dz\right] \qquad (10.64)$$

Numerical results are given by Croxton and Ferrier (1971*a*) to whose paper the reader should refer for further details.

10.6 Relation between isothermal compressibility and surface tension

There are at least two routes to understanding how a bulk property, the isothermal compressibility K_T, can be related to a surface property, the surface tension σ. Both involve the idea that a 'bulk theory' can be used in the presence of the surface, provided we add density gradients. The first argument we give is generally referred to as the Cahn–Hilliard theory. It involves a number of physical assumptions which are difficult to justify from first principles. But it does lead to insight into the basic origin of such a relation between K_T and σ.

The second argument is specific to liquid metals, and focuses attention on the behaviour of the conduction electrons (Brown and March, 1973).

10.6.1 Cahn–Hilliard theory

The argument of Cahn and Hilliard is perhaps already at least implicit in the early work of van der Waals. Essentially, the argument goes as follows. If there is a (statistical) fluctuation in density, say $\delta\rho$ in a volume v of a bulk liquid, then we can associate with it a free energy F_1 which is given by

$$F_1 = c_1 v(\delta\rho)^2/\rho^2 K_T \tag{10.65}$$

K_T being the isothermal compressibility. Although, strictly speaking, it is extrapolating this formula outside its proper range of validity, we now assume it can be used in the liquid surface, where there is an inhomogeneity in the density ρ of atoms per unit volume.

But due to the density gradient there must also be an additional free energy F_2, directly related to the density gradient by

$$F_2 = c_2 v(\delta\rho/L)^2 \tag{10.66}$$

where L is the surface thickness. Clearly, we can now construct the total surface tension

$$\sigma = \sigma_1 + \sigma_2 \tag{10.67}$$

from the contributions due to F_1 and F_2, namely

$$\sigma_1 = c_1 L(\delta\rho)^2/\rho^2 K_T \tag{10.68}$$

and

$$\sigma_2 = c_2 L(\delta\rho/L)^2 \tag{10.69}$$

the lengths L simply arising from v divided by a surface area.

To determine the surface thickness L, we now minimise σ with respect to L, when we find

$$\frac{d\sigma}{dL} = 0 = c_1 \frac{(\delta\rho)^2}{\rho^2 K_T} - c_2 \frac{(\delta\rho)^2}{L^2} \tag{10.70}$$

We see immediately that this equilibrium condition implies $F_1 = F_2$, i.e. equal contributions to the free energy from the 'density fluctuations' and from the density gradient. Hence it follows that

$$\sigma = 2c_1 L \frac{(\delta\rho)^2}{\rho^2 K_T} \tag{10.71}$$

It is clearly reasonable at this point to take as an estimate of $\delta\rho$ the difference between the bulk liquid density ρ and the vapour density, ρ_v say. But near the triple point $\rho \gg \rho_v$ and hence it follows that

$$\sigma \sim 2c_1 L/K_T \tag{10.72}$$

or

$$K_T \sigma \sim L \tag{10.73}$$

That there is such a correlation between K_T and σ was known to Frenkel (1942) and has been rediscovered by Egelstaff and Widom (1970). *Table 10.2* has been taken from data collected in their paper for liquids near the triple point. The product of $K_T\sigma$ is seen to be roughly constant (variations of a factor 2–3 only), the constant being of the order of 1 Å. This constancy is remarkable, bearing in mind (see *table 10.2*) that K_T and σ vary separately by more than 50 among the different liquids.

Table 10.2. Values of $K_T\sigma$ for liquids at or near their triple points (from Egelstaff and Widom, 1970)

	σ(dyne/cm)	$K_T(10^{-12}\ cm^2/dyne)$	$K_T\sigma$(Å)
Metals:			
Na	194	21	0.40
K	113	40	0.45
Rb	95	49	0.46
Cs	71	67	0.47
Cu	1280	1.45	0.19
Ag	940	1.86	0.18
Zn	785	2.4	0.19
Cd	666	3.2	0.21
Pb	470	3.5	0.17
Bi	395	4.3	0.17
Fe	1790	1.43	0.25
Molten salts:			
NaCl	116	29	0.34
KCl	97	38	0.37
NaBr	99	34	0.34
KBr	90	44	0.39
NaI	86	40	0.35
KI	78	50	0.39
Insulating liquids:			
Argon	13.1	212	0.28
Krypton	16.3	172	0.28
Xenon	18.7	166	0.31
Nitrogen	11.8	211	0.25
Oxygen	18.4	120	0.22
Bromine	41.5	63	0.26

10.6.2 Liquid metals

Among the entries in *table 10.2* are quite a lot of liquid metals. Modern treatments of compressibility and surface tension must involve, in an essential way, the conduction electron behaviour. An argument specific to such liquid metals has been given by Brown and March (1973) and we record the gist of it below, although we must refer the reader to the original paper for a quantitative discussion. As we shall see, there is a similarity of approach to that of the Cahn–Hilliard theory, although the density gradients which are now important are those in the electron density ρ. Nevertheless, charge neutrality and screening arguments appropriate in liquid metals must ensure that the electron density and the atomic density follow one another closely in the liquid surface, the former being slightly more diffuse than the latter.

The most elementary argument starts from the compressibility K_0 of the free electron gas, that we used in section 7.4 to derive the so-called Bohm-Staver formula for the velocity of sound in metals. The result for K_0 is simply

$$K_0 = \frac{3}{2\rho_0 \varepsilon_F} \tag{10.74}$$

ρ_0 being the constant electron density in the bulk metal, related to the Fermi energy ε_F and the Fermi wave number k_F by $\rho_0 = k_F^3/(3\pi^2)$ and $\varepsilon_F = \hbar^2 k_F^2/2m$, m being the electron mass.

We can next argue that the surface energy is coming essentially from the inhomogeneity in the electron density near the liquid metal surface. Thus, let us calculate the kinetic energy associated with this inhomogeneity. If we have only a single electron, with wave function ψ, then

$$\mathrm{KE} = -\frac{\hbar^2}{2m}\int \psi^* \nabla^2 \psi \, \mathrm{d}\boldsymbol{r} = \frac{\hbar^2}{2m}\int (\nabla\psi)^2 \, \mathrm{d}\boldsymbol{r} \tag{10.75}$$

But for this case of one electron, $\psi = \rho^{1/2}$ and hence

$$\mathrm{KE} = \frac{\hbar^2}{8m}\int \frac{(\nabla\rho)^2}{\rho} \, \mathrm{d}\boldsymbol{r} \tag{10.76}$$

When the density gradients are small, such a formula can be justified in a dense Fermi gas, although the coefficient is crucially altered to yield

$$\mathrm{KE} = \frac{\hbar^2}{72m}\int \frac{(\nabla\rho)^2}{\rho} \, \mathrm{d}\boldsymbol{r} \tag{10.77}$$

This formula is due, in its essential form, to von Weizsacker (1935), though the correct coefficient was first given by Kirznits (1957).

Paralleling the Cahn–Hilliard discussion, let us write, if we have a planar surface, with the z axis perpendicular to the surface,

$$\nabla\rho(\boldsymbol{r}) = \frac{\mathrm{d}\rho}{\mathrm{d}z} \simeq \frac{\rho_0}{L} \tag{10.78}$$

where L is the surface thickness. Inserting this we can write

$$\mathrm{KE} \sim \frac{\hbar^2}{72m} \frac{\rho_0^2}{L^2} \frac{AL}{\rho_0} \tag{10.79}$$

since $\nabla\rho$ is non-zero only over a range of $z \sim L$, A being the surface area. Hence

$$\sigma \sim \frac{\hbar^2}{72m} \frac{\rho_0}{L} \tag{10.80}$$

and it follows from eqns (10.74) and (10.80) that the product $K_0\sigma$ is given by

$$K_0\sigma \sim \frac{\hbar^2}{72m}\frac{\rho_0}{L}\frac{3}{2\rho_0\varepsilon_F} \qquad (10.81)$$

But a measure of the surface thickness is π/k_F, directly related to the de Broglie wavelength of an electron at the Fermi level.† Thus $L \propto \varepsilon_F^{-1/2}$ and again

$$K_T\sigma \sim L \qquad (10.82)$$

The constant in this relation is discussed quantitatively by Brown and March. The conclusion is that it is no surprise that liquid metals fit into *Table 10.2*, with the other liquids, since the thickness of the electron density at the liquid metal surface is nearly the same as the atomic density thickness.

10.7 Surface tension of solutions

In an ideal solution the chemical potentials are of the form

$$\mu_i' = \mu_i^{0\prime}(T, p) + RT \ln x_i' \qquad (10.83)$$

The prime refers to a specific phase, say the liquid phase, and the subscript refers to the *i*th component in the solution, with concentration x_i' in phase $'$. The superscript zero refers to the pure phase *i*. The above equation follows from the definition of activity, this latter quantity being equal to the concentration only in an ideal solution.

The free energy of such an ideal solution can be written as

$$\begin{aligned} F' = G' - pV' &= \sum_i n_i'\mu_i' - p\sum_i n_i'v_i' \\ &= \sum_i n_i'[\mu_i^{0\prime}(T, p) - pv_i' + RT \ln x_i'] \end{aligned} \qquad (10.84)$$

If it is assumed that the surface phase can be treated as an ideal solution then we can write for the free energy, from the above expression,

$$F^m = \sum_1 n_i^m[\mu_i^{0m}(T, p) - pv_i^m + RT \ln x_i^m] \qquad (10.85)$$

This implies that F^m is independent of σ. For an ideal solution, μ^{0m} becomes μ_i^{0m}, the surface chemical potential of the pure component. Furthermore, we assume that the surface areas a_i of an ideal surface solution, like the molar volumes in an ideal solution, are independent of composition and dependent only on pressure and temperature. Similarly, it will be assumed that v_i^m depends only on T and p. Thus we can write

$$\mu_i^m = \mu_i^{0m}(T,p) + RT \ln x_i^m - \sigma a_i \qquad (10.86)$$

†The electron density ρ can be calculated, for example, explicitly for electrons in a one-dimensional box of length a. It is clearly $\rho(z) = \frac{2}{a}\sum_{n=1}^{N} \sin^2\frac{n\pi z}{a}$. As $a \to \infty$, the summation can be replaced by an integration, to yield $\rho(z) = \rho_0\left[1 - \frac{\sin(2k_f z)}{2k_f z}\right]$. The result that the surface thickness $L \sim \pi/k_f$ follows.

If equilibrium is established between surface and solution,

$$\mu_i^{m} = \mu_i' \tag{10.87}$$

and hence

$$\mu_i^{0\prime}(T, p) + RT \ln x_i' = \mu_i^{0m}(T, p) + RT \ln x_i^{m} - \sigma a_i \tag{10.88}$$

We now turn to discuss the surface tension of ideal solutions using these results.

To be specific we associate phase $'$ with the liquid phase (l). According to elementary theory, solutions only approach ideal behaviour when the molecules of the components are similar in size. Assume then that the surface areas a_i occupied by the molecules are equal. For a pure component, therefore

$$\mu_i^{0m}(T, p) - \mu_i^{0l}(T, p) = \sigma_i a \tag{10.89}$$

where σ_i is the surface tension of pure i at given values of T and p. Combining this with (10.88) gives the result of Butler (1932)

$$\sigma = \sigma_i + \frac{RT}{a} \ln \frac{x_i^{m}}{x_i^{l}} \tag{10.90}$$

For a two-component system, σ can be eliminated between the two equations got by putting $i = 1, 2$ to obtain

$$\frac{x_1^{m}}{x_2^{m}} = \frac{x_1^{l}}{x_2^{l}} \exp\left[(\sigma_2 - \sigma_1)a/RT\right] \tag{10.91}$$

Thus

$$x_1^{m} = x_1^{l} \frac{\exp(-\sigma_1 a/RT)}{x_1^{l} \exp(-\sigma_1 a/RT) + x_2^{l} \exp(-\sigma_2 a/RT)} \tag{10.92}$$

and

$$x_2^{m} = x_2^{l} \frac{\exp(-\sigma_2 a/RT)}{x_1^{l} \exp(-\sigma_1 a/RT) + x_2^{l} \exp(-\sigma_2 a/RT)} \tag{10.93}$$

equations which appear to be due to Schuchowitzky (1944). These results demonstrate that the composition of the surface of an ideal solution is determined both by the composition of the solution and by the surface tensions of the two pure components. The surface is richer in the component which has the lower surface tension.

10.7.1 Surface tension of an ideal binary solution

Thus, using the identity $x_1 + x_2 = 1$, we obtain (Belton and Evans, 1945)

$$\exp(-\sigma a/RT) = x_1^{l} \exp(-\sigma_1 a/RT) + x_2^{l} \exp(-\sigma_2 a/RT) \tag{10.94}$$

This can be rearranged to take the form

$$\sigma_1 - \sigma = \frac{RT}{a} \ln\left[1 + x_2^{l}\{\exp\left[(\sigma_1 - \sigma_2)a/RT\right] - 1\}\right] \tag{10.95}$$

This has the form of an empirical equation proposed by von Szyszkowski (1968) namely

$$\sigma_1 - \sigma = \mathrm{b} \ln (1 + c_2/\mathrm{a}) \tag{10.96}$$

where a and b are constants independent of concentration.

Assuming that σ_1 and σ_2 are sufficiently similar so that

$$|(\sigma_1 - \sigma_2)a/RT| \ll 1 \tag{10.97}$$

the above expression evidently reduces to

$$\sigma = x_1{}^l\sigma_1 + x_2{}^l\sigma_2 \tag{10.98}$$

showing that there is then a simple additivity law for the surface tension of an ideal solution. Belton and Evans (1945) have found good agreement between the above theory and experiment for a number of equimolar mixtures, chosen to approximate to ideal behaviour.

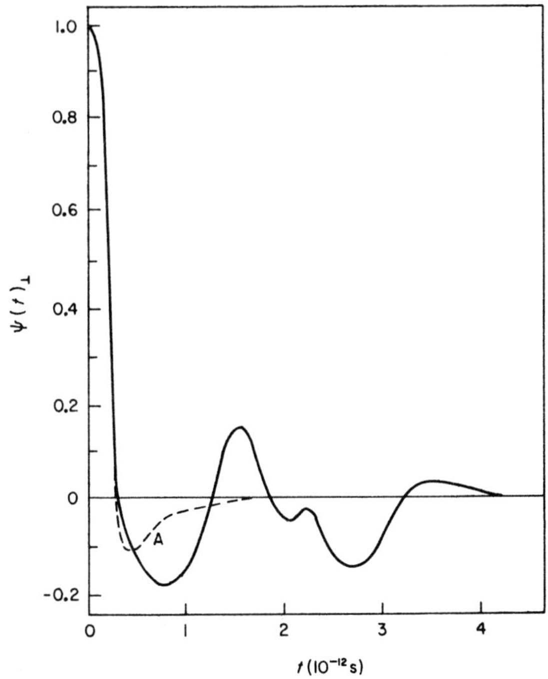

Figure 10.3 Velocity autocorrelation function in a two-dimensional argon liquid, for motions normal to the liquid surface (from Croxton, 1974). The broken curve reports the results of Rahman for bulk liquid argon

10.7.2 Regular solutions

The above arguments can be generalised to regular solutions (closely related to conformal solutions, as discussed earlier). In particular, Butler's equations are replaced by (α being a measure of the difference in interactions between $\frac{1}{2}(11+22)$ and 12)

$$\sigma=\sigma_1+\frac{RT}{a}\ln\frac{x_1^{\mathrm{m}}}{x_1^{l}}+\frac{\alpha_l}{a}\left[(x_2^{\mathrm{m}})^2-(x_2^{l})^2\right]-\frac{\alpha_{\mathrm{m}}}{a}(x_2^{l})^2$$

$$=\sigma_2+\frac{RT}{a}\ln\frac{x_2^{\mathrm{m}}}{x_2^{l}}+\frac{\alpha_l}{a}\left[(x_1^{\mathrm{m}})^2-(x_1^{l})^2\right]-\frac{\alpha_{\mathrm{m}}}{a}(x_1^{l})^2 \qquad (10.99)$$

By eliminating σ again, the relationship between x_2^{m} and $x_1^{\mathrm{m}}(=1-x_2^{\mathrm{m}})$ and the composition can be found. Hence the composition of a surface phase in equilibrium with a regular solution of known composition can be found.

When α and a are known, the above treatment can be used to calculate the surface tension of a regular solution in terms of the surface tension of either of the components.

10.8 Atomic dynamics in the liquid surface

Not a great deal of work has been done on atomic dynamics in the liquid surface at the time of writing. However, Croxton and Ferrier (1971*b*; 1971*c*)

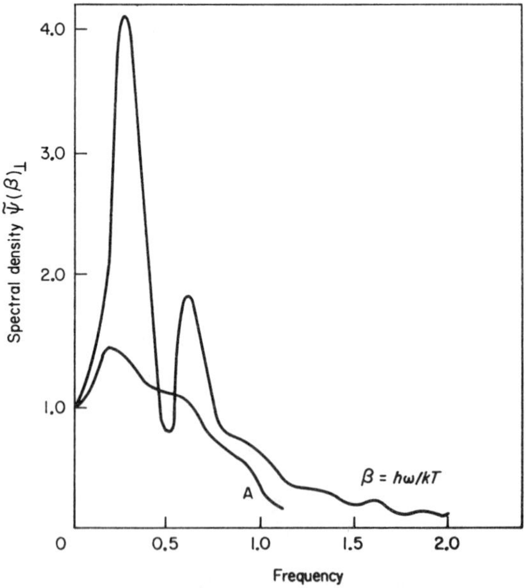

Figure 10.4 Fourier transform of the velocity autocorrelation functions reported in *figure 10.3* (from Croxton, 1974)

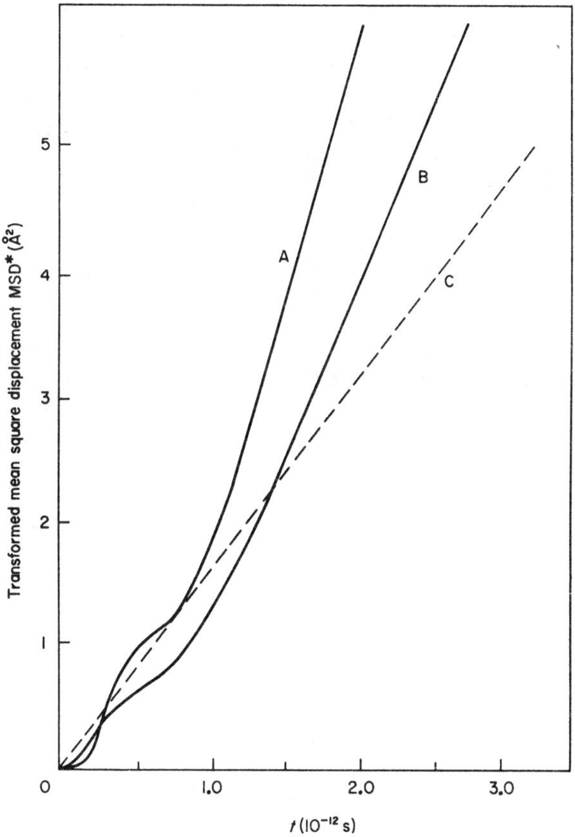

Figure 10.5 Mean square displacement in a two-dimensional argon liquid, for motions normal (*A*) and parallel (*B*) to the surface (from Croxton 1974)

have carried out some computer experiments on a two-dimensional assembly and we shall briefly summarise their main findings here (for a fuller account, see Croxton, 1974).

The velocity autocorrelation function for motions normal to the surface, namely

$$\psi_{\perp}(t) = \langle \dot{z}(0)\dot{z}(t)\rangle / \langle \dot{z}^2 \rangle \tag{10.100}$$

has the pronounced oscillatory nature shown in *figure 10.3*. It should be noted that the amplitude remains significant out to times of around 4×10^{-12} sec. Bulk studies, which we discussed in chapter 4, lead to no such exaggerated characteristics. Croxton and Ferrier interpret the pronounced structure in terms of a quasi-crystalline or structured interphase.

These workers have also calculated the Fourier transform of the velocity autocorrelation function (10.100). This is the frequency spectrum for motions normal to the surface, and highly resolved peaks are evident from *figure 10.4*. For comparison, the frequency spectrum calculated by Rahman (1964) for bulk argon is also shown in the figure. The strong peaks are attributed by Croxton and Ferrier, as remarked above, to a structured interphase.

Finally, in *figure 10.5* we show the results for the mean square displacement normal and parallel to the surface. Again for direct comparison we plot the results for the bulk liquid. The long time behaviour, linear in the time, has a slope which yields the diffusion constant (see eqn (4.4)). The two diffusion coefficients obtained from Croxton and Ferrier's computer experiments, designed to simulate liquid argon, are $D_{\perp} = 1 \times 10^{-4}$ cm^2/sec and $D \ = 5 \times 10^{-5}$ cm^2/sec at 94 K.

APPENDIX 2.1

HIERARCHY OF RELATIONS BETWEEN *s* AND (*s*+1)-PARTICLE CORRELATIONS

The general expression for the s-particle correlation function in the grand ensemble, given in eqn (2.6), can be differentiated with respect to either the coordinate of one particle or density or temperature in order to find relations between these correlations and higher order correlations. As immediate extension of the derivation of the force equation given in section 2.2 we find

$$\nabla_{r_1} g^{(s)}(\boldsymbol{r}_1,\ldots,\boldsymbol{r}_s) = -\frac{g^{(s)}(\boldsymbol{r}_1,\ldots,\boldsymbol{r}_s)}{k_B T}\sum_{i=2}^{s}\nabla_{r_1}\phi(r_{1i}) - \frac{\rho}{k_B T} \times \int \mathrm{d}\boldsymbol{r}_{s+1} g^{(s+1)}(\boldsymbol{r}_1,\ldots,\boldsymbol{r}_{s+1})\nabla_{r_1}\phi(|\boldsymbol{r}_1-\boldsymbol{r}_{s+1}|) \qquad \text{(A2.1.1)}$$

which is usually known as the hierarchy of equations for the distribution functions. It implies, in particular, that approximate theories for the triplet function could be developed on the basis of approximate expressions for the quadruplet function. Similarly, by the procedure followed in section 2.3 we find

$$k_B T\left(\frac{\partial[\rho^s g^{(s)}(\boldsymbol{r}_1,\ldots,\boldsymbol{r}_s)]}{\partial p}\right)_T = s\rho^{s-1}g^{(s)}(\boldsymbol{r}_1,\ldots,\boldsymbol{r}_s) + \rho^s\int \mathrm{d}\boldsymbol{r}_{s+1}[g^{(s+1)}(\boldsymbol{r}_1,\ldots,\boldsymbol{r}_{s+1}) - g^{(s)}(\boldsymbol{r}_1,\ldots,\boldsymbol{r}_s)] \qquad \text{(A.2.1.2)}$$

Again, as a point of principle, this relation implies the possibility of obtaining information on the quadruplet function from a study of the pressure dependence of the triplet function.

Finally, we consider the temperature derivative of $g^{(s)}$ at constant density. The expression (2.6) for $g^{(s)}$ depends on T both explicitly and implicitly through z. We can write then

$$\left(\frac{\partial[\rho^s g^{(s)}]}{\partial T}\right)_\rho = \left(\frac{\partial[\rho^s g^{(s)}]}{\partial T}\right)_z - \left(\frac{\partial[\rho^s g^{(s)}]}{\partial z}\right)_T\left(\frac{\partial z}{\partial\langle N\rangle}\right)_T\left(\frac{\partial\langle N\rangle}{\partial T}\right)_z = \left(\frac{\partial[\rho^s g^{(s)}]}{\partial T}\right)_z - \frac{1}{V}\left(\frac{\partial[\rho^s g^{(s)}]}{\partial\rho}\right)_T\left(\frac{\partial\langle N\rangle}{\partial T}\right)_z \qquad \text{(A2.1.3)}$$

where

$$\left(\frac{\partial\langle N\rangle}{\partial T}\right)_z=(k_BT^2)^{-1}[\langle NV_N\rangle-\langle N\rangle\langle V_N\rangle]$$

$$=\frac{\rho^2V}{k_BT^2}\int \mathrm{d}\boldsymbol{r}\,g(r)\phi(r)+\frac{\rho^3V^2}{2k_BT^2}\int\int \mathrm{d}\boldsymbol{r}\,\mathrm{d}\boldsymbol{s}\,[g^{(3)}(\boldsymbol{r},\boldsymbol{s})-g(r)]\phi(r) \quad \text{(A2.1.4)}$$

and

$$\left(\frac{\partial[\rho^s g^{(s)}]}{\partial T}\right)_z=(k_BT^2)^{-1}[\langle\rho^{(s)}V_N\rangle-\langle\rho^{(s)}\rangle\langle V_N\rangle]$$

$$=\frac{\rho^s}{k_BT^2}\left\{\sum_{i<j=1}^{s}\phi(r_{ij})g^{(s)}(\boldsymbol{r}_1,\ldots,\boldsymbol{r}_s)+\rho\int \mathrm{d}\boldsymbol{r}_{s+1}\,g^{(s+1)}(\boldsymbol{r}_1,\ldots,\boldsymbol{r}_{s+1})\right.$$

$$\times\sum_{i=1}^{s}\phi(|\boldsymbol{r}_i-\boldsymbol{r}_{s+1}|)+\tfrac{1}{2}\rho^2\int\int \mathrm{d}\boldsymbol{r}_{s+1}\,\mathrm{d}\boldsymbol{r}_{s+2}$$

$$\times[g^{(s+2)}(\boldsymbol{r}_1,\ldots,\boldsymbol{r}_{s+2})-g^{(s)}(\boldsymbol{r}_1,\ldots,\boldsymbol{r}_s)g(|\boldsymbol{r}_{s+1}-\boldsymbol{r}_{s+2}|)]$$

$$\times\phi(|\boldsymbol{r}_{s+1}-\boldsymbol{r}_{s+2}|) \quad \text{(A2.1.5)}$$

From these equations we then find

$$k_BT^2\left(\frac{\partial g^{(s)}}{\partial T}\right)_\rho=g^{(s)}\sum_{i<j=1}^{s}\phi(r_{ij})+\rho\int \mathrm{d}\boldsymbol{r}_{s+1}\,g^{(s+1)}\sum_{i=1}^{s}\phi(|\boldsymbol{r}_i-\boldsymbol{r}_{s+1}|)$$

$$+\tfrac{1}{2}\rho^2\int\int \mathrm{d}\boldsymbol{r}_{s+1}\,\mathrm{d}\boldsymbol{r}_{s+2}[g^{(s+2)}-g^{(s)}g(|\boldsymbol{r}_{s+1}-\boldsymbol{r}_{s+2}|)]$$

$$\times\phi(|\boldsymbol{r}_{s+1}-\boldsymbol{r}_{s+2}|)-\left\{\rho^2\int \mathrm{d}\boldsymbol{r}\,g(r)\phi(r)+\tfrac{1}{2}\rho^3\int\int \mathrm{d}\boldsymbol{r}\,\mathrm{d}\boldsymbol{s}\right.$$

$$\left.\times[g^{(3)}(\boldsymbol{r},\boldsymbol{s})-g(r)]\phi(r)\right\}\frac{1}{\rho^s}\left(\frac{\partial[\rho^s g^{(s)}]}{\partial p}\right)_T \quad \text{(A2.1.6)}$$

as reported by Schofield (1966). From this equation the expression (5.23) for the heat capacity c_V discussed in section 5.2.2 is easily recovered, using the thermodynamic definition $c_V=(\partial E/\partial T)_V$ and the expression (1.11) for the internal energy. Schofield shows that such an expression for c_V can also be obtained from a fluctuation-theory formula,

$$c_V=\frac{1}{\rho k_BT^2}\left(\langle EE\rangle-\frac{\langle E_\rho\rangle\langle\rho E\rangle}{\langle\rho\rho\rangle}\right) \quad \text{(A2.1.7)}$$

APPENDIX 3.1

FOURTH MOMENT THEOREM

We now turn to extract the coefficient of t^4 in (3.52) for $F_s(k, t)$. After a straightforward calculation, we find the real part (the imaginary part is readily shown to average to zero) as

$$[\tfrac{1}{24}k^4\langle\{\dot{x}_1(0)\}^4\rangle - \tfrac{1}{6}k^2\langle\dot{x}_1(0)\dddot{x}_1(0)\rangle - \tfrac{1}{8}\langle\{\ddot{x}_1(0)\}^2\rangle]$$

As a consequence of the fact that for a fluid the various time-dependent correlation functions must be independent of the origin of time, it is easily shown that $\langle\dot{x}_1(0)\dddot{x}_1(0)\rangle = -\langle\{\ddot{x}_1(0)\}^2\rangle$ and hence from (3.52) it follows that

$$\left.\frac{\partial^4 F_s(k, t)}{\partial t^4}\right|_{t=0} = k^4\langle\{\dot{x}_1(0)\}^4\rangle + k^2\langle\{\ddot{x}_1(0)\}^2\rangle \tag{A3.1.1}$$

The first term can be immediately calculated to yield $3k^4(k_BT)^2/m^2$ while in the second we use the Newtonian equation of motion,

$$m\ddot{x}_1 = -\frac{\partial\Phi_N}{\partial x_1}$$

Carrying out the averaging with respect to $\exp(-\Phi_N/k_BT)$ we note that

$$\frac{\partial}{\partial x_1}\{\exp(-\Phi_N/k_BT)\} = -\frac{1}{k_BT}\frac{\partial\Phi_N}{\partial x_1}\exp(-\Phi_N/k_BT)$$

and using this to integrate by parts we obtain

$$\left.\frac{\partial^4 F_s(k, t)}{\partial t^4}\right|_{t=0} = \frac{3k^4(k_BT)^2}{m^2} + \frac{\rho k^2 k_BT}{m^2}\int d\mathbf{r}\, g(r)\frac{\partial^2\Phi}{\partial x^2} \tag{A3.1.2}$$

Hence from (3.50) we have

$$\int_{-\infty}^{\infty}\frac{d\omega}{2\pi}\omega^4 S_s(k, \omega) = \frac{3k^4(k_BT)^2}{m^2} + \frac{\rho k^2 k_BT}{3m^2}\int d\mathbf{r}\, g(r)\nabla^2\Phi(r) \tag{A3.1.3}$$

A similar calculation leads to the fourth moment for $S(k, \omega)$ as

$$\int_{-\infty}^{\infty}\frac{d\omega}{2\pi}\omega^4 S(k, \omega) = \frac{3k^4(k_BT)^2}{m^2} + \frac{\rho k^2 k_BT}{m^2}\int d\mathbf{r}\, g(r)(1-\cos kx)\frac{\partial^2\Phi}{\partial x^2} \tag{A3.1.4}$$

APPENDIX 3.2

SOLUTION OF THE BLOCH EQUATION

If we have a single-particle Schrödinger equation, with eigenvalues E_i and eigenfunctions ψ_i, then the canonical or Bloch density matrix C, defined by

$$C(\mathbf{r}, \mathbf{r}_0, \beta) = \sum_{\text{all i}} \psi_i^*(\mathbf{r})\psi_i(\mathbf{r}_0) \exp(-\beta E_i) \tag{A3.2.1}$$

is such that the Hamiltonian H_r acting on $C(\mathbf{r}, \mathbf{r}_0, \beta)$ yields $\sum_i E_i \psi_i^*(\mathbf{r})\psi_i(\mathbf{r}_0) \exp(-\beta E_i)$. Evidently, this latter quantity is the same as $-\partial C/\partial\beta$, from (A3.2.1) and hence C satisfies the Bloch equation

$$H_r C = -\frac{\partial C}{\partial \beta} \tag{A3.2.2}$$

It is now clear that, for free particles, when

$$H_r = -\tfrac{1}{2}\nabla_r^2 \tag{A3.2.3}$$

eqn (A3.2.2) is identical with the diffusion equation, (3.65), provided we replace $\beta/2$ by Dt. Furthermore, as $\beta \to 0$, it follows from (A3.2.1) that

$$\begin{aligned} C(\mathbf{r}, \mathbf{r}_0, 0) &= \sum_{\text{all i}} \psi_i^*(\mathbf{r})\psi_i(\mathbf{r}_0) \\ &= \delta(\mathbf{r} - \mathbf{r}_0) \end{aligned} \tag{A3.2.4}$$

which is a statement of the well-known completeness theorem for eigenfunctions. Thus, the desired boundary condition (3.66) is also obeyed.

Hence, putting $\psi_i = V^{-1/2} \exp(i\mathbf{k} \cdot \mathbf{r})$, $E_i = k^2/2$ which are evidently appropriate solutions of the Schrödinger equation with the free-particle Hamiltonian (A3.2.3), into eqn (A3.2.1) and integrating over $\mathbf{k}$, remembering that the density of states in $\mathbf{k}$ space is $V/(8\pi^3)$ we find

$$\begin{aligned} C(\mathbf{r}, \mathbf{r}_0, \beta) &= \frac{1}{8\pi^3} \int d\mathbf{k} \exp[i\mathbf{k} \cdot (\mathbf{r} - \mathbf{r}_0)] \exp(-\beta k^2/2) \\ &= (2\pi\beta)^{-3/2} \exp(-|\mathbf{r} - \mathbf{r}_0|^2/2\beta) \end{aligned} \tag{A3.2.5}$$

Replacing $|\mathbf{r} - \mathbf{r}_0|$ by r, and β by $2D|t|$, the desired result (3.68) follows.

APPENDIX 4.1

MEAN SQUARE DISPLACEMENT AND VELOCITY AUTOCORRELATIONS

The mean square distance travelled by a particle over a time interval t is obviously given by

$$u(t) = \langle [\boldsymbol{R}(t) - \boldsymbol{R}(0)]^2 \rangle \tag{A4.1.1}$$

By using $\boldsymbol{R}(t) - \boldsymbol{R}(0) = \int_0^t \mathrm{d}t_1\, v(t_1)$ and the invariance under time reversal of the velocity autocorrelation function, we find

$$\begin{aligned} u(t) &= \int_0^t \mathrm{d}t_1 \int_0^t \mathrm{d}t^2 \langle v(t_1) \cdot v(t_2) \rangle \\ &= \int_0^t \mathrm{d}t_1 \int_0^t \mathrm{d}t_2 \phi(|t_2 - t_1|) \\ &= 2\int_0^t \mathrm{d}t_1 \int_0^{t-t_1} \mathrm{d}\tau\, \phi(\tau) \end{aligned} \tag{A4.1.2}$$

The time interval $(t - t_1)$ varies from t to 0 as t_1 varies from 0 to t, and if we interchange the order of the two integrations we have

$$\begin{aligned} u(t) &= 2 \int_0^t \mathrm{d}\tau\, \phi(\tau) \int_0^{t-\tau} \mathrm{d}t_1 \\ &= 2 \int_0^t \mathrm{d}\tau\, (t - \tau)\phi(\tau) \end{aligned} \tag{A4.1.3}$$

which is eqn (4.2). From this equation we also have

$$2\phi(t) = \ddot{u}(t) \tag{A4.1.4}$$

APPENDIX 4.2

KINETIC CALCULATION OF DIFFUSION COEFFICIENT FOR HARD SPHERES

Following Longuet-Higgins and Pople (1956), we evaluate the decay time γ^{-1} of the velocity autocorrelation function in a hard-sphere fluid as

$$\gamma = -\lim_{t\to 0}\left\{\frac{\langle v\cdot\Delta v\rangle}{\langle v^2\rangle\Delta t}\right\} = -\frac{1}{3mk_{\mathrm{B}}T}\lim_{t\to 0}\frac{\langle \boldsymbol{p}\cdot\Delta\boldsymbol{p}\rangle}{\Delta t} \tag{A4.2.1}$$

where Δv and $\Delta\boldsymbol{p}$ are the vector increments in the velocity and the momentum of a particle in a positive time interval Δt. If $\boldsymbol{p}_1$ and $\boldsymbol{p}_2$ are the momenta of two colliding particles, and p_1 and p_2 are their components along the interparticle separation vector, then the momentum gained by sphere 1 can be written

$$-\boldsymbol{p}_1\cdot\Delta\boldsymbol{p}_1 = (p_1-p_2)p_1 \tag{A4.2.2}$$

The probability of sphere 1 experiencing such a collision per unit (short) time is $2\pi\rho a^2 g_0 f(p_2)\,\mathrm{d}p_2/m$ where $f(p_2)$ is the Maxwell distribution and g_0 is the value of the pair correlation function at contact. Thus we write

$$-\lim_{\Delta t\to 0}\frac{\langle \boldsymbol{p}_1\cdot\Delta\boldsymbol{p}_1\rangle}{\Delta t} = \frac{2\pi\rho a^2 g_0}{m}\iint_{(p_1-p_2)>0}\mathrm{d}p_1\,\mathrm{d}p_2 p_1(p_1-p_2)^2 f(p_1)f(p_2)$$
$$= 4\pi\rho a^2 g_0 k_{\mathrm{B}}T\,(mk_{\mathrm{B}}T/\pi)^{1/2} \tag{A4.2.3}$$

The value g_0 of the pair correlation at contact can be related to the pressure p in the fluid, using eqn (1.15) with $\phi'(r) = -k_{\mathrm{B}}T\delta(r-R)$ where R is the diameter of the spheres:

$$g_0 = \frac{3}{2\pi\rho a^3}\left(\frac{p}{\rho k_{\mathrm{B}}T}-1\right) \tag{A4.2.4}$$

The final result is

$$\gamma = \frac{2}{ma}\left(\frac{mk_{\mathrm{B}}T}{\pi}\right)^{1/2}\left(\frac{p}{\rho k_{\mathrm{B}}T}-1\right) \tag{A4.2.5}$$

which immediately leads to the expression (4.41) for the self-diffusion coefficient.

APPENDIX 5.1

BOLTZMANN TRANSPORT EQUATION

Having made contact between the collective approach of Zwanzig and the collisionless Boltzmann equation of Vlasov, it will be convenient to present a brief introduction to the Boltzmann transport equation. We deal with a system of particles that is in dynamic equilibrium under external forces and define a distribution function such that $f(\boldsymbol{r}, v)\,d\boldsymbol{r}\,dv$ is the number of particles having positional coordinates in the range between x and $x+dx$, etc., and velocities between v_x and v_x+dv_x, etc.

In the absence of interparticle collisions, the *total* rate of change of the one-particle distribution function with time would be zero by Liouville's theorem. In a real fluid, instead, we write

$$\frac{df}{dt}\equiv\frac{\partial f}{\partial t}+\left.\frac{df}{dt}\right|_{\text{drift}}=\left.\frac{df}{dt}\right|_{\text{coll}} \tag{A5.1.1}$$

where the drift term arises because particles are moving from one part of the system to another and are accelerated by the external field during this process, while the term $(df/dt)|_{\text{coll}}$ represents changes that accompany collisions between the particles. In the present steady-state situation, of course, the distribution function does not depend explicitly on time, namely $(\partial f/\partial t)=0$.

The drift process can be discussed by noting that the number of particles that at time $t+dt$ have drifted to the region of phase space corresponding to the coordinates x, y, z and v_x, v_y, v_z must be equal to the number that were in the cell of phase space at $x-v_x\,dt, y-v_y\,dt, z-v_z\,dt$ and $v_x-a_x\,dt, v_y-a_y\,dt, v_z-a_z\,dt$ at time t. Here the *a's* denote the Cartesian components of the acceleration. This relation holds provided the time interval dt is sufficiently short, so that collisions do not have an appreciable effect on the distribution. Thus the change in f due to the drift process is

$$\Delta f=f(x, y, z, v_x, v_y, v_z)-f(x-v_x\,dt, y-v_y\,dt, z-v_z\,dt, v_x-a_x\,dt, v_y-a_y\,dt,$$
$$v_z-a_z\,dt)=\left(\frac{\partial f}{\partial x}v_x+\frac{\partial f}{\partial y}v_y+\frac{\partial f}{\partial z}v_z+\frac{\partial f}{\partial v_x}a_x+\frac{\partial f}{\partial v_y}a_y+\frac{\partial f}{\partial v_z}a_z\right)dt \tag{A5.1.2}$$

whence

$$\left.\frac{df}{dt}\right|_{\text{drift}}=\frac{\partial f}{\partial x}v_x+\frac{\partial f}{\partial y}v_y+\frac{\partial f}{\partial z}v_z+\frac{\partial f}{\partial v_x}a_x+\frac{\partial f}{\partial v_y}a_y+\frac{\partial f}{\partial v_z}a_z \tag{A5.1.3}$$

The collisional process can be tackled as follows. If $P(v_x, v_y, v_z; v'_x, v'_y, v'_z)$ $dv'_x\, dv'_y\, dv'_z$ is the probability per unit time that a particle will change its velocity from v to v', then the total number of particles which have velocities changing from v to some other value is

$$A = f(x, y, z, v_x, v_y, v_z) \int dv'_x\, dv'_y\, dv'_z\, P(v, v') \tag{A5.1.4}$$

Similarly, the total number of particles having velocities which change to v from other values is

$$B = \int dv'_x\, dv'_y\, dv'_z\, f(\mathbf{r}, v')P(v', v) \tag{A5.1.5}$$

Hence the rate of change of f caused by collisions is

$$\left.\frac{df}{dt}\right|_{\text{coll}} = B - A \tag{A5.1.6}$$

The condition for dynamic equilibrium is that the drift term equal the collision term, that is

$$\frac{\partial f}{\partial x} v_x + \frac{\partial f}{\partial y} v_y + \frac{\partial f}{\partial z} v_z + \frac{\partial f}{\partial v_x} a_x + \frac{\partial f}{\partial v_y} a_y + \frac{\partial f}{\partial v_z} a_z = \left.\frac{df}{dt}\right|_{\text{coll}} \tag{A5.1.7}$$

This is the famous Boltzmann transport equation for a fluid in steady state. If, instead, the external forces vary with time, the distribution depends explicitly on time and the term $\partial f/\partial t$ must be added on the left-hand side of the eqn (A5.1.7).

APPENDIX 5.2

CALCULATION OF $S(k, \omega)$ BY THE VLASOV EQUATION

The Boltzmann equation for the one-particle distribution function $f(\boldsymbol{r}, v, t)$ in a non-stationary situation is obtained from the results of appendix 5.1 by setting

$$\frac{\partial f}{\partial t}+\left.\frac{\mathrm{d}f}{\mathrm{d}t}\right|_{\mathrm{drift}}=\left.\frac{\mathrm{d}f}{\mathrm{d}t}\right|_{\mathrm{coll}} \tag{A5.2.1}$$

where the last two terms are given by eqns (A5.1.3) and (A5.1.6), respectively. In the Vlasov equation, the collision term above is omitted but the interactions between the particles are included to some extent through a mean field approximation, namely, by relating the particle acceleration entering the drift term not to the external force but to a suitably chosen internal force. This procedure becomes correct in certain situations, when the fluid can sustain collisionless collective motions (e.g., zero sound in liquid He^3 or the plasmon excitation in the electron gas, for which $\omega\tau \gg 1$), whereas the opposite condition, $\omega\tau \ll 1$, applies for ordinary sound in gases.

In particular, in the Zwanzig-type approach presented in section 5.3 we write the particle acceleration as

$$m\boldsymbol{a}(\boldsymbol{r}, t) = -\nabla[V_e(\boldsymbol{r}, t)+V(\boldsymbol{r}, t)] \tag{A5.2.2}$$

where $V_e(\boldsymbol{r}, t)$ is the external potential applied to the fluid and $V(\boldsymbol{r}, t)$ is chosen to be given in Fourier transform by

$$V(\boldsymbol{k}, t) = -\frac{k_B T}{\rho}\tilde{c}(k)\rho(\boldsymbol{k}, t) \tag{A5.2.3}$$

Writing

$$f(\boldsymbol{r}, v, t) = \phi_B(v) + f^{(1)}(\boldsymbol{r}, v, t) \tag{A5.2.4}$$

and assuming $V_e(\boldsymbol{r}, t)$ and the deviation from equilibrium, $f^{(1)}(\boldsymbol{r}, v, t)$, to be small, we can now write the Vlasov equation in Fourier transform in the form

$$(\omega+\boldsymbol{k}\cdot v)f^{(1)}(\boldsymbol{k}, v, \omega) = \frac{1}{m}\boldsymbol{k}\cdot\frac{\partial\phi_B(v)}{\partial v}\left[V_e(\boldsymbol{k}, \omega) - \frac{k_B T}{\rho}\tilde{c}(k)\rho(\boldsymbol{k}, \omega)\right] \tag{A5.2.5}$$

Since $\rho(\boldsymbol{k}, \omega) = \int \mathrm{d}v\, f^{(1)}(\boldsymbol{k}, v, \omega)$, this equation has the same form as eqn (5.52) when we take $V_e = 0$.

To calculate $S(k, \omega)$ from eqn (A5.2.5) we may either solve it for $V_e = 0$ under given initial conditions on $f^{(1)}(\boldsymbol{k}, v, t=0)$ to find the time evolution of the particle density from a given initial value, or directly find from it the density response to the external potential $V_e(\boldsymbol{k}, \omega)$ and then use the fluctuation–dissipation theorem. Following the second route yields

$$f^{(1)}(\boldsymbol{k}, v, \omega)=\frac{\dfrac{1}{m}\boldsymbol{k}\cdot\dfrac{\partial\phi_B(v)}{\partial v}}{\omega+\boldsymbol{k}\cdot v+i\eta}\left[V_e(\boldsymbol{k}, \omega)-\frac{k_BT}{\rho}\tilde{c}(k)\rho(\boldsymbol{k}, \omega)\right] \tag{A5.2.6}$$

where we have inserted an infinitesimal imaginary part ($\eta=0^+$) in the denominator to account for the causal nature of the response.

This is, in fact, equivalent to inserting a collision term of the type

$$\left.\frac{\mathrm{d}f}{\mathrm{d}t}\right|_{\mathrm{coll}}=-f^{(1)}/\tau \tag{A5.2.7}$$

with $\tau\to\infty$.

Integration of eqn (A5.2.6) over v yields

$$\rho(\boldsymbol{k}, \omega)=\chi_0(k, \omega)\left[V_e(\boldsymbol{k}, \omega)-\frac{k_BT}{\rho}\tilde{c}(k)\rho(\boldsymbol{k}, \omega)\right] \tag{A5.2.8}$$

or

$$\rho(\boldsymbol{k}, \omega)=\frac{\chi_0(k, \omega)}{1+\dfrac{k_BT}{\rho}\tilde{c}(k)\chi_0(k, \omega)}V_e(\boldsymbol{k}, \omega) \tag{A5.2.9}$$

where

$$\chi_0(k, \omega)=\int \mathrm{d}v\,\frac{\dfrac{1}{m}\boldsymbol{k}\cdot\dfrac{\partial\phi_B(v)}{\partial v}}{\omega+\boldsymbol{k}\cdot v+i\eta} \tag{A5.2.10}$$

This latter function can clearly be identified with the density response function of a fluid of non-interacting particles, whose dynamic structure factor we have already evaluated in section 3.4.1. Using the fluctuation–dissipation theorem on that result, eqn (3.63), or by direct evaluation of eqn (A5.2.10), we can write

$$\mathrm{Im}\,\chi_0(k, \omega)=-\frac{\rho\pi^{1/2}}{k_BT}x\exp(-x^2) \tag{A5.2.11}$$

where $x=[(\omega/k)(m/2k_BT)^{1/2}]$. By the Kramers–Kronig relation we find

$$\mathrm{Re}\,\chi_0(k, \omega)=-\frac{\rho}{k_BT}\left[1-2x\exp(-x^2)\int_0^x\exp(t^2)\,\mathrm{d}t\right] \tag{A5.2.12}$$

Finally, using the fluctuation–dissipation theorem to derive $S(k, \omega)$ from the response function given in eqn (A5.2.9), we have

$$S(k,\omega)=\frac{1}{k}\left(\frac{2\pi m}{k_{\mathrm{B}}T}\right)^{1/2}$$
$$\times\frac{\exp(-x^2)}{\left[\frac{1}{S(k)}+2\tilde{c}(k)x\exp(-x^2)\int_0^x\exp(t^2)\,\mathrm{d}t\right]^2+[\sqrt{\pi}\tilde{c}(k)x\exp(-x^2)]^2} \quad \text{(A5.2.13)}$$

in agreement with eqn (5.53) of the text.

By comparison with the generalised hydrodynamic result reported in eqn (4.89) we note that the Vlasov equation approach effectively fixes the generalised viscosity function $D_{\mathrm{L}}(k,\omega)$ from the properties of the perfect gas. Thus, this approach assumes that the liquid can sustain collective oscillations through the mean field given by eqn (A5.2.3), while the damping of these oscillations arises from the disordered motions of the atoms treated as free particles.

APPENDIX 6.1

PARTIAL STRUCTURE FACTORS AND THERMODYNAMICS OF MULTI-COMPONENT FLUIDS

The statistical mechanical theory of solutions, developed in a classic paper by Kirkwood and Buff (1951), extends the Ornstein–Zernike relation between the structure factor at long wavelength in a one-component fluid and its compressibility to the case of multi-component systems. The thermodynamic quantities of interest evidently are, in addition to the compressibility which measures the response of the fluid to the application of pressure, the 'osmotic compressibilities' which describe the separability of the components.

By the procedure already followed in section 1.3 we consider for a multi-component fluid an open region of volume V containing $N_1, \ldots, N$ particles of species $1, \ldots, \nu$. Equation (1.21), relating the structure factor to the particle number fluctuations, is immediately extended to the multi-component fluid to read

$$\lim_{k\to 0} a_{\alpha\beta}(k) = 1 + \langle N \rangle \left[\frac{\langle N_\alpha N_\beta \rangle - \langle N_\alpha \rangle \langle N_\beta \rangle}{\langle N_\alpha \rangle \langle N_\beta \rangle} - \frac{\delta_{\alpha\beta}}{\langle N_\alpha \rangle} \right] \tag{A6.1.1}$$

Thus the partial structure factors in the long wavelength limit are related to the composition fluctuations in the fluid mixture.

We now use the theory of the grand ensemble to relate the composition fluctuations to thermodynamics quantities. The grand partition function is given by

$$\exp(-\Omega/k_{\mathrm{B}}T) = \sum_{N_1, \cdots, N_v = 0}^{\infty} \exp\left([N_1\mu_1 + \cdots + N_v\mu_v - F(N_1, \ldots, N_v, T, V)]/k_{\mathrm{B}}T\right) \tag{A6.1.2}$$

where $F(N_1, \ldots, N_v, T, V)$ is the Helmholtz free energy of a member of the grand ensemble with given numbers of particles of each species, and the chemical potentials $\mu_1, \ldots, \mu_v$ are to be determined from the average number of particles of each species,

$$\langle N_\alpha \rangle = \sum_{N_1, \cdots, N_v = 0}^{\infty} N_\alpha \exp\left([\Omega + N_1\mu_1 + \cdots + N_v\mu_v - F(N_1, \ldots, N_v, T, V)]/k_{\mathrm{B}}T\right) \tag{A6.1.3}$$

We immediately find

$$\left(\frac{\partial\Omega}{\partial\mu_\alpha}\right)_{T,V,\mu_{\bar{\alpha}}}=-\langle N_\alpha\rangle \tag{A6.1.4}$$

and

$$\begin{aligned} k_BT\left(\frac{\partial N_\alpha}{\partial\mu_\beta}\right)_{T,V,\mu_{\bar{\beta}}} &= \langle N_\alpha\rangle\left(\frac{\partial\Omega}{\partial\mu_\beta}\right)_{T,V,\mu_{\bar{\beta}}}+\langle N_\alpha N_\beta\rangle \\ &= \langle N_\alpha N_\beta\rangle-\langle N_\alpha\rangle\langle N_\beta\rangle \end{aligned} \tag{A6.1.5}$$

where the index $\bar{\alpha}$ denotes all species different from α. Since the thermodynamic derivatives $(\partial N_\alpha/\partial\mu_\beta)_{T,V,\mu_{\bar{\beta}}}$ are the elements of the inverse of the matrix whose elements are $(\partial\mu_\alpha/\partial N_\beta)_{T,V,N_{\bar{\beta}}}$, we can invert the above relation to read

$$\frac{V}{k_BT}\left(\frac{\partial\mu_\alpha}{\partial N_\beta}\right)_{T,V,N_{\bar{\beta}}}=|A|_{\alpha\beta}/|A| \tag{A6.1.6}$$

where

$$A_{\alpha\beta}=\frac{1}{V}\left[\langle N_\alpha N_\beta\rangle-\langle N_\alpha\rangle\langle N_\beta\rangle\right] \tag{A6.1.7}$$

$|A|=|A_{\alpha\beta}|$, and $|A|_{\alpha\beta}$ represents the cofactor of $A_{\alpha\beta}$ in the determinant $|A|$. We have thus expressed thermodynamic derivatives in terms of the composition fluctuations, namely, of the partial structure factors at long wavelength.

To make contact with experimental quantities we now transform the thermodynamic derivatives at constant volume into derivatives at constant pressure, by means of the identity

$$\left(\frac{\partial\mu_\alpha}{\partial N_\beta}\right)_{T,V,N_{\bar{\beta}}}=\left(\frac{\partial\mu_\alpha}{\partial N_\beta}\right)_{T,p,N_{\bar{\beta}}}+\frac{v_\alpha v_\beta}{VK_T} \tag{A.6.1.8}$$

where v_α is the partial molar volume, per molecule, of species α. By making use of the Gibbs–Duhem equation,

$$\sum_{\alpha=1}^{\nu}\langle N_\alpha\rangle\left(\frac{\partial\mu_\alpha}{\partial N_\beta}\right)_{T,p,N_{\bar{\beta}}}=0 \tag{A6.1.9}$$

one readily finds from eqns (A6.1.6) and (A6.1.8)

$$\rho v_\alpha=\sum_{\beta=1}^{\nu}c_\beta|A|_{\alpha\beta}\Big/\sum_{\beta,\gamma=1}^{\nu}c_\beta c_\gamma|A|_{\beta\gamma} \tag{A6.1.10}$$

and

$$\rho^2k_BTK_T=|A|\Big/\sum_{\alpha,\beta=1}^{\nu}c_\alpha c_\beta|A|_{\alpha\beta} \tag{A6.1.11}$$

where $c_\alpha \equiv \langle N_\alpha \rangle \Big/ \sum_{\beta=1}^{\nu} \langle N_\beta \rangle$ is the number concentration of species α, and the quantities $A_{\alpha\beta}$ are given by eqn (A6.1.7), namely

$$A_{\alpha\beta} = \rho[c_\alpha \delta_{\alpha\beta} + c_\alpha c_\beta (a_{\alpha\beta} - 1)] \tag{A6.1.12}$$

Finally, eqns (A6.1.6) and (A6.1.8) are rewritten as expressions for the derivatives of the chemical potentials with respect to concentrations of the various species,

$$(k_B T)^{-1} \left(\frac{\partial \mu_\alpha}{\partial c_\beta} \right)_{T,p,c_{\bar{\beta}}} = \frac{\rho V}{(1-c_\beta) k_B T} \left(\frac{\partial \mu_\alpha}{\partial N_\beta} \right)_{T,p,N_{\bar{\beta}}}$$

$$= (1-c_\beta)^{-1} \left\{ \frac{\rho |A|_{\alpha\beta}}{|A|} - \frac{\rho v_\alpha v_\beta}{k_B T K_T} \right\} \tag{A6.1.13}$$

In the special case of a two-component fluid, writing $c_2 = c$ and $c_1 = 1 - c$, the above results take the following forms:

$$\rho v_1 = \frac{1 + c(a_{22} - a_{12})}{1 + c(1-c)(a_{11} + a_{22} - 2a_{12})} \tag{A6.1.14}$$

$$(1-c) v_1 + c v_2 = 1/\rho \tag{A6.1.15}$$

$$\rho k_B T K_T = \frac{[(1-c)a_{11} + c][c a_{22} + (1-c)] - c(1-c)(a_{12} - 1)^2}{1 + c(1-c)(a_{11} + a_{22} - 2a_{12})} \tag{A6.1.16}$$

and

$$\frac{1}{k_B T} \left(\frac{\partial \mu_2}{\partial c} \right)_{T,p} = \frac{1}{c[1 + c(1-c)(a_{11} + a_{22} - 2a_{12})]} \tag{A6.1.17}$$

This completes the deviation of the equations given in section 6.2. Notice that the thermodynamic quantities there introduced are related to the present quantities by

$$\delta = \rho v_1 - \rho v_2 \tag{A6.1.18}$$

and

$$S_{cc} = (1-c) k_B T \Big/ \left(\frac{\partial \mu_2}{\partial c} \right)_{T,p} \tag{A6.1.19}$$

APPENDIX 6.2

BASIC EQUATIONS OF THEORY OF CONFORMAL SOLUTIONS

The idea is to relate the thermodynamic properties of a multi-component solution to those of a reference species L_0 say. The basic equation derived by Longuet-Higgins (1951) for the molar Gibbs free energy in terms of the mole fractions x_r and x_s of species L_r and L_s in the solution is

$$G=\sum_{r} x_r[G_r+RT\ln x_r]+\sum_{r<s}\sum x_r x_s E_0 d_{rs} \tag{A6.2.1}$$

Here G_r is the molar free energy of species L_r at given T and p, while E_0 is the molar configurational energy of the reference species and is explicitly $RT-Q_0$, where Q_0 is the latent heat of vaporisation of L_0 at temperature T and pressure p. d_{rs} is an interaction parameter for each pair of components.

The dependence of volume on composition follows from eqn (A6.2.1) as

$$V=\sum_{r} x_r V_r+V_0(p\beta_0-T\alpha_0)\sum_{r<s}\sum x_r x_s d_{rs} \tag{A6.2.2}$$

where α_0 and β_0 are, respectively, the thermal expansion coefficient and the isothermal compressibility of the reference liquid L_0.

These forms, (A6.2.1) and (A6.2.2), express the thermodynamic properties of a conformal solution in terms of those of its components, together with a single interaction parameter for each pair of components. It is clear that when this model is applicable, therefore, the thermodynamic properties of a solution of more than two components can be determined from a study of the appropriate possible binary systems.

The theory is closely related to the theory of regular solutions. In fact, the only assumption made in the model of Longuet-Higgins that is not invoked in the theory of regular solutions is that the intermolecular forces are approximately equal in magnitude for different pairs of species.

A generalisation of this model has been effected by Byers Brown (see Rowlinson, 1969 for a summary and further references). However, with the present dearth of experimental data we do not feel that to proceed to such refinements, with their associated complications, would be useful at the present time.

APPENDIX 6.3

WAVE NUMBER-DEPENDENT GENERALISATIONS OF COMPRESSIBILITY AND SIZE FACTORS IN ALLOYS

We record here some useful relations between the number–concentration correlation functions S_{NN}, S_{Nc} and S_{cc} and the Pearson–Rushbrooke (1957) partial direct correlation functions $c_{ij}(k)$. First, as follows from the definitions

$$S_{cc}(k)=c_1c_2+\rho c_1{}^2c_2{}^2 \times\frac{c_{11}(k)+c_{22}(k)-2c_{12}(k)-\rho[c_{11}(k)c_{22}(k)-c_{12}^2(k)]}{1-\rho[c_1c_{11}(k)+c_2c_{22}(k)]+\rho^2c_1c_2[c_{11}(k)c_{22}(k)-c_{12}^2(k)]} \tag{A6.3.1}$$

Introducing a generalisation of the compressibility through

$$\vartheta(k)=S_{NN}(k)-\Delta^2(k)S_{cc}(k) \tag{A6.3.2}$$

which tends to $\rho k_B T K_T$ as $k\to 0$, $\Delta(k)$ being simply defined by

$$\Delta(k)=\frac{S_{Nc}(k)}{S_{cc}(k)}\to -\delta \qquad \text{as } k\to 0 \tag{A6.3.3}$$

$S_{NN}(k)$ can be written as

$$S_{NN}(k)=\vartheta(k)\left[1+\frac{c_1c_2\rho^2\{c_1[c_{11}(k)-c_{12}(k)]-c_2[c_{22}(k)-c_{12}(k)]\}^2}{1-\rho[c_1c_{11}(k)+c_2c_{22}(k)]+\rho^2c_1c_2[c_{11}(k)c_{22}(k)-c_{12}^2(k)]}\right] \tag{A6.3.4}$$

In terms of the direct correlation functions, we can write $\vartheta(k)$ explicitly as

$$\vartheta(k)=\{1-\rho[c_1{}^2c_{11}(k)+c_2{}^2c_{22}(k)+2c_1c_2c_{12}(k)]\}^{-1} \tag{A6.3.5}$$

and in terms of $\vartheta(k)$, $\Delta(k)$ can be written

$$\Delta(k)=\rho\vartheta(k)[c_1\{c_{11}(k)-c_{12}(k)\}-c_2\{c_{22}(k)-c_{12}(k)\}] \tag{A6.3.6}$$

This latter equation allows $S_{Nc}(k)$ to be written explicitly in terms of the direct correlation functions.

Finally, we note that the partial structure factors can be expressed in terms of $\vartheta(k)$, $\Delta(k)$ and $S_{cc}(k)$ as

$$\left.\begin{aligned} a_{11}(k)&=\vartheta(k)+S_{cc}(k)\left[\Delta(k)+\frac{1}{c_1}\right]^2-\frac{c_2}{c_1}\\ a_{22}(k)&=\vartheta(k)+S_{cc}(k)\left[\Delta(k)-\frac{1}{c_2}\right]^2-\frac{c_1}{c_2}\\ a_{12}(k)&=\vartheta(k)+S_{cc}(k)\left[\Delta(k)+\frac{1}{c_1}\right]\left[\Delta(k)-\frac{1}{c_2}\right]+1 \end{aligned}\right\} \qquad \text{(A6.3.7)}$$

These are readily seen to be k-dependent generalisations of eqns (6.19)–(6.21).

APPENDIX 6.4

PERTURBATION THEORY FOR CLASSICAL FLUIDS

In some circumstances the calculation of thermodynamic functions and also pair functions can be effected to useful accuracy by perturbation theory (Barker and Henderson, 1967; 1971). Here the derivation of expressions to second order for thermodynamic functions and first order for radial distribution functions will be given. The argument parallels that of Henderson, Barker and Smith (1972). It will be convenient to derive the results in a form applicable to a multi-component fluid.

Force law assumed

Consider a system of N molecules of species $1, \dots, m$. The potential energy V_N of the molecules is taken to depend on a parameter γ in such a way that $(\partial V_N/\partial\gamma)$ is a sum of terms each depending only on the coordinates of two molecules:

$$V_N = V_N(\gamma; \boldsymbol{r}_1, \dots, \boldsymbol{r}_N) \tag{A6.4.1}$$

$$\frac{\partial V_N}{\partial \gamma} = \sum_{i>j=1}^{N} w_{\alpha_i \alpha_j}(\gamma; R_{ij}) \tag{A6.4.2}$$

Here $\alpha_i = 1, \dots, m$ if molecule i is of species $1, \dots, m$, respectively and $R_{ij} = |\boldsymbol{r}_i - \boldsymbol{r}_j|$.

Derivatives of partition function with respect to γ

The aim of the calculation below is to expand the thermodynamic functions to second order in a Taylor series about an arbitrary point $\gamma = \gamma_0$. To this end we shall evaluate the first and second derivatives of the partition function of the system with respect to the parameter γ. These derivatives are best found in the grand canonical ensemble. Results for the canonical ensemble may then be obtained by appropriate transformation.

In the grand canonical ensemble, in which the independent variables are T, V and the chemical potentials μ_i, the probability that there are n molecules of species $j_1, j_2, \dots, j_n$ in the elements $d\boldsymbol{r}_1, \dots, d\boldsymbol{r}_n$ is $n^{(n)}_{j_1,\dots,j_n}(\boldsymbol{r}_1, \dots, \boldsymbol{r}_n)\, d\boldsymbol{r}_1 \dots d\boldsymbol{r}_n$, where

$$n^{(n)}_{j_1,\dots,j_n}(\boldsymbol{r}_1, \dots, \boldsymbol{r}_n) = \Xi^{-1} \prod_{l=1}^{m} \sum_{N_l \geq i_l} \frac{\exp(N_l v_l)}{(N_l - i_l)!} \lambda_l^{3N_l} \int \exp(-\beta V_N)\, d\boldsymbol{r}_{n+1} \dots d\boldsymbol{r}_N \tag{A6.4.3}$$

As usual, $\beta=(k_BT)^{-1}$, while $\nu_l=\mu_l/(k_BT)$, $\lambda_l=(2\pi m_l k_BT/h^2)^{1/2}$, and m_l and N_l are, respectively, the molecular mass and the number of molecules of species l. Furthermore, we have $i_k=\sum_{l=1}^{n}\delta_{j_lk}$, $n=\sum_{l=1}^{m} i_l$ and $N=\sum_{l=1}^{m} N_l$. Finally, Ξ is the grand partition function defined by

$$\Xi=\prod_{l=1}^{m}\sum_{N_l>0}\frac{\exp(N_l\nu_l)}{N_l!}\lambda_l^{3N_l}\int\exp(-\beta V_N)\,d\boldsymbol{r}_1\ldots d\boldsymbol{r}_N \tag{A6.4.4}$$

The integrations in the above equations are taken throughout the volume V, and the pressure p in the system is given by

$$\frac{pV}{k_BT}=\ln\Xi \tag{A6.4.5}$$

Although it is not exhibited explicitly, $n^{(n)}$ and Ξ are to be regarded as functions of the parameter γ.

It is then a straightforward matter to show that

$$\left[\frac{\partial\ln\Xi}{\partial\gamma}\right]_{\nu_l,T,V}=-\tfrac{1}{2}\beta\sum_{i,j=1}^{m}\int n_{ij}^{(2)}(12)w_{ij}(12)\,d\boldsymbol{r}_1\,d\boldsymbol{r}_2 \tag{A6.4.6}$$

and

$$\begin{aligned}\left[\frac{\partial^2\ln\Xi}{\partial\gamma^2}\right]_{\nu_l,T,V}&=\tfrac{1}{4}\beta^2\sum_{i,j,k,l=1}^{m}\int\left[n_{ijkl}^{(4)}(1234)-n_{ij}^{(2)}(12)n_{kl}^{(2)}(34)\right]\\&\quad\times w_{ij}(12)w_{kl}(34)\,d\boldsymbol{r}_1\,d\boldsymbol{r}_2\,d\boldsymbol{r}_3\,d\boldsymbol{r}_4\\&\quad+\beta^2\sum_{i,j,k=1}^{m}\int n_{ijk}^{(3)}(123)w_{ij}(12)w_{jk}(23)d\boldsymbol{r}_1\,d\boldsymbol{r}_2\,d\boldsymbol{r}_3\\&\quad+\tfrac{1}{2}\beta^2\sum_{i,j=1}^{m}\int n_{ij}^{(2)}(12)[w_{ij}(12)]^2\,d\boldsymbol{r}_1\,d\boldsymbol{r}_2\\&\quad-\tfrac{1}{2}\beta\sum_{i,j=1}^{m}\int n_{ij}^{(2)}(12)\frac{\partial w_{ij}(12)}{\partial\gamma}\,d\boldsymbol{r}_1\,d\boldsymbol{r}_2\end{aligned} \tag{A6.4.7}$$

Here the obvious notation $n^{(2)}(12)\equiv n^{(2)}(\boldsymbol{r}_1,\boldsymbol{r}_2)$, etc., has been used.

The above results allow the Taylor expansion of the pressure to second order in $(\gamma-\gamma_0)$, at constant chemical potential, to be obtained. At this point it should be noted that the expression for the second derivative is in a suitable form for numerical evaluation in the thermodynamic limit since $[n_{ijkl}^{(4)}(1234)-n_{ij}^{(2)}(12)\,n_{kl}^{(2)}(34)]$ approaches zero exactly when the pairs 12 and 34 are widely separated.

Derivatives of canonical partition function

The derivatives of the canonical partition function, for which the independent variables are T, V and $\bar{N}_i$, can now be obtained from these results using thermodynamic arguments.

The canonical partition function Z_N is defined by

$$Z_N = \prod_{l=1}^{m} \frac{\lambda_l^{3N_l}}{N_l!} \int \exp(-\beta V_N)\, d\boldsymbol{r}_1 \ldots d\boldsymbol{r}_N \tag{A6.4.8}$$

and the Helmholtz free energy F is then

$$F = -k_B T \ln Z_N \tag{A6.4.9}$$

However, F can be related to the grand partition function through

$$F = \sum_{l=1}^{m} N_l \mu_l - k_B T \ln \Xi \tag{A6.4.10}$$

From this result it is easy to show that

$$\frac{1}{k_B T}\left(\frac{\partial F}{\partial \gamma}\right)_{\bar{N}_i} = \sum_{l=1}^{m} N_l \left(\frac{\partial \nu_l}{\partial \gamma}\right)_{\bar{N}_i} - \left(\frac{\partial \ln \Xi}{\partial \gamma}\right)_{\bar{N}_i} \tag{A6.4.11}$$

Thus

$$\left(\frac{\partial F}{\partial \gamma}\right)_{\bar{N}_i} = -k_B T \left(\frac{\partial \ln \Xi}{\partial \gamma}\right)_{\nu_i} \tag{A6.4.12}$$

since

$$\bar{N}_i = k_B T \left(\frac{\partial \ln \Xi}{\partial \mu_i}\right)_{\gamma} = \left(\frac{\partial \ln \Xi}{\partial \nu_i}\right)_{\gamma} \tag{A6.4.13}$$

Hence, using eqn (A6.4.6), we have the expression we were seeking for the first derivative of the Helmholtz free energy.

Proceeding to the calculation of the second derivative, we find

$$\frac{1}{k_B T}\left(\frac{\partial^2 F}{\partial \gamma^2}\right)_{\bar{N}_i} = -\left(\frac{\partial^2 \ln \Xi}{\partial \gamma^2}\right)_{\nu_i} - \sum_{l=1}^{m} \left(\frac{\partial^2 \ln \Xi}{\partial \nu_l \partial \gamma}\right)\left(\frac{\partial \nu_l}{\partial \gamma}\right)_{\bar{N}_i} \tag{A6.4.14}$$

If we write in the last term

$$Q_l = -\frac{\rho k_B T}{\bar{N}} \frac{\partial^2 \ln \Xi}{\partial \nu_l \partial \gamma} \tag{A6.4.15}$$

where $\rho = \bar{N}/V$, then, from eqn (A6.4.6), this can be expressed as

$$Q_l = \frac{1}{V}\frac{\partial}{\partial \nu_l}\left[\tfrac{1}{2} \sum_{i,j=1}^{m} \int n_{ij}^{(2)}(12) w_{ij}(12)\, d\boldsymbol{r}_1\, d\boldsymbol{r}_2\right] \tag{A6.4.16}$$

Also it is convenient to write

$$R_l = k_B T \left(\frac{\partial v_l}{\partial \gamma}\right)_{N_i} \tag{A6.4.17}$$

Now $\mu_j = \partial F / \partial \bar{N}_j$ and hence

$$\begin{aligned} R_l &= \frac{\partial^2 F}{\partial \gamma \partial \bar{N}_l} = -k_B T \frac{\partial^2 \ln \Xi}{\partial \bar{N}_l \partial \gamma} \\ &= \frac{\partial}{\partial \bar{N}_l}\left[\tfrac{1}{2} \sum_{i,j=1}^{m} \int n_{ij}^{(2)}(12) w_{ij}(12)\, d\boldsymbol{r}_1\, d\boldsymbol{r}_2\right] \end{aligned} \tag{A6.4.18}$$

In terms of normalised distribution functions defined by

$$g_{j_1,\cdots,j_n}(1 \cdots n) = n^{(n)}_{j_1,\cdots,j_n}(1 \cdots n)/[\rho_1{}^{j_1} \cdots \rho_n{}^{j_n}] \tag{A6.4.19}$$

where $\rho_k = \bar{N}_k / V$, it follows that

$$Q_l = \frac{\partial}{\partial v_l}\left[\tfrac{1}{2} \sum_{i,j=1}^{m} \rho_i \rho_j \int g_{ij}(12) w_{ij}(12)\, d\boldsymbol{r}_2\right] \tag{A6.4.20}$$

and

$$R_l = \frac{\partial}{\partial \rho_l}\left[\tfrac{1}{2} \sum_{i,j=1}^{m} \rho_i \rho_j \int g_{ij}(12) w_{ij}(12)\, d\boldsymbol{r}_2\right] \tag{A6.4.21}$$

Thus the second derivative of the Helmholtz free energy with respect to γ takes the form

$$\left(\frac{\partial^2 F}{\partial \gamma^2}\right)_{N_i} = -k_B T \left(\frac{\partial^2 \ln \Xi}{\partial \gamma^2}\right)_{v_j, T, V} + \beta \bar{N} \rho^{-1} \psi \tag{A6.4.22}$$

where

$$\psi = \sum_{l=1}^{m} Q_l R_l \tag{A6.4.23}$$

The vectors $\boldsymbol{Q}$ and $\boldsymbol{R}$ are not, in fact, independent. The relation between them is

$$\begin{aligned} Q_l &= -\frac{\rho k_B T}{\bar{N}} \frac{\partial^2 \ln \Xi}{\partial v_l \partial \gamma} \\ &= -\frac{\rho k_B T}{\bar{N}} \sum_{j=1}^{m} \left[\frac{\partial}{\partial \bar{N}_j}\left(\frac{\partial \ln \Xi}{\partial \gamma}\right)\right]\left(\frac{\partial \bar{N}_j}{\partial v_l}\right) \\ &= \sum_{j=1}^{m} A_{lj} R_j \end{aligned} \tag{A6.4.24}$$

In the last expression (Kirkwood and Buff, 1951; see appendix 6.1)

$$A_{lj}=A_{jl}=(\bar{N})^{-1}\rho\left(\frac{\partial\bar{N}_j}{\partial v_l}\right)$$

$$=\left(\frac{\partial\rho_j}{\partial v_l}\right)_\gamma=\rho_j\delta_{lj}+\rho_l\rho_j\int[g_{lj}(12)-1]\,d\boldsymbol{r}_2 \tag{A6.4.25}$$

It is now straightforward to show that

$$\psi=\sum_{i,j=1}^{m} R_i A_{ij} R_j \tag{A6.4.26}$$

Hence

$$\left(\frac{\partial F}{\partial\gamma}\right)_{\bar{N}_l}=\tfrac{1}{2}V\sum_{i,j=1}^{m}\rho_i\rho_j\int g_{ij}(12)w_{ij}(12)\,d\boldsymbol{r}_2 \tag{A6.4.27}$$

and

$$\left(\frac{\partial^2 F}{\partial\gamma^2}\right)_{N_l}=-\beta V\Bigg\{\tfrac{1}{4}\sum_{i,j,k,l=1}^{m}\rho_i\rho_j\rho_k\rho_l\int[g_{ijkl}(1234)-g_{ij}(12)g_{kl}(34)]$$
$$\times w_{ij}(12)w_{kl}(34)\,d\boldsymbol{r}_2\,d\boldsymbol{r}_3\,d\boldsymbol{r}_4+\sum_{i,j,k=1}^{m}\rho_i\rho_j\rho_k\int g_{ijk}(123)$$
$$\times w_{ij}(12)w_{jk}(23)\,d\boldsymbol{r}_2\,d\boldsymbol{r}_3+\tfrac{1}{2}\sum_{i,j=1}^{m}\rho_i\rho_j\int g_{ij}(12)[w_{ij}(12)]^2\,d\boldsymbol{r}_2$$
$$-\tfrac{1}{2}k_BT\sum_{i,j=1}^{m}\rho_i\rho_j\int g_{ij}(12)\frac{\partial w_{ij}(12)}{\partial\gamma}\,d\boldsymbol{r}_2-\psi\Bigg\} \tag{A6.4.28}$$

First derivative of radial distribution function

Since the perturbation $w_{ij}(12)$ is arbitrary, the above results can be used to find the first derivative of the radial distribution function with respect to γ. The result is

$$\left(\frac{\partial g_{ij}(12)}{\partial\gamma}\right)_{N_l}=-\beta\Bigg\{\tfrac{1}{2}\sum_{k,l=1}^{m}\rho_k\rho_l\int[g_{ijkl}(1234)-g_{ij}(12)g_{kl}(34)]w_{kl}(34)\,d\boldsymbol{r}_3\,d\boldsymbol{r}_4$$
$$+\sum_{k=1}^{m}\rho_k\int g_{ijk}(123)[w_{ik}(13)+w_{jk}(23)]\,d\boldsymbol{r}_3$$
$$+g_{ij}(12)w_{ij}(12)-2(\rho_i\rho_j)^{-1}\phi_{ij}\Bigg\} \tag{A6.4.29}$$

where

$$\phi_{ij}=\sum_{k=1}^{m}\frac{\partial}{\partial v_k}[\tfrac{1}{2}\rho_i\rho_j g_{ij}(12)]R_k$$
$$=\sum_{l=1}^{m}\frac{\partial}{\partial\rho_l}[\tfrac{1}{2}\rho_i\rho_j g_{ij}(12)]Q_l \tag{A6.4.30}$$

Buff and his coworkers earlier found similar results to those of Henderson, Barker and Smith. However, Buff's results were derived in the grand canonical and constant pressure ensemble. The present results are more useful for perturbation theory calculations (Henderson and Barker, 1969).

The case of pure fluid

Let us now specialise to the case of a pure fluid, $m=1$. Then the derivatives of the free energy take the form

$$\left(\frac{\partial F}{\partial \gamma}\right)_\rho = \tfrac{1}{2}\bar{N}\rho \int g(12)w(12)\,\mathrm{d}\boldsymbol{r}_2 \tag{A6.4.31}$$

and

$$\begin{aligned}\left(\frac{\partial^2 F}{\partial \gamma^2}\right)_\rho = &-\beta\bar{N}\left\{\tfrac{1}{4}\rho^3 \int [g(1234)-g(12)g(34)]w(12)w(34)\,\mathrm{d}\boldsymbol{r}_2\,\mathrm{d}\boldsymbol{r}_3\,\mathrm{d}\boldsymbol{r}_4\right.\\ &\left.+\rho^2 \int g(123)w(12)w(23)\,\mathrm{d}\boldsymbol{r}_2\,\mathrm{d}\boldsymbol{r}_3 + \tfrac{1}{2}\rho \int g(12)[w(12)]^2\,\mathrm{d}\boldsymbol{r}_2\right\}\\ &+\tfrac{1}{2}\rho\bar{N}\int g(12)\frac{\partial w(12)}{\partial \gamma}\,\mathrm{d}\boldsymbol{r}_2\\ &+\bar{N}\left(\frac{\partial \rho}{\partial p}\right)_\gamma \left\{\frac{\partial}{\partial \rho}\left[\tfrac{1}{2}\rho^2 \int g(12)w(12)\,\mathrm{d}\boldsymbol{r}_2\right]\right\}^2\end{aligned} \tag{A6.4.32}$$

Similarly, for the derivative of the radial distribution function $g(12)$, the following result is obtained

$$\begin{aligned}\left(\frac{\partial g(12)}{\partial \gamma}\right)_\rho = &-\beta\left\{\tfrac{1}{2}\rho^2 \int [g(1234)-g(12)g(34)]w(34)\,\mathrm{d}\boldsymbol{r}_3\,\mathrm{d}\boldsymbol{r}_4\right.\\ &\left.+2\rho \int g(123)w(23)\,\mathrm{d}\boldsymbol{r}_3 + g(12)w(12)\right\}\\ &+\frac{2}{\rho}\left(\frac{\partial \rho}{\partial p}\right)_\gamma \left\{\frac{\partial}{\partial \rho}[\tfrac{1}{2}\rho^2 g(12)]\right\}\left\{\frac{\partial}{\partial \rho}\left[\tfrac{1}{2}\rho^2 \int g(34)w(34)\,\mathrm{d}\boldsymbol{r}_4\right]\right\}\end{aligned} \tag{A6.4.33}$$

In obtaining these results, use has been made of the fact that, for a single-component system

$$\begin{aligned}A_{11} &= \frac{\partial \rho}{\partial v} = \rho k_\mathrm{B} T\left(\frac{\partial \rho}{\partial p}\right)_\gamma\\ &= \rho\left\{1+\rho \int [g(12)-1]\,\mathrm{d}\boldsymbol{r}_2\right\}\end{aligned} \tag{A6.4.34}$$

These results for single-component systems were derived by Henderson and Barker (1967).

APPENDIX 6.5

HYDRODYNAMIC CORRELATION FUNCTIONS IN A BINARY MIXTURE

The linearised hydrodynamic equations for a two-component neutral fluid (Landau and Lifshitz, 1959) comprise the mass continuity equation,

$$\frac{\partial m(\boldsymbol{r}, t)}{\partial t} + \nabla \cdot \boldsymbol{p}(\boldsymbol{r}, t) = 0 \tag{A6.5.1}$$

the Navier–Stokes equation, from which we retain only the longitudinal part,

$$\left(\frac{\partial}{\partial t} - D_L \nabla^2\right) \boldsymbol{p}(\boldsymbol{r}, t) = -\nabla P(\boldsymbol{r}, t) \tag{A6.5.2}$$

$P(\boldsymbol{r}, t)$ being the local pressure; the heat diffusion equation,

$$\frac{\partial Q(\boldsymbol{r}, t)}{\partial t} = \kappa \nabla^2 T(\boldsymbol{r}, t) - m\rho \left[k_T \left(\frac{\partial \mu}{\partial x}\right)_{p,T} - T\left(\frac{\partial \mu}{\partial T}\right)_{p,x} \right] \nabla \cdot j(\boldsymbol{r}, t) \tag{A6.5.3}$$

and the continuity equation for the mass concentration $x(\boldsymbol{r}, t)$,

$$\frac{\partial x(\boldsymbol{r}, t)}{\partial \mathrm{t}} + \nabla \cdot j(\boldsymbol{r}, t) = 0 \tag{A6.5.4}$$

The first two equations are the same as those for a one-component fluid, discussed in section 4.3.1, and the intervening quantities are the same as those defined there; in particular, $D_L = (\frac{4}{3}\eta + \zeta)/m\rho$ with $m = c_1 m_1 + c_2 m_2$. The heat diffusion equation contains, instead, an additional term determined by the diffusion flux $j(\boldsymbol{r}, t)$, with a coefficient involving the thermal diffusion ratio, k_T, and thermodynamic derivatives of $\mu = (\mu_1/m_1) - (\mu_2/m_2)$, the difference in chemical potential per unit mass of the two components. Finally, the continuity equation for the mass concentration leads to the diffusion equation when we write the diffusion flux as

$$j(\boldsymbol{r}, t) = -D\left[\nabla x(\boldsymbol{r}, t) + \frac{k_T}{T} \nabla T(\boldsymbol{r}, t) + \frac{k_p}{P} \nabla P(\boldsymbol{r}, t)\right] \tag{A6.5.5}$$

in terms of the interdiffusion coefficient D, the thermal diffusion ratio and the thermodynamic quantity $k_p = P(\partial\mu/\partial P)_{T,x}/(\partial\mu/\partial x)_{p,T}$.

The solution of the above equations to determine the time-dependent correlation functions in the mixture, to terms linear in the transport coefficients, has been given by Cohen, Sutherland and Deutch (1971). Formally, one can rewrite the hydrodynamic equations as

$$\frac{\partial N_i(\boldsymbol{r}, t)}{\partial t} = -\sum_j \int \mathrm{d}\boldsymbol{r}' L_{ij}(\boldsymbol{r}-\boldsymbol{r}') N_j(\boldsymbol{r}', t) \tag{A6.5.6}$$

where $L_{ij}(\boldsymbol{r}-\boldsymbol{r}')$ is a suitable matrix and $N_i(\boldsymbol{r}, t)$ are a set of four dynamic variables, which are conveniently chosen as the velocity potential $\psi(\boldsymbol{r}, t) = (1/m)\nabla \cdot \boldsymbol{p}(\boldsymbol{r}, t)$, the pressure $P(\boldsymbol{r}, t)$, the mass concentration $x(\boldsymbol{r}, t)$, and the quantity $\phi(\boldsymbol{r}, t) = T(\boldsymbol{r}, t) - (T\alpha_T/\rho m C_p)P(\boldsymbol{r}, t)$, where α_T is the thermal expansion coefficient and C_p the heat capacity. The formal solution of the above set of equations has the form

$$\langle \tilde{N}_i(\boldsymbol{k}, z) N_j(-\boldsymbol{k})\rangle = [\det M(\boldsymbol{k}, z)]^{-1} \sum_l P_{il}(\boldsymbol{k}, z) \langle N_l(\boldsymbol{k}) N_j(-\boldsymbol{k})\rangle \tag{A6.5.7}$$

where $\tilde{N}_i(\boldsymbol{k}, z)$ is the Laplace–Fourier transform of $N_i(\boldsymbol{r}, t)$, $M_{ij}(\boldsymbol{k}, z) = z\delta_{ij} + L(\boldsymbol{k})$, and $P_{ij}(\boldsymbol{k}, z)$ are suitable algebraic functions. The static correlation functions in the thermodynamic limit which enter the right-hand side of the above equation are determined by fluctuation theory, which for the above choice of dynamical variables gives the entropy change per unit mass in a fluctuation as

$$\Delta s = -\frac{1}{2T_0}\left[\frac{C_p}{T}(\delta\phi)^2 + \frac{K_T}{m\rho\gamma}(\delta P)^2 + \left(\frac{\partial \mu}{\partial x}\right)_{p,T}(\delta x)^2\right] \tag{A6.5.8}$$

Explicitly, one finds for the concentration–concentration correlation function (Bhatia, Thornton and March, 1974) the expression

$$S_{cc}(k, \omega) = \frac{Nk_B T}{Z}\left\{\frac{2A_7 Xk^2}{\omega^2 + X^2k^4} + \frac{2A_8 Yk^2}{\omega^2 + Y^2k^4}\right\} \tag{A6.5.9}$$

where

$$X = \tfrac{1}{2}\{\chi + \mathscr{D} + [(\chi + \mathscr{D})^2 - 4\chi D]^{1/2}\} \tag{A6.5.10}$$

$$Y = \tfrac{1}{2}\{\chi + \mathscr{D} - [(\chi + \mathscr{D})^2 - 4\chi D]^{1/2}\} \tag{A6.5.11}$$

$$A_7 = (Y - D)/(Y - X) \tag{A6.5.12}$$

$$A_8 = (X - D)/(X - Y) \tag{A6.5.13}$$

and $\chi = V\kappa/C_p$, $\mathscr{D} = D[1 + Zk_T{}^2/TC_p]$, and $Z = (\partial^2 G/\partial c_2{}^2)_{p,T,N}$, $Z_x = (\partial^2 G/\partial x^2)_{p,T,N}$

Similarly, the number–concentration correlation function is given by

$$S_{Nc}(k, \omega) = Nk_B T\left\{\frac{2A_4 Xk^2}{\omega^2 + X^2k^4} + \frac{2A_5 Yk^2}{\omega^2 + Y^2k^4} + A_6 \frac{k}{c_0}\left[\frac{\omega + c_0 k}{(\omega + c_0 k)^2 + \Gamma^2 k^4} - \frac{\omega - c_0 k}{(\omega - c_0 k)^2 + \Gamma^2 k^4}\right]\right\} \tag{A6.5.14}$$

where the new parameters are the adiabatic speed of sound, $c_0 = (\gamma/m\rho K_T)^{3/2}$, and

$$\Gamma = \tfrac{1}{2}\left[D_L + (\gamma-1)\chi + \frac{\gamma DVZ}{K_T}\Sigma^2\right] \quad \text{(A6.5.15)}$$

$$A_4 = (Y-X)^{-1}\left[(D-Y)\frac{\delta}{Z} + \frac{Dk_T\alpha_T}{C_p}\right] \quad \text{(A6.5.16)}$$

$$A_5 = (X-Y)^{-1}\left[(D-X)\frac{\delta}{Z} + \frac{Dk_T\alpha_T}{C_p}\right] \quad \text{(A6.5.17)}$$

$$A_6 = -D\Sigma \quad \text{(A6.5.18)}$$

with $\Sigma = (\delta - \delta_m)/Z + k_T\alpha_T/C_p$, $\delta = (1/V)(\partial V/\partial c)_{p,T,N}$ and $\delta_m = (m_1 - m_2)/m$.

Finally, the number–number correlation function is given by

$$S_{NN}(k,\omega) = \frac{Nk_BTK_T}{V\gamma}\left\{\frac{2A_1Xk^2}{\omega^2+X^2k^4} + \frac{2A_2Yk^2}{\omega^2+Y^2k^4} + \left[\frac{\Gamma k^2}{(\omega+c_0k)^2+\Gamma^2k^4} + \frac{\Gamma k^2}{(\omega-c_0k)^2+\Gamma^2k^4}\right] + A_3\frac{k}{c_0}\left[\frac{\omega+c_0k}{(\omega+c_0k)^2+\Gamma^2k^4} - \frac{\omega-c_0k}{(\omega-c_0k)^2+\Gamma^2k^4}\right]\right\} \quad \text{(A6.5.19)}$$

where the new parameters are

$$A_1 = \frac{1-\gamma}{X-Y}\left[D - X - \frac{2Dk_T\delta}{T\alpha_T} + (Y-D)\frac{\delta^2C_p}{ZT\alpha_T{}^2}\right] \quad \text{(A6.5.20)}$$

$$A_2 = \frac{1-\gamma}{Y-X}\left[D - Y - \frac{2Dk_T\delta}{T\alpha_T} + (X-D)\frac{\delta^2C_p}{ZT\alpha_T{}^2}\right] \quad \text{(A6.5.21)}$$

and

$$A_3 = (3\Gamma - D_L) + \frac{2V\gamma D}{K_T}\delta_m\Sigma \quad \text{(A6.5.22)}$$

APPENDIX 6.6

INELASTIC SCATTERING FROM FLUIDS IN TERMS OF PARTIAL STRUCTURE FACTORS

In terms of the partial dynamic structure factors $S_{\alpha\beta}(k, \omega)$ for a mixture,

$$S_{\alpha\beta}(k, \omega) = (c_\alpha c_\beta)^{-1/2} \int dt \exp(-i\omega t) \langle \rho_\alpha(\mathbf{k}, t)\rho_\beta(-\mathbf{k}, 0)\rangle \qquad \text{(A6.6.1)}$$

we can rewrite the probability $P(\mathbf{k}, \omega)$ for scattering processes with momentum transfer $\hbar\mathbf{k}$ and energy transfer $\hbar\omega$ to the fluid, evaluated in section 3.2.3, as

$$P(\mathbf{k}, \omega) = \sum_{\alpha,\beta} (c_\alpha c_\beta)^{1/2} \mathscr{V}_\alpha(\mathbf{k}) \mathscr{V}_\beta(\mathbf{k}) S_{\alpha\beta}(k, \omega) \qquad \text{(A6.6.2)}$$

In a binary mixture this can be rewritten in terms of the number–concentration structure factors (Bhatia and Thornton, 1970) as

$$P(\mathbf{k}, \omega) = (\bar{\mathscr{V}})^2 S_{NN}(k, \omega) + (\mathscr{V}_1 - \mathscr{V}_2)^2 S_{cc}(k, \omega) + 2\bar{\mathscr{V}}(\mathscr{V}_1 - \mathscr{V}_2) S_{Nc}(k, \omega) \qquad \text{(A6.6.3)}$$

where

$$\bar{\mathscr{V}} = c_1 \mathscr{V}_1 + c_2 \mathscr{V}_2 \qquad \text{(A6.6.4)}$$

With reference to the discussion of section 3.2.3 for an isotopic mixture, this identifies the 'coherent' scattering factor as $S_{NN}(k, \omega)$. In reducing the other contributions to the self-function $S_s(k, \omega)$, we have thus omitted the interference term $S_{Nc}(k, \omega)$, which according to the discussion of section 6.7 is a very good approximation in an isotopic mixture for small mass differences, but we have also neglected possible 'coherent' effects in $S_{cc}(k, \omega)$.

APPENDIX 7.1

DENSITY RESPONSE FUNCTION FOR A FERMI GAS

For the evaluation of the density response function $\chi_0(k, \omega)$ of the ideal Fermi gas at zero temperature we use the general expression (3.43) of the density response function. The state $|\Psi_m\rangle$ is the ground state of the gas, which is a Slater determinant of plane waves $|\boldsymbol{k}'\rangle$ with $|\boldsymbol{k}'| < k_F$. The operator $\rho_k{}^\dagger$ creates a density fluctuation of wave vector $\boldsymbol{k}$, and for a free-particle system can be written as a superposition of all processes in which an electron is scattered with a change of momentum given by $\hbar\boldsymbol{k}$. The states $|\Psi_{m'}\rangle$ therefore differ from the ground state because of an electron having been taken from the state $|\boldsymbol{k}'\rangle$ to the state $|\boldsymbol{k}'+\boldsymbol{k}\rangle$: because of the statistics we must have $|\boldsymbol{k}'+\boldsymbol{k}| > k_F$, and the corresponding excitation energy is $\hbar\omega_{m'm} = \varepsilon_{k'+k} - \varepsilon_{k'}$. We can therefore write

$$\chi_0(k, \omega) = 2 \sum_{\substack{k' \\ (k' < k_F,\ |k'+k| > k_F)}} \left[\frac{1}{\hbar\omega - \varepsilon_{k'+k} + \varepsilon_{k'} + i\eta} - \frac{1}{\hbar\omega + \varepsilon_{k'+k} - \varepsilon_{k'} + i\eta} \right] \tag{A7.1.1}$$

where the factor 2 accounts for spin degeneracy of each plane wave state.

The evaluation of the integral (A7.1.1) was carried out by Lindhard (1954), with the following result:

$$\begin{aligned} \operatorname{Re} \chi_0(k, \omega) = -\frac{mk_F}{\pi^2\hbar^2} \Bigg\{ \frac{1}{2} &+ \frac{k_F}{4k} \left[\frac{(\omega + \hbar k^2/2m)^2}{(kv_F)^2} - 1 \right] \ln \left| \frac{\omega - kv_F + \hbar k^2/2m}{\omega + kv_F + \hbar k^2/2m} \right| \\ &- \frac{k_F}{4k} \left[\frac{(\omega - \hbar k^2/2m)^2}{(kv_F)^2} - 1 \right] \ln \left| \frac{\omega - kv_F - \hbar k^2/2m}{\omega + kv_F - \hbar k^2/2m} \right| \Bigg\} \end{aligned} \tag{A7.1.2}$$

with $v_F = \hbar k_F/m$, and

$$\operatorname{Im} \chi_0(k, \omega) = \begin{cases} -\dfrac{m^2\omega}{2\pi\hbar^3 k} & (0 \leqslant \omega \leqslant kv_F - \hbar k^2/2m) \\ -\dfrac{mk_F{}^2}{4\pi\hbar^2 k} \left[1 - \dfrac{(\omega - \hbar k^2/2m)^2}{(kv_F)^2} \right] & \left(\left| kv_F - \dfrac{\hbar k^2}{2m} \right| \leqslant \omega \leqslant kv_F + \dfrac{\hbar k^2}{2m} \right) \\ 0 & \left(\omega \geqslant kv_F + \dfrac{\hbar k^2}{2m} \right) \end{cases} \tag{A7.1.3}$$

In the static limit ($\omega = 0$) eqn (A7.1.2) gives

$$\chi_0(k, 0) = -\frac{mk_F}{2\pi^2\hbar^2}\left\{1 + \frac{k_F}{k}\left[1 - \left(\frac{k}{2k_F}\right)^2\right]\ln\left|\frac{1 + k/2k_F}{1 - k/2k_F}\right|\right\} \qquad \text{(A7.1.4)}$$

which corresponds to eqn (7.63) in the text.

APPENDIX 9.1

RENORMALISATION, LENGTH AND ASYMPTOTIC SCALING;— CALLAN–SYMANZIK EQUATION

One of the first suggestions that the renormalisation group would provide a microscopic foundation for length scaling was made by Di Castro and Jona-Lasinio (1969)—see also Hubbard (1972). They argue that the equations of scaling theory are just what the functional equations of the renormalisation group become near the critical point. Although their argument is couched in very general terms, and has been made much more specific by many later workers, we feel it is of sufficient interest still to give it, in essence, below. We shall then proceed to a more detailed treatment, involving asymptotic scaling and the Callan–Symanzik equation.

We consider a system described by a field ψ, and a two-body interaction $V = \alpha v(\mu, q^2)$, α being a dimensionless parameter measuring the strength of the interaction, while μ measures the range of the force. If we consider explicitly the two-point temperature-dependent Green function, the renormalisation group property (see Bogoliubov and Shirkov, Chapter VIII, 1959) leads to the exact functional relation

$$g(x, \omega_n, y_1, y_2, \alpha) = Z(\tau, y_1, y_2, \ldots, \alpha) \times g\left[\frac{x}{\tau}, \frac{\omega_n}{\tau}, \frac{y_1}{\tau}, \frac{y_2}{\tau}, \ldots, \alpha Z_V{}^{-1} Z^2(\tau, y_1, y_2, \ldots, \alpha)\right] \quad \text{(A9.1.1)}$$

Here g is a dimensionless function related to the Green function by

$$G = G^0 g$$

G^0 being the free two-point function. The rest of the notation is such that x is the square of the momentum, ω_n is the discrete frequency variable, while the y_1 represent the 'fixed' dimensional quantities (temperature, mass, range of forces, etc.). All the variables are expressed in terms of a reference momentum λ, while $\tau = \lambda'^2/\lambda^2$ is an arbitrary scaling factor. Z can be expressed in terms of g at $\omega_n = 0$ and satisfies the normalisation condition

$$Z(1, y_1, y_2 \cdots \alpha) = 1 \quad \text{(A9.1.2)}$$

It is important to choose the appropriate temperature-dependent Green function in each case, for not all choices of G are compatible with this normalisation. Z_v describes the vertex renormalisation. Equation (A9.1.1) is coupled to a similar equation for the vertex but there will be no need to invoke this explicitly in what follows.

It will be seen that eqn (A9.1.1) is an exact generalised homogeneity condition and it seems reasonable to anticipate that it will contain the equations of scaling theory as a special case. However, since as we have seen these equations are only valid in the immediate vicinity of the critical point, it is necessary to clarify how they come from the renormalisation group equation (A9.1.1), which is valid at all temperatures.

To make the connection, Di Castro and Jona-Lasinio consider the Ising model in the molecular field approximation. Applying the renormalisation group method to this case, they find that an equation of the form (A9.1.1) can be written for the static correlation function S and that this equation near the critical temperature T_c does indeed satisfy the scaling form

$$S(q, t) = Z_c\, S(qZ_c{}^{1/2}, tZ_c) \tag{A9.1.3}$$

where t as usual is $(T-T_c)/T_c$, while $Z_c = Z(\tau, T_c, \alpha)$. This form is just that considered by Kadanoff et al. (1967). If $\lambda \simeq \mu$, the range parameter, then the approximate equation $Z_c \simeq \tau^{-1}$ holds and the scaling factor reduces to the ratio of two lengths, in agreement with the physical arguments of Kadanoff et al.

Next consider the general case, the argument however still being restricted to static scaling. According to Halperin and Hohenberg (1967) static scaling is equivalent to the following two hypotheses:

(1) The order parameter correlation length ξ which diverges at $T = T_c$ for vanishing external field contains all the 'relevant' information on critical fluctuations and on the nature of the interaction.

(2) The order parameter static correlation function near the critical point has the form

$$S(q) \simeq q^z f(q\xi) \tag{A9.1.4}$$

Di Castro and Jona-Lasinio show that if the assumption (1) is combined with the renormalisation group equation (A9.1.1), then assumption (2) follows automatically.

Now the order parameter is assumed to coincide with $\langle\psi\rangle$ in which case the static correlation function S can be obtained from G by summing over the discrete frequency variable ω_n. According to the assumption (1), the entire dependence on the variables y_i and α near the critical point reduces to the dependence on the single variable $y = \xi^{-2}/\lambda^2$. Under this assumption the renormalisation group equation for the static correlation function S becomes

$$S(x, y) = Z(\tau, y) S(x/\tau, y/\tau) \tag{A9.1.5}$$

Differentiating this equation with respect to τ and setting $\tau=1$, it is straightforward to show that

$$x\frac{\partial S}{\partial x}+y\frac{\partial S}{\partial y}=\sigma(y)S; \qquad \sigma(y)=\left.\frac{\partial Z}{\partial \tau}\right|_{\tau=1} \tag{A9.1.6}$$

The solution of this equation, following Di Castro and Jona-Lasinio, requires the integration of the following ordinary differential equations

$$\mathrm{d}x/\mathrm{d}y=x/y; \qquad \mathrm{d}S/\mathrm{d}y=\sigma(y)S/y \tag{A9.1.7}$$

In the vicinity of $y=0$ (critical point) the second of these equations is dominated by the singularity of y^{-1}, and it is permissible if σ is regular to make the approximation $\sigma(y)\simeq\sigma(0)=\sigma_0$. The assumption that σ is regular turns out to be closely related to the idea of a fixed point, later introduced by Wilson, who thereby overcomes the need for the assumption. The general solution of eqn (A9.1.6) in such an approximation is given by

$$S(x, y)=f(x/y)y^{\sigma_0} \tag{A9.1.8}$$

with f arbitrary. This has the desired form (compare eqn (9.49) of main text).

Asymptotic scaling and Callan–Symanzik equation

After this introduction we shall develop the field theory approach a little more formally below. To motivate this, we can say that two approaches have been developed for calculating the critical exponents as functions of dimensionality and number of components of the order parameter n. These are, first, the Feynman graph method, pioneered by Wilson (see Wilson and Kogut, 1974) and the use of the Callan–Symanzik equation.

We remind the reader again that the conclusions which follow from the discussion in the main text are, first, that phase transitions can be expressed in terms of renormalisable field theory,† and secondly that the theory scales asymptotically following the Gell–Mann and Low argument.

We recall from the main text that the renormalised correlation function S_R can be written in terms of Z and the self-energy as

$$S_\mathrm{R}^{-1}=Z(q^2+\mu^2)-Z\Sigma(q^2)+Z(\mu_0{}^2-\mu^2) \tag{A9.1.9}$$

In this form, the effective unperturbed propagator is considered to be $Z^{-1}\times(q^2+\mu^2)^{-1}$ and the perturbation results in the presence of the second (self-energy) and third terms on the right-hand side. At $q=0$ the sum of these two terms is $(1-Z)\mu^2$ and one finds that the effect of these interaction terms is to give back $S_\mathrm{R}{}^{-1}(0)=\mu^2$.

The utility of this form is that it gives the relation between the un-renormalised

†Almost the whole of the rest of this book can be read with minimal knowledge of field theory. For the uninitiated, a good reference to renormalised field theory is Bogoliubov and Shirkov (1959).

Green function G_n and the renormalised object $G_n{}^R$ as

$$G_n{}^R = Z^{-n/2} G_n \tag{A9.1.10}$$

Where G_n is calculated with the renormalised coupling constant, then as with S_R it is finite.

We now define a modification Γ_4 to be

$$\Gamma_4(q_1 \ldots q_4) = \prod_{i=1}^{4} S^{-1}(q_i) G_4(q_1 \ldots q_4) \tag{A9.1.11}$$

In perturbation theory, Γ_4 is written as

$$\Gamma_4 = g_0 + \Lambda(\{q\}, g_0) \tag{A9.1.12}$$

where Λ is the sum of connected graphs which dress Γ_4. Similarly, $\Gamma_4{}^R$ is defined as the $G_4{}^R$, 'amputated' by cutting of the renormalised $S_R(q_i)$. One further expresses the interaction in terms of the renormalised coupling constant g defined by

$$\left.\begin{aligned} g &= \Gamma_4{}^R(0, g) \\ \Gamma_4{}^R(\{q\}, g) &= Z^2 \Gamma_4(\{q\}, g_0) \end{aligned}\right\} \tag{A9.1.13}$$

With these definitions, it can then be proved (compare Brout, 1974) that $\Gamma_4{}^R$ is finite in the limit.

Callan–Symanzik scaling condition

Following Brezin, Le Guillou and Zinn-Justin (1973; 1974) we get first the Callan–Symanzik equation for dimensionality $d=4$. It can be shown in this case that $g = \Gamma_4{}^R(0)$ is dimensionless. The graphs contributing to $\Gamma_n{}^R$ are built out of g and $[q^2+\mu^2]^{-1}$ and these facts lead to the scaling law

$$\Gamma_n{}^R(\{\lambda p_i\}; \mu, g) = \lambda^{4-\eta} \Gamma_n{}^R(\{p_i\}; \mu/\lambda; g) \tag{A9.1.14}$$

We now use this, together with the relations

$$\left.\begin{aligned} G_n{}^R &= Z^{-n/2} G_n \\ \Gamma_n{}^R &= Z^{n/2} \Gamma_n \end{aligned}\right\} \tag{A9.1.15}$$

and consider variations in μ at fixed g_0. Then g varies through changes in μ and we must write

$$\frac{d}{d\mu} = \frac{\partial}{\partial\mu} + \left(\frac{\partial g}{\partial\mu}\right)_{g_0} \frac{\partial}{\partial g} \tag{A9.1.16}$$

The right-hand side of (A9.1.14) is an explicit function of μ_0 and its derivative with respect to μ_0 can be calculated in terms of graphs. The essential step is then to switch to μ_0 on the right-hand side while using μ on the left-hand side. The

Callan–Symanzik equation follows (Callan, 1970; Symanzik, 1970; 1971), namely

$$\left[\mu\frac{\partial}{\partial\mu}+\beta\frac{\partial}{\partial g}-\tfrac{1}{2}n\eta\right]\Gamma_{\mathrm{n}}{}^{\mathrm{R}}=\Delta\Gamma_{\mathrm{n}} \tag{A9.1.17}$$

where $\beta=\mu(\partial g/\partial\mu)_{g_0}$, $\eta=(\mu/Z)(\partial Z/\partial\mu)^{\bullet}$

$$\Delta\Gamma_{\mathrm{n}}=Z^{n/2}\left(\frac{\mathrm{d}\mu_0}{\mathrm{d}\mu}\right)\frac{\partial\Gamma_{\mathrm{n}}}{\partial\mu_0}$$

In the notation, it has been anticipated that $\partial\ln Z/\partial\ln\mu$ is the 'anomalous dimension' η.

The merit of (A9.1.17) in the study of asymptotic behaviour is clear when it is recognised that the term $(\partial\Gamma_{\mathrm{n}}/\partial\mu_0)$ on the right-hand side falls off with one additional power of the squared momentum than the left-hand side. This is because the operation $(\partial/\partial\mu_0)$, graph by graph, introduces one extra power $S^{(0)}[=(p^2+\mu_0{}^2)^{-1}]$. Hence, in the asymptotic limit, the simple procedure of counting powers leads to an one extra power of p^{-2} in each graph of $\partial\Gamma_{\mathrm{n}}/\partial\mu_0$ as compared to Γ_{n}. Hence in the limit $p/\mu\to\infty$ the right-hand side is negligible (compare the argument with that corresponding to eqn (A9.1.6) above).

The general solution of the resulting homogeneous equation is given by Brezin et al. This is not necessary for our purpose; we impose asymptotic scaling, that is Γ_{n} varies like a power of the momentum scale λ of eqn (A9.1.14). Then we get from the Callan–Symanzik equation the asymptotic scaling condition

$$\left.\begin{array}{l}\beta(g)=0\\ \Gamma_{\mathrm{n}}{}^{\mathrm{R}}(\{p_{\mathrm{i}}\};\mu,g)\to\mu^{n\eta/2},\ p_{\mathrm{i}}\gg\mu\end{array}\right\} \tag{A9.1.18}$$

and hence from eqn (A9.1.14)

$$\Gamma_{\mathrm{n}}{}^{\mathrm{R}}(\{\lambda p_{\mathrm{i}}\};\mu;g)\to\lambda^{4-n-n\eta/2} \tag{A9.1.19}$$

(In particular, for $n=2$, $S_{\mathrm{R}}{}^{-1}(p)\to p^2(p/\mu)^{-\eta}$, which completes the identification of η referred to above). Thus from (A9.1.18), the renormalised coupling constant g is determined by the requirement of asymptotic scaling (a proof of this has been given by Brezin et al.). In particular, at $\mu=0$, the power law behaviour typifies all behaviour on a momentum scale infinitesimal with respect to Λ. Once g is known, η is to be determined through $\eta=(\mu/Z)\partial Z/\partial\mu$.

Brezin et al. have stressed the distinction between two possible approaches to asymptotic scaling. In one, there is a renormalised coupling constant g, g_λ which is a function of the renormalisation scale. Then $g_\lambda\to g$ as $\lambda\to\infty$. In the other, g_λ is equal to g once and for all. In the first case they show that $\mathrm{d}\beta/\mathrm{d}g<0$ and in the second case $\mathrm{d}\beta/\mathrm{d}g>0$. They demonstrate that for $d<4$, the second case is realised, this being related to the fact that $Z\to0$ and $\Lambda\to\infty$.

Case $d<4$

In this case, using dimensional analysis, we can arrive at the result that a non-dimensional coupling u is defined uniquely through

$$g=u\mu^{\varepsilon},\ \varepsilon=4-d \tag{A9.1.20}$$

Then eqn (A.9.1.14) is generalised to $d<4$ if u rather than g is used:

$$\Gamma_n^{R}(\{\lambda p_i\};\mu,u)=\lambda^{d-n(d-2)/2}\Gamma_n^{R}(\{p_i\},\mu/\Lambda,u) \tag{A9.1.21}$$

The entire argument then goes through as before, provided u is used rather than g. The scaling condition is

$$\left.\begin{array}{l}\beta(u)=\mu(\partial u/\partial\mu)_{g_0}=0\\ \Gamma_n^{R}(\{\lambda p_i\};\mu,u)\to\lambda^{[d-n(d-2)/2-n\eta/2]}\end{array}\right\} \tag{A9.1.22}$$

A more convenient form for calculation is

$$\beta=-\varepsilon[\partial\ln(uZ_1Z^{-2})/\partial\mu]_{g_0}^{-1} \tag{A9.1.23}$$

$$\eta=\beta(u)\frac{\partial\ln Z}{\partial\mu}$$

The use of the Callan–Symanzik method to calculate critical exponents in conjunction with the $4-d=\varepsilon$ or $1/n$ expansion has been referred to in the main text.

APPENDIX 10.1

SURFACE STRESS AT A PLANAR LIQUID SURFACE

Following Kirkwood and Buff (1949) we calculate the pressure $p^{t}(z)$ normal to area element in the y–z plane at height z, for a liquid surface in the x–y plane, by noting that the x component of the force exerted on the molecules in a volume element $\mathrm{d}\boldsymbol{r}_1$ by the molecules in a volume element $\mathrm{d}\boldsymbol{r}_2$ is

$$\frac{x_2-x_1}{R}\,\phi'(R)\rho^{(2)}(z_1,\boldsymbol{R})\,\mathrm{d}\boldsymbol{r}_1\,\mathrm{d}\boldsymbol{r}_2 \qquad \text{(A10.1.1)}$$

where $\boldsymbol{R}=\boldsymbol{r}_2-\boldsymbol{r}_1$. Integration over all the pairs gives the force per unit area transmitted across the y–z plane as

$$F_x(z_1)=\int_{-\infty}^{0}\mathrm{d}x_1\int_{(x_2>0)}\mathrm{d}\boldsymbol{R}\,\frac{x_2-x_1}{R}\,\phi'(R)\rho^{(2)}(z_1,\boldsymbol{R}) \qquad \text{(A10.1.2)}$$

The integration over x_1 is easily carried out and, using the fact that the integrand becomes an even function of x_2-x_1, we can write

$$F_x(z_1)=\tfrac{1}{2}\int\mathrm{d}\boldsymbol{R}\,\frac{(x_2-x_1)^2}{R}\,\phi'(R)\rho^{(2)}(z_1,\boldsymbol{R}) \qquad \text{(A10.1.3)}$$

To this we must add the kinetic part of the stress, given by the momentum transport $k_BT\rho(z_1)$ across the unit area, thus finding eqn (10.1):

$$p^{t}(z)=k_BT\rho(z)-\tfrac{1}{2}\int\mathrm{d}\boldsymbol{R}\,\frac{(x_2-x_1)^2}{R}\,\phi'(R)\rho^{(2)}(z,\boldsymbol{R}) \qquad \text{(A10.1.4)}$$

It is easily checked that at large distance from the surface, when $\rho(z)$ and $\rho^{(2)}(z,\boldsymbol{R})$ become independent of z, this reduces to the expression (1.15).

References

Abe, R. (1958). *Prog. Theor. Phys.*, **19**, 57 and 407
Abel, W. R., Anderson, A. C. and Wheatley, J. C. (1966). *Phys. Rev. Letts*, **17**, 74
Abowitz, G. and Gordon, R. B. (1962). *J. Chem. Phys.*, **37**, 125
Abramo, M. C., Parrinello, M. and Tosi, M. P. (1973*a*). *J. Non-Metals*, **2**, 57
—, —, — (1973*b*). *J. Non-Metals*, **2**, 67
—, —, — (1974). *J. Phys.* **C8**, 4201
Abramo, M. C. and Tosi, M. P. (1972). *Nuovo Cimento*, **B10**, 21
—, — (1974). *Nuovo Cimento*, **B21**, 363
Abrikosov, A. A. and Khalatnikov, I. M. (1959). *Rep. Prog. Phys.*, **22**, 329
Ackasu, A. Z. and Daniels, E. (1970). *Phys. Rev.*, **A2**, 962
Adams, D. J. and McDonald, I. R. (1974). *J. Phys.*, **C7**, 2761
Adelman, S. A. and Deutch, J. M. (1974). *J. Chem. Phys.*, **60**, 3935
Ailawadi, N. K., Rahman, A. and Zwanzig, R. (1971). *Phys. Rev.*, **A4**, 1616
Alder, B. J. (1964). *Phys. Rev. Letts*, **12**, 317
— and Hecht, C. E. (1969). *J. Chem. Phys.*, **50**, 2032
— and Wainwright, T. E. (1970). *Phys. Rev.*, **A1**, 18
Allen, G. and Higgins, J. S. (1973). *Rep. Prog. Phys.*, **36**, 1073
Alvesalo, T. A., Anufriyev, Yu. D., Collan, H. K., Lounasmaa, O. V. and Wennerstrom, P. (1973). *Phys. Rev. Letts*, **30**, 962
Amit, D. J., Kane, J. W. and Wagner, H. (1968). *Phys. Rev.*, **175**, 313
Andersen, H. C. and Chandler, D. (1971). *J. Chem. Phys.*, **55**, 1497
—, Weeks, J. D. and Chandler, D. (1971). *Phys. Rev.*, **A4**, 1597
Anderson, A. C., Roach, W. R., Sarwinski, R. E. and Wheatley, J. C. (1966). *Phys. Rev. Letts*, **16**, 263
Anderson, P. W. and Brinkman, W. F. (1973). *Phys. Rev. Letts*, **30**, 1108
Arcovito, G., Faloci, C., Roberti, M. and Mistura, L. (1969). *Phys. Rev. Letts*, **22**, 1040
Ascarelli, P., Paskin, A. and Harrison, R. (1967). *1st International Conference on Liquid Metals*. Taylor and Francis, London
Ashcroft, N. W. and Langreth, D. C. (1967*a*). *Phys. Rev.* **156**, 685
—, — (1967*b*). *Phys. Rev.*, **155**, 682
— and March, N. H. (1967). *Proc. Roy. Soc.*, **A297**, 336
Asnin, V. M., Rogachev, A. A. and Sablina, N. I. (1970). *JETP Letts*, **11**, 99

Balian, R. and Werthamer, N. R. (1963). *Phys. Rev.*, **131**, 1553
Bardasis, A., Falk, D. S. and Simkin, D. A. (1965). *J. Phys. Chem. Solids*, **26**, 1269
Bardeen, J., Baym, G. and Pines, D. (1966). *Phys. Rev. Letts*, **17**, 372
—, —, — (1967). *Phys. Rev.*, **156**, 207
—, Cooper, L. M. and Schrieffer, J. R. (1957). *Phys. Rev.*, **108**, 1175
Barker, J. and Henderson, D. (1967). *J. Chem. Phys.*, **47**, 4714
—, — (1971). *Adv. Chem. Revs*, **4**, 303
Barker, M. I., Johnson, M. W., March, N. H. and Page, D. I. (1973). *2nd International Conference on Liquid Metals*, p. 99. Taylor and Francis, London
Baym, G. (1964). *Phys. Rev.*, **135**, A1691
— and Saam, W. F. (1968). *Phys. Rev.*, **171**, 172
Beeby, J. L. (1973). *J. Phys.*, **C6**, 2262
Belton, J. W. and Evans, M. G. (1945). *Trans. Far. Soc.*, **41**, 1
Benedek, G. (1966). *Brandeis University Summer Institute in Theoretical Physics*. Gordon and Breach, New York

Ben-Naim, A. and Stillinger, F. H. (1972). In *Structure and Transport Processes in Water and Aqueous Solutions* (ed. R. A. Horne). Wiley–Interscience, New York
Benoît à la Guillame, C., Voos, M., Salvan, F., Laurant, J. M. and Bonnot, A. (1971). *C. R. Acad. Sci.*, **B272**, 236
Berge, P. and Dubois, M. (1971). *Phys. Rev. Letts*, **27**, 1125
Berk, N. F. and Schrieffer, J. R. (1966). *Phys. Rev. Letts*, **17**, 433
Bernal, J. D. and Fowler, R. H. (1933). *J. Chem. Phys.*, **1**, 515
Berry, M. V., Durrans, R. F. and Evans, R. (1972). *J. Phys.*, **A5**, 166
—and Resnek, S. R. (1971). *J. Phys.*, **A4**, 77
Bhatia, A. B., Hargrove, W. H. and March, N. H. (1973). *J. Phys.*, **C6**, 621
—and March, N. H. (1972). *Phys.Letts*, **41A**, 397
—, — (1975). *J. Phys.*, **F5**, 1100
— and Thornton, D. E. (1970). *Phys. Rev.* **B2**, 3004
—, — and March, N. H. (1974). *Phys. Chem. Liquids*, **4**, 93
Blum, L. (1972). *J. Chem. Phys.*, **57**, 1862
— (1973*a*). *J. Chem. Phys.*, **58**, 3295
— (1974). *Chem. Phys. Letts*, **26**, 200
— and Torruella, A. J. (1972). *J. Chem. Phys.*, **56**, 303
Bogoliubov, N. and Shirkov, D. (1959). *Introduction to the Theory of Quantized Fields*. Interscience, New York
Bohm, D. and Pines, D. (1953). *Phys. Rev.*, **92**, 609
— and Staver, T. (1952). *Phys. Rev.*, **84**, 836
Bolzer, H. M., Bernier, M. E. R., Gully, W. J., Richardson, R. C. and Lee, D. M. (1974). *Phys. Rev. Letts*, **32**, 875
Born, M. and Green, H. S. (1946). *Proc. Roy. Soc.*, **A188**, 10
Bratby, P., Gaskell, T. and March, N. H. (1970). *Phys. Chem.Liquids*, **2**, 53
Brezin, E. and Wallace, D. J. (1973). *Phys. Rev.*, **B7**, 1967
—, Le Guillou, J. C. and Zinn-Justin (1973). *Phys. Rev.*, **D8**, 434 and 2418
—, —, —(1974). *Phys. Rev.*, **D9**, 1121
Brillouin, L. (1922). *Ann. Physique*, **17**, 88
Brinkman, W. F., Rice, T. M., Anderson, P. W. and Chui, S. T. (1972). *Phys. Rev. Letts*, **28**, 961
Brout, R. (1974). *Physics Reports*, **10C**, 1
Brown, R. C. and March, N. H. (1968). *Phys. Chem.Liquids*, **1**, 141
—, —(1973). *J. Phys.*, **C5**, L363
Brueckner, K. A., Soda, T., Anderson, P. W. and Morel, F. (1960). *Phys. Rev.*, **118**, 1442
Brumberger, H., Alexandropoulos, N. G. and Claffey, W. (1967). *Phys. Rev. Letts*, **19**, 555
Brumer, P. (1974). *Phys. Rev.*, **A10**, 1
— and Karplus, M. (1973). *J. Chem. Phys.*, **58**, 3903
Brush, S. G., Sahlin, H. L. and Teller, E. (1966). *J. Chem. Phys.*, **45**, 2102
Buckingham, A. D. (1967). *Disc. Far. Soc.*, No. 43, The structure and properties of liquids, p. 205
Buff, F. P. and Lovett, R. A. (1968). In *Simple Dense Fluids* (eds H. L. Frisch and Z. W. Salsburg) Academic Press, New York
Butler, J. A. V. (1932). *Proc. Roy. Soc.*, **A135**, 348

Callan, C. G. (1970). *Phys. Rev.*, **D2**, 1541
Callen, H. B. and Welton, E. A. (1951). *Phys. Rev.*, **83**, 34
Care, C. M. and March, N. H. (1975). *Adv. Phys.*, **24**, 101
Carley, D. D. (1963). *Phys. Rev.*, **131**, 1406
Chandler, D. and Weeks, J. D. (1970). *Phys. Rev. Letts*, **25**, 149
Chapman, S. and Cowling, T. G. (1939). *The Mathematical Theory of Non-Uniform Gases*. University Press, Cambridge
Chester, G. V., Metz, R. and Reatto, L. (1968). *Phys. Rev.*, **175**, 275
Chihara, J. (1973). *2nd International Conference on Liquid Metals.* Taylor and Francis, London
Chung, C. H. and Yip, S. (1969). *Phys. Rev.*, **182**, 323
Cochran, W. (1963). *Rep. Prog. Phys.*, **26**, 1
Cohen, C., Sutherland, J. W. H. and Deutch, J. M. (1971). *Phys. Chem. Liquids*, **2**, 213

Cole, G. H. A. (1967). *An Introduction to the Statistical Theory of Classical Simple Dense Fluids*. Pergamon Press, Oxford
Cole, R. H. (1938). *J. Chem. Phys.*, **6**, 385
Combescot, M. and Nozières, P. (1972). *J. Phys.*, **C5**, 2369
Cooper, M. S. (1973). *Phys. Rev.*, **A7**, 1
Copley, J. R. D. and Lovesey, S. L. (1975). *Rep. Prog. Phys.*, **38**, 461
– and Rowe, J. M. (1974*a*). *Phys. Rev. Letts*, **32**, 49
–, –(1974*b*). *Phys. Rev.*, **A9**, 1656
Cowley, R. A. and Woods, A. D. B. (1971). *Can. J. Phys.*, **49**, 177
Croxton, C. A. (1974). *Liquid State Physics—A Statistical Mechanical Introduction*. University Press, Cambridge
– and Ferrier, R. D. (1971*a*). *J. Phys.*, **C4**, 1921; *Phil. Mag.*, **24**, 489
–, –(1971*b*). *J. Phys.*, **C4**, 2447
–, –(1971*c*). *Phys.Letts*, **35A**, 330

Davison, T. B. and Feenberg, E. (1969). *Phys. Rev.*, **178**, 306
de Boer, J. (1957). In *Progress in Low Temperature Physics* (ed. C. J. Gorter), vol. 2, p. 23. North-Holland, Amsterdam
Debye, P. (1915). *Ann. d. Phys.*, **46**, 809
–and Hückel, E. (1923). *Z. Phys.*, **24**, 185
de Gennes, P. G. (1959). *Physica*, **25**, 825
Delbene, J. and Pople, J. A. (1970). *J. Chem. Phys.*, **52**, 4848
Derrien, J. Y. and Dupuy, J. (1975). *J. de Physique*, **36**, 191
Di Castro, C. (1972). *Lettere, Nuovo Cimento*, **5**, 69
—and Jona-Lasinio, G. (1969) *Phys. Lett*, **29A**, 322
Dick, B. G. and Overhauser, A. W. (1958). *Phys. Rev.*, **112**, 90
Dietrich, O. W., Graf, E. H., Huang, C. H. and Passell, L. (1972). *Phys. Rev.*, **A5**, 1377
Domb, C. and Green, M. S. (1972 and 1974). *Critical Phenomena*, Academic Press, London
Domb, C. and Hunter, D. L. (1965). *Proc. Phys. Soc.*, **86**, 117
Doniach, S. and Engelsberg, S. (1966). *Phys. Rev. Letts*, **17**, 750
Dorfman, J. R. D. and Cohen, E. G. D. (1970). *Phys. Rev. Letts*, **25**, 1257
–, –(1972). *Phys. Rev.*, **A6**, 776
Dubois, D. F. (1959). *Ann. Phys.*, **8**, 24
– and Kivelson, M. G. (1969). *Phys. Rev.*, **186**, 409
Dundon, J. M., Stolfa, D. L. and Goodkind, J. M. (1973). *Phys. Rev. Letts*, **30**, 843

Ebbsjö, I., Schofield, P., Sköld, K. and Waller, I. (1974). *J. Phys.*, **C8**, 3891
Edwards, D. O., Brewer, D. F., Seligman, P., Skertic, M. and Yaqub, M. (1965). *Phys. Rev. Letts*, **15**, 773
Egelstaff, P. A. (1967). *An Introduction to the Liquid State*. Interscience, New York
–, March, N. H. and McGill, N. C. (1974). *Can. J. Phys.*, **52**, 1651
–, Page, D. I. and Heard, C. R. T. (1969). *Phys.Letts*, **30A**, 376
–, –, –(1971). *J. Phys.*, **C4**, 1453
–, – and Powles, J. G. (1971). *Molecular Phys.*, **20**, 881
– and Ring, J. (1968). In *Physics of Simple Fluids* (eds H. N. V. Temperley, G. S. Rushbrooke and J. S. Rowlinson), Chap. 7. North-Holland, Amsterdam
– and Widom, B. (1970). *J. Chem. Phys.*, **53**, 2667
– and Wignall, G. D. (1970). *J. Phys.*, **C3**, 1673
Ehrenreich, H. and Cohen, M. H. (1959). *Phys. Rev.*, **115**, 786
Eisenberg, D. and Kauzmann, W. (1969). *The Structure and Properties of Water*. University Press, Oxford
Eisenschitz, R. and Wilford, M. J. (1962). *Proc. Phys. Soc.*, **80**, 1078
Eliezer, I. and Krindel, P. (1972). *J. Chem. Phys.*, **57**, 1884
Emery, V. J. (1964). *Ann. Phys.*, **28**, 1
– (1966). *Phys. Rev.*, **148**, 138
Enderby, J. E., Gaskell, T. and March, N. H. (1965). *Proc. Phys. Soc.*, **85**, 217

— and March, N. H. (1966). In *Phase Stability in Metals and Alloys* (eds P. S. Rudman, J. Stringer and R. T. Jaffee). McGraw-Hill, New York
— and North, D. M. (1968). *Phys. Chem. Liquids*, **1**, 1
— and Egelstaff, P. A. (1966). *Phil. Mag.*, **14**, 961
Ernst, M. H., Hauge, E. H. and van Leeuwen, J. M. J. (1970). *Phys. Rev.Letts*, **25**, 1254
—, —, —(1971). *Phys. Rev.*, **A4**, 2055

Faber, T. E. (1973). *An Introduction to the Theory of Liquid Metals*. University Press, Cambridge
Fatuzzo, E. and Mason, P. R. (1967). *Proc. Phys. Soc.*, **90**, 729
Feenberg, E. (1969). *Theory of Quantum Fluids*. Academic Press, New York
Felderhof, B.U. (1970) *Physica*, **48**, 451
Fermi, E. (1936) *Ricerca Scient.*, **7**, 13
Fetter, A. L. (1965) *Phys. Rev.*, **138**, A429
Feynman, R. P. (1954). *Phys. Rev.*, **94**, 264
—(1955). In *Progress in Low Temperature Physics* (ed. C. J. Gorter), vol, 1, chap. 2. North-Holland, Amsterdam
—and Cohen, M. (1956). *Phys. Rev.*, **102**, 1189
Fisher, I. Z. (1964). *Statistical Theory of Liquids*, p. 159. University Press, Chicago
—(1972). *Soviet Phys. JETP*, **35**, 811
—, and Bokut, B. V. (1956). *Zh. Fiz. Khim.*, **30**, 2547
Fisher, M. E. (1964). *J. Math. Phys.*, **5**, 944
—(1967). *Rep. Prog. Phys.*, **30**, 615
—(1973). In *Collective Properties of Physical Systems* (eds B. Lundqvist and S. Lundqvist), p. 16. Nobel Foundation, Stockholm
Fisk, S. and Widom, B. (1969). *J. Chem. Phys.*, **50**, 3219
Flory, P. J. (1942). *J. Chem. Phys.*, **10**, 51
Forster, D. and Martin, P. C. (1970). *Phys. Rev.*, **A2**, 1575
Fowler, R. H. (1937). *Proc. Roy. Soc.*, **A159**, 229
Frank, H. S. (1966). In *Chemical Physics of Ionic Solutions* (eds B. F. Conway and R. G. Barradas). Wiley, New York
Franks, F. (1973). *Treatise on Water*, Plenum Press, New York
Frenkel, J. (1942). *Kinetic Theory of Liquids*. University Press, Oxford
Friedel, J. (1958). *Nuovo Cimento*, Suppl. **7**, 287
Frisch, H. L. and Lebowitz, J. L. (1964). *Equilibrium Theory of Classical Fluids*. Benjamin, New York
Fröhlich, H. (1958). *Theory of Dielectrics*. Clarendon Press, Oxford
Fumi, F. G. and Tosi, M. P. (1964). *J. Phys. Chem. Solids*, **25**, 31 and 45

Gaskell, T. (1965). *Proc. Phys. Soc.*, **86**, 693
—(1966). *Proc. Phys. Soc.*, **89**, 231
— and March, N. H. (1970). *Phys. Letts* **33A**, 460
Gavoret, J. and Nozières, P. (1964). *Ann. Phys.*, **28**, 349
Geldart, D. J. W. and Taylor, R. (1970). *Can. J. Phys.*, **48**, 167
Gell-Mann, M. and Low, F. E. (1951). *Phys. Rev.*, **84**, 350
Giaquinta, P. V. (1974). *Doctoral thesis*, unpublished (University of Messina)
Giglio, M. and Benedek, G. B. (1969). *Phys. Rev. Letts*, **23**, 1145
Gillan, M., 1974, *J. Phys.*, **C7**, L1
Gillan, M., Larsen, B., Tosi, M. P. and March, N. H. (1976). *J. Phys.*, **C9**,889
Gingrich, N. S. (1943). *Rev. Mod. Phys.*, **15**, 90
Glick, A. and Ferrell, R. A. (1960). *Ann. Phys. (NY)*, **11**, 359
— and Long, W. E. (1971). *Phys. Rev.*, **B4**, 3455
Gray, P. (1973). *J. Phys.*, **F3**, L43
Green, H. S. (1952). *Molecular Theory of Fluids*. North-Holland, Amsterdam
—(1960). *Handbuch der Physik*, **10**, 79
Green, M. S. (1952). *J. Chem. Phys.*, **20**, 1281
—(1954). *J. Chem. Phys.*, **22**, 398

Greytak, T. J., Johnson, R. T., Paulson, D. N. and Wheatley, J. C. (1973). *Phys. Rev. Letts*, **31**, 452
—, Woerner, R., Yan, J. and Benjamin, R. (1970). *Phys. Rev. Letts*, **25**, 1547
Gross, E. P. (1961). *Nuovo Cimento*, **20**, 454
Guggenheim, E. A. (1945). *J. Chem. Phys.*, **13**, 253
—(1952). *Mixtures*. University Press, Oxford
Gurney, R. N. (1953). *Ionic Processes in Solution*. Dover, New York
Gyorffy, B. L. and March, N. H. (1971). *Phys. Chem. Liquids*, **2**, 197

Hall, H. E. (1960). *Adv. Phys.*, **9**, 89
Halperin, B. I. (1973). In *Collective Properties of Physical Systems* (eds B. Lundqvist and S. Lundqvist). Nobel Foundation, Stockholm
— and Hohenberg, P. C. (1967). *Phys. Rev. Letts*, **19**, 700
—, — and Ma, S. (1972). *Phys. Rev. Letts*, **29**, 1548
Hankins, D., Moscowitz, J. W. and Stillinger, F. H. (1970). *J. Chem. Phys.*, **53**, 4544
Hansen, J. P. (1970). *Phys. Rev.*, **A2**, 221
—(1973). *Phys. Rev.*, **A8**, 3096
—, Pollock, E. L. and McDonald, I. R. (1974). *Phys. Rev. Letts*, **32**, 277
Hartmann, W. M. (1971). *Phys. Rev. Letts*, **26**, 1640
Hedin, L. (1965). *Phys. Rev.*, **139**, A796
— and Lundqvist, S. (1969). *Solid State Physics*, vol. 23 (eds F. Seitz, D. Turnbull and H. Ehrenreich). Academic Press, New York
Heine, V. (1970). *Solid State Physics*, vol. 24 (eds F. Seitz, D. Turnbull and H. Ehrenreich). Academic Press, New York
Heller, P. (1967). *Rep. Prog. Phys.*, **30**, 731
Henderson, D. and Barker, J. A. (1969). *Aust. J. Chem.*, **22**, 2263
—, — (1967). *J. Chem. Phys.*, **47**, 2856
—, — and Smith, W. R. (1972). *Utilitas Mathematica*, **1**, 211; (cf. W. R. Smith, *Statistical Mechanics*, **1**, Chemical Society, London, 1973, p. 71)
Hill, T. L. (1955). *Statistical Mechanics*. McGraw-Hill, New York
Hinton, J. F. and Amis, E. S. (1971). *Chem. Revs.*, **71**, 627
Hirt, C. W. (1967). *Physics of Fluids*, **10**, 565
Ho, J. T. and Litster, J. D. (1969). *Phys. Rev. Letts*, **22**, 603
Hohenberg, P. C. and Platzman, P. M. (1966). *Phys. Rev.*, **152**, 198
Hoover, W. G., Ross, M., Johnson, K. W., Henderson, D., Barker,J. A. and Brown, B. C. (1970). *J. Chem. Phys.*, **52**, 4931
Howe, R. A., Howells, W. S. and Enderby, J. E. (1974). *J. Phys.*, **C7**, L111
Hubbard, J. (1957). *Proc. Roy, Soc.*, **A243**, 336
—(1972). *Phys. Letts*, **40A**, 111
— and Beeby, J. L. (1969). *J. Phys.*, **C2**, 556
— and Schofield, P. (1972). *Phys. Letts*, **40A**, 245
Hultgren, R., Orr, R. L., Anderson, P. D. and Kelley, K. K. (1963). *Selected Values of Thermodynamic Properties of Metals and Alloys*. Wiley, New York
Hutchinson, P. (1967). *Proc. Phys. Soc.*, **91**, 506

Ichikawa, K. and Thompson, J. C. (1974). *J. Phys.*, **F4**, 9
Ivanov, V. A., Makarenko, I. N., Nikolaenko, A. M. and Stishov, S. M. (1974). *Phys. Letts*, **47A**, 75

Jackson, H. W. and Feenberg, E. (1962). *Rev. Mod. Phys.*, **34**, 686
Jacucci, G. and McDonald, I. R. (1976) (In the press)
Jasnow, D., Moore, M. A. and Wortes, M. (1969). *Phys. Rev. Letts*, **22**, 940
Jepsen, D. W. (1966). *J. Chem. Phys.*, **44**, 774
— (1966). *J. Chem. Phys.*, **45**, 709
— and Friedman, H. L. (1963). *J. Chem. Phys.*, **38**, 846
Johnson, M. W., March, N. H. Page, D. I. Parrinello, M. and Tosi, M. P. (1974). *J. Phys.*, **C8**, 751
Jones, W. and March, N. H. (1973). *Theoretical Solid State Physics*. Wiley-Interscience, London

Kadanoff, L. P. (1966). *Physics*, **2**, 263
— (1970). *Comments on Solids*, **2**,
—, Götze, W., Hamblen, D., Hecht, R., Lewis, E. A. S., Pakiauskas, V. V., Rayl, M., Swift, J., Aspnes, D. and Kane, J. (1967). *Rev. Mod. Phys.*, **39**, 395
– and Martin, P. C. (1963). *Ann. Phys. (NY)*, **24**, 419
Kawasaki, K. (1970). *Ann. Phys. (NY)*, **61**, 1
Keldysh, L. V. (1968). In *Proceedings of the 9th International Conference on the Physics of Semiconductors*, p. 1303 (eds S. M. Ryvkin and V. V. Shmastser). Nauka, Leningrad
Kerr, W. C. (1968). *Phys. Rev.*, **174**, 316
–, Pathak, K. N. and Singwi, K. S. (1970). *Phys. Rev.*, **A2**, 2416
Khalatnikov, I. M. and Abrikosov, A. A. (1958). *Sov. Phys. JETP*, **6**, 84
Kirkwood, J. G. (1935). *J. Chem. Phys.*, **3**, 300
– and Buff, F. (1949). *J. Chem. Phys.*, **17**, 338
–, –(1951). *J. Chem. Phys.*, **19**, 774
Kirznits, D. A. (1957). *Sov. Phys. JETP*, **5**, 64
Kittel, C. (1963). *Quantum Theory of Solids*. Wiley, New York
Kohn, W. (1959). *Phys. Rev. Letts*, **2**, 393
– and Vosko, S. H. (1960). *Phys. Rev.*, **119**, 912
Kojima, H., Paulson, D. N. and Wheatley, J. C. (1974). *Phys. Rev. Letts*, **32**, 141
Komarov, L. I. and Fisher, I. Z. (1963). *Sov. Phys. JETP*, **16**, 1358
Krishnan, K. S. and Bhatia, A. B. (1945). *Nature*, **156**, 503
Krumhansl, J. A. and Wang, S. (1972). *J. Chem. Phys.*, **56**, 2034 and 2179
Kubo, R. (1957). *J. Phys. Soc. Japan*, **12**, 570
–(1966). *Rep. Progr. Phys.*, **29**, 255

Landau, L. D. (1957). *Sov. Phys. JETP*, **3**, 920
–and Lifshitz, E. M. (1959). *Fluid Mechanics*. Pergamon Press, Oxford
– and Placzek, G. (1934). *Phys. Z. Sovjetunion*, **5**, 172
– and Pomeranchuk, I. (1948). *Dokl. Akad. Nauk SSSR*, **59**, 669
Langer, J. S. and Vosko, S. H. (1959). *J. Phys. Chem. Solids*, **12**, 196
Larsen, B. (1974). *CECAM Workshop on Ionic Liquids* (to be published)
Lawson, D. T., Gully, W. J., Goldstein, S., Richardson, R. C. and Lee, D. M. (1973). *Phys. Rev. Letts*, **30**, 541
Lebowitz, J. L. (1964). *Phys. Rev.*, **133**, A895
— and Percus, J. K. (1963). *J. Math. Phys.*, **4**, 248
Leggett, A. J. (1973*a*). *Phys. Rev. Letts*, **31**, 352
–(1973*b*). In *Collective Properties of Physical Systems*, (eds B. Lundqvist and S. Lundqvist). Nobel Foundation, Stockholm
Levesque, D. and Ashurst, W. T. (1974). *Phys. Rev. Letts*, **33**, 277
–, Verlet, L. and Kürkijarvi, J. (1973). *Phys. Rev.*, **A7**, 1690
Lewis, J. W., Singer, K. and Woodcock, L. V. (1975). *J. Chem. Soc., Faraday Trans.* II, **71**, 308
Lighthill, M. J. (1958). *An Introduction to Fourier Analysis and Generalized Functions*, University Press, Cambridge
Lindhard, J. (1954). *Kgl. Danske Mat-Fys Medd.*, **28**, No. 8
Lobo, R., Robinson, J. E. and Rodriguez, S. (1973). *J. Chem. Phys.*, **59**, 5992
–, Rodriguez, S. and Robinson, J. E. (1967). *Phys. Rev.*, **161**, 513
Lomer, W. M. and Low, G. C. (1965). In *Thermal Neutron Scattering* (ed P. A. Egelstaff). Academic Press, New York
Longuet-Higgins, H. C. (1951). *Proc. Roy. Soc.*, **A205**, 247
– and Pople, J. A. (1956). *J. Chem. Phys.*, **25**, 884
– and Widom, B. (1964). *Molec. Phys.*, **8**, 549
Lovesey, S. L. (1971). *J. Phys.*, **C4**, 3057
Luttinger, J. M. (1964). *Phys. Rev.*, **135**, A1505

Ma, S. (1973). *Phys. Letts*, **43A**, 475
McAlister, S. P. and Turner, R. (1972). *J. Phys.*, **F2**, L51

McDonald, I. R. and Freeman, K. S. C. (1973). *Molec. Phys.*, **26**, 529
McMillan, W. L. (1965). *Phys. Rev.*, **138**, A442
Malomuzh, N. P., Oleinik, V. P. and Fisher, I. Z. (1973). *Soviet Phys. JETP*, **36**, 1233
Mandel, F., Bearman, R. J. and Bearman, M. Y. (1970). *J. Chem. Phys.*, **52**, 3315
March, N. H. (1968). *Liquid Metals*. Pergamon Press, Oxford
– (1974). *Self-consistent Fields in Atoms*. Pergamon Press, Oxford
– and Tosi, M. P. (1974). *Phys. Letts*, **50A**, 224
–, – and Bhatia, A. B. (1973). *J. Phys.*, **C6**, L59
–, Young, W. H. and Sampanthar, S. (1967). *The Many-Body Problem in Quantum Mechanics*. University Press, Cambridge
Marshall, W. C. and Lovesey, S. L. (1973). *Thermal Neutron Scattering*. University Press, Oxford
Martin, P. C. (1967). *Phys. Rev.*, **161**, 143
Mermin, N. D. and Ambegaokar, V. (1973). In *Collective Properties of Physical Systems*, p. 97 (eds B. Lundqvist and S. Lundqvist). Nobel Foundation, Stockholm
Mikolaj, P. G. and Pings, C. J. (1967). *J. Chem. Phys.*, **46**, 1401
Miller, A., Pines, D. and Nozières, P. (1962). *Phys. Rev.*, **127**, 1452
Mo, K. C., Gubbins, K. E., Jacucci, G. and McDonald, I. R. (1974). *Molec. Phys.*, **27**, 1173
Montgomery, D. C. and Tidman, D. A. (1964). *Plasma Kinetic Theory*. McGraw-Hill, New York
Mori, H. (1958). *Phys. Rev.*, **112**, 1829
Mori, H. (1965*a*). *Prog. Theor. Phys.*, **33**, 423
– (1965*b*). *Prog. Theor. Phys.*, **34**, 399
Mott, B. W. (1957). *Phil. Mag.*, **2**, 259
Mountain, R. D. (1966). *Rev. Mod. Phys.*, **38**, 205; also *Advances in Molecular Relaxation Processes*, Vol. 9 (1976)
Murase, C. (1970). *J. Phys. Soc. Japan*, **29**, 549

Nee, T. W. and Zwanzig, R. (1970). *J. Chem. Phys.*, **52**, 6353
Nelkin, M. and Ranganathan, S. (1967). *Phys. Rev.*, **164**, 221
Nienhuis, G. and Deutch, J. M. (1971*a*). *J. Chem. Phys.*, **55**, 4213
–, – (1971*b*). *J. Chem. Phys.*, **56**, 235
Nijboer, B. R. A. and Rahman, A. (1966). *Physica*, **32**, 415
– and van Hove, L. (1952). *Phys. Rev.*, **85**, 777
Nozières, P. (1963). *The Theory of Interacting Fermi Systems*. Benjamin, New York
– and Pines., D. (1958). *Nuovo Cimento*, **9**, 470
Nyquist, H. (1928). *Phys. Rev.*, **32**, 110

O'Neil, T. and Rostoker, N. (1965). *Physics of Fluids*, **8**, 1109
Onsager, L. (1936). *J. Amer. Chem. Soc.*, **58**, 1486
– (1944). *Phys. Rev.*, **65**, 117
Ornstein, L. S. and Zernike, F. (1914). *Proc. Acad. Sci. Amsterdam*, **17**, 793
–, – (1918). *Physik Z.*, **19**, 134
Osaka, Y. (1962). *J. Phys. Soc. Japan*, **17**, 547
Osheroff, D. D. (1974). *Phys. Rev. Letts*, **33**, 1009
– and Brinkman, W. F. (1974). *Phys. Rev. Letts*, **32**, 584
–, Gully, W. J., Richardson, R. C. and Lee, D. M. (1972). *Phys. Rev. Letts*, **29**, 920
–, Richardson, R. C. and Lee, D. M. (1972). *Phys. Rev. Letts*, **28**, 885
Ostgaard, E. (1970). *Phys. Rev.*, **A1**, 1048

Paalman, H. H. and Pings, C. J. (1963). *Rev. Mod. Phys.*, **35**, 389
Page, D. I. and Mika, K. (1971). *J. Phys.*, **C4**, 3034
– and Powles, J. G. (1971). *Molec. Phys.*, **21**, 901
Parrinello, M. and Tosi, M. P. (1973). *J. Phys.*, **C6**, L254
–, – and March, N. H. (1974*a*). *J. Phys.*, **C7**, 2577
–, –, – (1974*b*). *Proc. Roy. Soc.*, **A341**, 91
Patey, G. N. and Valleau, J. P. (1973). *Chem. Phys. Letts*, **21**, 297
Pathak, K. N. and Bansal, R. (1973). *J. Phys.*, **C6**, 1989
– and Singwi, K. S. (1970). *Phys. Rev.*, **A2**, 2427

Paulson, D. N., Johnson, P. T. and Wheatley, J. C. (1973). *Phys. Rev. Letts*, **30**, 829
—, Kojima, H., and Wheatley, J. C. (1974). *Phys. Rev. Letts*, **32**, 1098
Pearson, F. J. and Rushbrooke, G. S. (1957). *Proc. Roy. Soc. Edinburgh*, **A64**, 305
Percus, J. K. (1964). In *Equilibrium Theory of Classical Fluids* (eds H. L. Frisch and J. L. Lebowitz). Benjamin Inc., New York
— and Yevick, G. J. (1958). *Phys. Rev.*, **110**, 1
Pethick, C. J. (1970). *Phys. Rev.*, **B2**, 1789
Pines, D. (1963*a*). In *Liquid Helium*. Academic Press, New York
—(1963*b*). *Elementary Excitations in Solids*. Benjamin Inc., New York
—(1966) In *Quantum Fluids*, p. 257. (ed D. F. Brewer). North-Holland, Amsterdam
— and Nozières, P. (1966). *Theory of Quantum Fluids*. Benjamin Inc., New York
Pitaevskii, L. P. (1959). *Soviet Phys. JETP*, **9**, 830
—(1960). *Soviet Phys. JETP*, **37**, 1267
—(1961). *Soviet Phys. JETP*, **13**, 451
Pizzimenti, G., Tosi, M. P. and Villari, A. (1971). *Lett. Nuovo Cimento*, **1**, 743
Platzman, P. M. and Eisenberger, P. (1974). *Phys. Rev. Letts*, **33**, 152
Pokrovsky, V. E. and Svistunova, K. I. (1971). *JETP Letts*, **13**, 212
Polyakov, A. M. (1970). *Sov. Phys. JETP Letts*, **12**, 381
Pomeau, Y. (1972). *Phys. Rev.*, **A5**, 2569
Pomeranchuk, I. (1949). *Zh. Eksp. Theor. Fiz.*, **19**, 42
Powles, J. G., Dore, J. C. and Page, D. I. (1972). *Molec. Phys.*, **24**, 1025
Price, D. L. (1971). *Phys. Rev.*, **A4**, 358
Prigogine, I. (1958). *The Molecular Theory of Solutions*. North-Holland, Amsterdam
—, Bingen, R. and Cohen, E. G. D. (1962). In *Proc. 8th Int. Conf. on Low Temperature Physics* (ed. R. O. Davies). Butterworths, London, 1963

Raether, H. (1965). *Springer Tracts*, **38**, Springer, Berlin
Rahman, A., (1964). *Phys. Rev.*, **136**, A405
—(1967*a*). In *Neutron Inelastic Scattering*, vol. 1, p. 561. International Atomic Energy Agency, Vienna
—(1967*b*). *Phys. Rev. Letts*, **19**, 420
—(1974*a*). *Phys. Rev. Letts*, **32**, 52
—(1974*b*). *Phys. Rev.*, **A9**, 1667
—(1974*c*). *CECAM Workshop on Ionic Liquids* (to be published)
— and Copley, J. R. D. (1974). (to be published)
— and Paskin, A. (1966). *Phys. Rev. Letts*, **16**, 300
—, Singwi, K. S. and Sjolander, A. (1962). *Phys. Rev.*, **126**, 986
— and Stillinger, F. H. (1971). *J. Chem. Phys.*, **55**, 3336
—, —(1972). *J. Chem. Phys.*, **57**, 1281
Ramshaw, J. D. (1972). *J. Chem. Phys.*, **57**, 2684
Randolph, P. D. (1964). *Phys. Rev.*, **134**, 1483
Rayfield, G. W. and Reif, F. (1964). *Phys. Rev.*, **136**, A1194
Reatto, L. and Chester, G. V. (1967). *Phys. Rev.*, **155**, 88
Reiss, H., Frisch, H. L. and Lebowitz, J. L. (1959). *J. Chem. Phys.*, **31**, 369
Rice, S. A. and Gray, P. (1965). *The Statistical Mechanics of Simple Liquids*. Wiley, New York
Rice, T. M. (1968). *Phys. Rev.*, **175**, 858
Richardson, R. C. and Lee, D. M. (1973). In *Collective Properties of Physical Systems*, p. 84. (eds B. Lundqvist and S. Lundqvist). Nobel Foundation, Stockholm
Robinson, G. and March, N. H. (1972). *J. Phys.*, **C5**, 2553
Romano, F. and Margheritis, C. (1974). *Physica*, **77**, 557
Ross, M. and Schofield, P. (1971). *J. Phys.*, **C4**, L305
Rowland, T. (1960). *Phys. Rev.*, **119**, 900
Rowlinson, J. S. (1951). *Trans. Far. Soc.*, **47**, 120
— 1969). *Mixtures*, University Press, Cambridge
Ruppersberg, H. (1973). *Phys. Letts*, **46A**, 75
Rushbrooke, G. S. (1960). *Physica*, **26**, 259

Schiff, L. I. (1959). *Quantum Mechanics*. McGraw-Hill, New York
Schofield, P. (1966). *Proc. Phys. Soc.*, **88**, 149
—(1973). *Comp. Phys. Commun.*, **5**, 17
— (1974). *CECAM Workshop on Ionic Liquids* (to be published); (cf. *Statistical Mechanics*, **2**, The Chemical Society, 1976, p. 1)
Schommers, W. (1973). *Thesis*, University of Karlsruhe (unpublished)
Schuchowitzky, A. A. (1944). *Acta Physicochem. URSS*, **19**, 176
Sears, V. F. (1969). *Can. J. Phys.*, **47**, 199
Singwi, K. S., Sjölander, A., Tosi, M. P. and Land, R. H. (1970). *Phys. Rev.*, **B1**, 1044
—, Sköld, K. and Tosi, M. P. (1968). *Phys. Rev. Letts*, **21**, 881
—, —, —(1970). *Phys. Rev.*, **A1**, 454
—, Tosi, M. P., Land, R. H. and Sjölander, A. (1968). *Phys. Rev.*, **176**, 589
Sköld, K. (1967). *Phys. Rev. Letts*, **19**, 1023
— and Larsson, K. E. (1967). *Phys. Rev.*, **161**, 102
—, Rowe, J. M., Ostrowski, G. and Randolph, P. D. (1972). *Phys. Rev.*, **A6**, 1107
Stanley, H. E. (1971). *Introduction to Phase Transitions and Critical Phenomena*. Clarendon Press, Oxford
Stell, G. (1973). *J. Chem. Phys.*, **59**, 3926
Swift, J. and Kadanoff, L. P. (1968). *Ann. Phys. (NY)*, **50**, 312
Symanzik, K. (1970). *Comm. Math. Phys.*, **18**, 227
—(1971). *Comm. Math. Phys.*, **23**, 49

ter Haar, D. (1954). *Elements of Statistical Mechanics*. Rinehart, New York
Thiele, E. (1963). *J. Chem. Phys.*, **39**, 474
Toombs, G. (1965). *Proc. Phys. Soc.*, **86**, 273
Tosi, M. P. (1964). *Solid State Physics*, vol. 16, (eds F. Seitz and D. Turnbull). Academic Press, New York
— and Doyama, M. (1967). *Phys. Rev.*, **160**, 716
— and March, N. H. (1973). *Nuovo Cimento*, **B15**, 308
—, Parrinello, M. and March, N. H. (1974). *Nuovo Cimento*, **B23**, 135
Toulouse, G. (1973). In *Collective Properties of Physical Systems*, p. 45. (eds B. Lundqvist and S. Lundqvist). Nobel Foundation, Stockholm
Triezenberg, D. G. and Zwanzig, R. (1972). *Phys. Rev. Letts*, **28**, 1183

van Hove, L. (1954*a*). *Phys. Rev.*, **95**, 249
—(1954*b*). *Phys. Rev.*, **93**, 1374
van Leeuwen, J. M. J. and Cohen, E. G. D. (1962). In *Proc. 8th Int. Conf. on Low-Temperature Physics* (ed R. O. Davies). Butterworths, London, 1963
Vashishta, P., Bhattacharyya, P. and Singwi, K. S. (1974). *Phys. Rev.*, **B10**, 5108
—, Das, S. G. and Singwi, K. S. (1974). *Phys. Rev. Letts*, **33**, 911
— and Singwi, K. S. (1972). *Phys. Rev.*, **B6**, 875
Verlet, L. (1968). *Phys. Rev.*, **165**, 201
Verlet, L. and Weis, J. J. (1974). *Molec. Phys.*, **28**, 665
Vineyard, G. H. (1958). *Phys. Rev.*, **110**, 999
von Szyszkowski, B. (1968). *Z. Phys. Chem.*, **64**, 385
von Weizsäcker, C. F. (1935). *Z. Phys.*, **96**, 431

Waisman, A. and Lebowitz, J. L. (1970). *J. Chem. Phys.*, **52**, 4307
—, —(1972). *J. Chem. Phys.*, **56**, 3086
Wannier, G. H. (1966). *Statistical Physics*. Wiley, New York
Watabe, M. and Hasegawa, H. (1973). *2nd Int. Conf. on Liquid Metals*. Taylor and Francis, London
— and Young, W. H. (1974). *J. Phys. F.*, **4**, 129
Watts, R. O. (1974). *Molec. Phys.*, **28**, 1069
Webb, R. A., Greytak, T. J. Johnson, R. T. and Wheatley, J. C. (1973). *Phys. Rev. Letts*, **30**, 210
—, Kleinberg, R. L. and Wheatley, J. C. (1974). *Phys. Rev. Letts*, **33**, 145
Wertheim, M. S. (1963). *Phys. Rev. Letts*, **10**, 321
—(1971). *J. Chem. Phys.*, **55**, 4291
—(1973). *Molec. Phys.*, **26**, 1425

Wheatley, J. C. (1966). In *Quantum Fluids*, (ed D. F. Brewer). North-Holland, Amsterdam
Widom, B. (1965). *J. Chem. Phys.*, **43**, 3892
Wignall, G. D. and Egelstaff, P. A. (1968). *J. Phys.*, **C1**, 1088
Wigner, E. P. (1938). *Trans. Faraday Soc.*, **34**, 678
Wilson, K. (1972). *Phys. Rev. Letts*, **28**, 548
Wilson, K. G. and Fisher, M. E. (1972). *Phys. Rev. Letts*, **28**, 240
— and Kogut, J. (1974). *Physics Reports*, **12C**, 75
Woodcock, L. V. and Singer, K. (1971). *Trans. Faraday Soc.*, **67**, 12
Woodhead-Galloway, J. and Gaskell, T. (1968). *J. Phys.*, **C1**, 1472
—, — and March, N. H. (1968). *J. Phys.*, **C1**, 271
Woods, A. D. B. (1965). In *Inelastic Scattering of Neutrons*, p. 191 International Atomic Energy Agency, Vienna
— and Cowley, R. A. (1973). *Rep. Progr. Phys.*, **36**, 1135
—, Svensson, E. C. and Martel, P. (1972). In *Neutron Inelastic Scattering* p. 359 International Atomic Energy Agency, Vienna
Wu, F. Y. and Feenberg, E. (1961). *Phys. Rev.*, **122**, 739

Zawadowski, A., Ruvalds, J. and Solana, J. (1972). *Phys. Rev.*, **A5**, 399
Zemansky, M. W. (1951). *Heat and Thermodynamics*. McGraw-Hill, New York
Zernike, F. and Prins, J. A. (1927). *Z. Phys.*, **41**, 184
Ziman, J. M. (1961). *Phil. Mag.*, **6**, 1013
— (1972). *Elements of Advanced Quantum Theory*, University Press, Cambridge
Zolleweg, J., Hawkins, G. and Benedek, G. B. (1971). *Phys. Rev. Letts*, **27**, 1182
Zwanzig, R. (1966). *Phys. Rev.*, **144**, 170

Index

A CATALOG OF SELECTED

DOVER BOOKS

IN SCIENCE AND MATHEMATICS

A CATALOG OF SELECTED
DOVER BOOKS
IN SCIENCE AND MATHEMATICS

ROTARY-WING AERODYNAMICS, W.Z. Stepniewski. Clear, concise text covers aerodynamic phenomena of the rotor and offers guidelines for helicopter performance evaluation. Originally prepared for NASA. 537 figures. 640pp. 6⅛ × 9¼.
64647-5 Pa. $14.95

DIFFERENTIAL GEOMETRY, Heinrich W. Guggenheimer. Local differential geometry as an application of advanced calculus and linear algebra. Curvature, transformation groups, surfaces, more. Exercises. 62 figures. 378pp. 5⅜ × 8½.
63433-7 Pa. $7.95

INTRODUCTION TO SPACE DYNAMICS, William Tyrrell Thomson. Comprehensive, classic introduction to space-flight engineering for advanced undergraduate and graduate students. Includes vector algebra, kinematics, transformation of coordinates. Bibliography. Index. 352pp. 5⅜ × 8½. 65113-4 Pa. $8.00

A SURVEY OF MINIMAL SURFACES, Robert Osserman. Up-to-date, in-depth discussion of the field for advanced students. Corrected and enlarged edition covers new developments. Includes numerous problems. 192pp. 5⅜ × 8½.
64998-9 Pa. $8.00

ANALYTICAL MECHANICS OF GEARS, Earle Buckingham. Indispensable reference for modern gear manufacture covers conjugate gear-tooth action, gear-tooth profiles of various gears, many other topics. 263 figures. 102 tables. 546pp. 5⅜ × 8½. 65712-4 Pa. $11.95

SET THEORY AND LOGIC, Robert R. Stoll. Lucid introduction to unified theory of mathematical concepts. Set theory and logic seen as tools for conceptual understanding of real number system. 496pp. 5⅜ × 8¼. 63829-4 Pa. $8.95

A HISTORY OF MECHANICS, René Dugas. Monumental study of mechanical principles from antiquity to quantum mechanics. Contributions of ancient Greeks, Galileo, Leonardo, Kepler, Lagrange, many others. 671pp. 5⅜ × 8½.
65632-2 Pa. $14.95

FAMOUS PROBLEMS OF GEOMETRY AND HOW TO SOLVE THEM, Benjamin Bold. Squaring the circle, trisecting the angle, duplicating the cube: learn their history, why they are impossible to solve, then solve them yourself. 128pp. 5⅜ × 8½. 24297-8 Pa. $3.95

MECHANICAL VIBRATIONS, J.P. Den Hartog. Classic textbook offers lucid explanations and illustrative models, applying theories of vibrations to a variety of practical industrial engineering problems. Numerous figures. 233 problems, solutions. Appendix. Index. Preface. 436pp. 5⅜ × 8½. 64785-4 Pa. $8.95

CURVATURE AND HOMOLOGY, Samuel I. Goldberg. Thorough treatment of specialized branch of differential geometry. Covers Riemannian manifolds, topology of differentiable manifolds, compact Lie groups, other topics. Exercises. 315pp. 5⅜ × 8½. 64314-X Pa. $6.95

HISTORY OF STRENGTH OF MATERIALS, Stephen P. Timoshenko. Excellent historical survey of the strength of materials with many references to the theories of elasticity and structure. 245 figures. 452pp. 5⅜ × 8½. 61187-6 Pa. $9.95

CATALOG OF DOVER BOOKS

TENSOR CALCULUS, J.L. Synge and A. Schild. Widely used introductory text covers spaces and tensors, basic operations in Riemannian space, non-Riemannian spaces, etc. 324pp. 5⅜ × 8¼. 63612-7 Pa. $7.00

A CONCISE HISTORY OF MATHEMATICS, Dirk J. Struik. The best brief history of mathematics. Stresses origins and covers every major figure from ancient Near East to 19th century. 41 illustrations. 195pp. 5⅜ × 8½. 60255-9 Pa. $7.95

A SHORT ACCOUNT OF THE HISTORY OF MATHEMATICS, W.W. Rouse Ball. One of clearest, most authoritative surveys from the Egyptians and Phoenicians through 19th-century figures such as Grassman, Galois, Riemann. Fourth edition. 522pp. 5⅜ × 8½. 20630-0 Pa. $9.95

HISTORY OF MATHEMATICS, David E. Smith. Non-technical survey from ancient Greece and Orient to late 19th century; evolution of arithmetic, geometry, trigonometry, calculating devices, algebra, the calculus. 362 illustrations. 1,355pp. 5⅜ × 8½. 20429-4, 20430-8 Pa., Two-vol. set $21.90

THE GEOMETRY OF RENÉ DESCARTES, René Descartes. The great work founded analytical geometry. Original French text, Descartes' own diagrams, together with definitive Smith-Latham translation. 244pp. 5⅜ × 8½. 60068-8 Pa. $6.00

THE ORIGINS OF THE INFINITESIMAL CALCULUS, Margaret E. Baron. Only fully detailed and documented account of crucial discipline: origins; development by Galileo, Kepler, Cavalieri; contributions of Newton, Leibniz, more. 304pp. 5⅜ × 8½. (Available in U.S. and Canada only) 65371-4 Pa. $7.95

THE HISTORY OF THE CALCULUS AND ITS CONCEPTUAL DEVELOPMENT, Carl B. Boyer. Origins in antiquity, medieval contributions, work of Newton, Leibniz, rigorous formulation. Treatment is verbal. 346pp. 5⅜ × 8½. 60509-4 Pa. $6.95

THE THIRTEEN BOOKS OF EUCLID'S ELEMENTS, translated with introduction and commentary by Sir Thomas L. Heath. Definitive edition. Textual and linguistic notes, mathematical analysis. 2500 years of critical commentary. Not abridged. 1,414pp. 5⅜ × 8½. 60088-2, 60089-0, 60090-4 Pa., Three-vol. set $26.85

A HISTORY OF VECTOR ANALYSIS: The Evolution of the Idea of a Vectorial System, Michael J. Crowe. The first large-scale study of the history of vector analysis, now the standard on the subject. Unabridged republication of the edition published by University of Notre Dame Press, 1967, with second preface by Michael C. Crowe. Index. 278pp. 5⅜ × 8½. 64955-5 Pa. $7.00

THE HISTORICAL ROOTS OF ELEMENTARY MATHEMATICS, Lucas N.H. Bunt, Phillip S. Jones, and Jack D. Bedient. Fundamental underpinnings of modern arithmetic, algebra, geometry and number systems derived from ancient civilizations. 320pp. 5⅜ × 8½. 25563-8 Pa. $7.95

CALCULUS REFRESHER FOR TECHNICAL PEOPLE, A. Albert Klaf. Covers important aspects of integral and differential calculus via 756 questions. 566 problems, most answered. 431pp. 5⅜ × 8½. 20370-0 Pa. $7.95